Lecture Notes in Computer Science 8879

Commenced Publication in 1973
Founding and Former Series Editors:
Gerhard Goos, Juris Hartmanis, and Jan van Leeuwen

Wei Tsang Ooi   Cees G.M. Snoek
Hung Khoon Tan   Chin-Kuan Ho
Benoit Huet   Chong-Wah Ngo (Eds.)

Advances in Multimedia Information Processing – PCM 2014

15th Pacific-Rim Conference on Multimedia
Kuching, Malaysia, December 1-4, 2014
Proceedings

 Springer

Volume Editors

Wei Tsang Ooi
National University of Singapore, Singapore
E-mail: ooiwt@comp.nus.edu.sg

Cees G.M. Snoek
University of Amsterdam, The Netherlands
E-mail: cgmsnoek@uva.nl

Hung Khoon Tan
Universiti Tunku Abdul Rahman, Kampar, Perak, Malaysia
E-mail: thkhoon@utar.edu.my

Chin-Kuan Ho
Multimedia University, Cyberjaya, Selangor, Malaysia
E-mail: ckho@mmu.edu.my

Benoit Huet
EURECOM, Sophia Antipolis, France
E-mail: benoit.huet@eurecom.fr

Chong-Wah Ngo
City University of Hong Kong, Kowloon, Hong Kong, China
E-mail: cscwngo@cityu.edu.hk

ISSN 0302-9743   e-ISSN 1611-3349
ISBN 978-3-319-13167-2   e-ISBN 978-3-319-13168-9
DOI 10.1007/978-3-319-13168-9
Springer Cham Heidelberg New York Dordrecht London

Library of Congress Control Number: 2014953543

LNCS Sublibrary: SL 3 – Information Systems and Application, incl. Internet/Web
and HCI

Typesetting: Camera-ready by author, data conversion by Scientific Publishing Services, Chennai, India

Printed on acid-free paper

Springer is part of Springer Science+Business Media (www.springer.com)

Preface

Welcome to the proceedings of the Pacific-Rim Conference on Multimedia (PCM 2014), held December 1–4, 2014, in Kuching, Malaysia. The Pacific-Rim Conference on Multimedia is a leading international conference for researchers and industry practitioners to share and showcase their new ideas, original research results, and engineering development experiences from areas related to multimedia data types. The 2014 edition of PCM marked its 15th anniversary. The longevity of the conference would not be possible without the strong support of the research community, and we take this opportunity to thank everyone who has contributed to the growth of the conference in one way or another in the last 15 years.

This year's conference was held in the beautiful and historical city of Kuching, the capital of the state of Sarawak, Malaysia. A city of great history and legends, Kuching is known as the Camelot of the *White Rajahs*, referring to the English Brooke family that ruled the Kingdom of Sarawak for slightly over a century, until 1946. Located on Borneo, the third largest island in the world and home to one of the oldest rainforests in the world, Kuching and its region are filled with tourist attractions such as the Bako National Park, Semenggoh Wildlife Center, Sarawak Museum, and the Sarawak Cultural Village.

PCM 2014 featured a comprehensive program including a workshop on "Multimedia Big Data Analytics," two tutorials on "Context-Aware Multimedia Retrieval and Delivery" and "Depth-Based 3D Video Techniques and Systems," ten oral presentation sessions, a best paper candidates session, a poster and demonstration session, and a special session on "Image Analysis and Processing Under Extra Help." The conference received 84 submissions, covering topics in multimedia content analysis, multimedia signal processing and communications, as well as multimedia applications and services. The submissions were reviewed by a strong technical Program Committee, consisting of 82 reviewers. Each submission was reviewed by at least three reviewers. The Program Chairs carefully considered the input and feedback from the reviewers and accepted 41 papers for presentation at the conference. Out of these accepted papers, 36 were presented as oral, three were presented as posters, and two as demos. Moreover, we had four papers presented in the special session. The technical program is an important aspect but only provides its full impact if introduced by a challenging keynote. We are pleased with and thankful to our keynote speaker, Roger Zimmermann, for having accepted to present his work at the Pacific-Rim Conference on Multimedia 2014.

We are heavily indebted to many individuals for their significant contribution. We thank the PCM Steering Committee for their invaluable input and guidance on crucial decisions. We wish to acknowledge and express our deepest appreciation to the Organizing Committee who worked hard to ensure that all aspects of the conference went smoothly. The various PCM 2014 programs were organized by the special session chairs, Lexing Xie and Jianfei Cai, the workshop chairs, Liang-Tien Chia and Ralph Ewerth, the tutorial chairs, Winston Hsu and Roger Zimmermann, the panel chairs, Pablo Cesar and Shin'ichi Satoh, and the demo chairs, Christian Timmerer, and Xueliang Liu. Klaus Schoeffmann, Jun Wang, and Jinhui Tang helped to publicize the conference. The conference Web and social media presence were championed by Xiao-Yong Wei and Lai-Kuan Wong. Finally, a special nod goes to Kok-Why Ng, Choo-Yee Ting, C. Eswaran, Noramiza Hashim, and Poo Kuan Hoong who made all the necessary local arrangements and logistics for PCM 2014 to take place smoothly.

We gratefully thank the Sarawak Convention Bureau for their generous support of PCM 2014, which made several key aspects of the conference possible.

Finally, we wish to thank all committee members, reviewers, session chairs, student volunteers, and supporters. Their contributions are much appreciated.

We hope you all enjoy the proceedings of the 2014 Conference on Multimedia.

December 2014

Wei Tsang Ooi
Cees G.M. Snoek
Hung Khoon Tan
Chin Kuan Ho
Benoit Huet
Chong-Wah Ngo

Organization

Program Committee

Hezerul Abdul Karim	Multimedia University, Malaysia
Pradeep Atrey	State University of New York at Albany, USA
Marco Bertini	Università degli Studi di Firenze, Italy
Yoong Choon Chang	Multimedia University, Malaysia
Kuan-Ta Chen	Academia Sinica, Taiwan
Xiangyu Chen	Institute for Infocomm Research, Singapore
Wen-Huang Cheng	Academia Sinica, Taiwan
Lai-Tee Cheok	Singapore Management University, Singapore
Ngai-Man Cheung	Singapore University of Technology and Design, Singapore
Chen-Kuo Chiang	National Chung Cheng University, Taiwan
Wei-Ta Chu	National Chung Cheng University, Taiwan
Gianfranco Doretto	West Virginia University, USA
Lingyu Duan	Peking University, China
Abdulmotaleb El Saddik	University of Ottawa, Canada
Ralph Ewerth	Jena University of Applied Sciences, Germany
Jianping Fan	University of North Carolina at Charlotte, USA
Chiou-Shann Fuh	National Taiwan University, Taiwan
Margrit Gelautz	Vienna University of Technology, Austria
Bernard Ghanem	Advanced Digital Sciences Center, Singapore
Richang Hong	Hefei University of Technology, China
Jian Hou	Bohai University, China
Changbo Hu	University of Texas, USA
Min-Chun Hu	National Cheng Kung University, Taiwan
Chun-Rong Huang	National Chung Hsing University, Taiwan
Naoyuki Ichimura	National Institute of Advanced Industrial Science and Technology (AIST), Japan
Bogdan Ionescu	University Politehnica of Bucharest, Romania
Daisuke Iwai	Osaka University, Japan
Yoshio Iwai	Tottori University, Japan
Rongrong Ji	Xiamen Unviersity, China
Jinyuan Jia	Tongji University, China
Shen Jialie	Singapore Management University, Singapore
Lu Jiang	Carnegie Mellon University, USA
Yu-Gang Jiang	Fudan University, China

Xin Jin	Yahoo Inc., USA
Xin Jin	Tsinghua University, China
Li-Wei Kang	National Yunlin University of Science and Technology, Taiwan
Mario Koeppen	Kyushu Institute of Technology, Japan
Shang-Hong Lai	National Tsing Hua University, Taiwan
Congyan Lang	Beijing Jiaotong University, China
Duy-Dinh Le	National Institute of Informatics, Japan
Tung-Ying Lee	National Tsing Hua University, Taiwan
Haojie Li	Dalian University of Technology, China
Xirong Li	Renmin University of China, China
Ke Liang	Advanced Digital Sciences Center, Singapore
Guo-Shiang Lin	Da-Yeh University, Taiwan
Damon Shing-Min Liu	National Chung Cheng University, Taiwan
Dong Liu	Columbia University, USA
Xueliang Liu	Hefei University of Technology, China
Hong Lu	Fudan University, China
Mathias Lux	Klagenfurt University, Austria
Yasushi Makihara	Osaka University, Japan
Vasileios Mezaris	Information Technologies Institute/ Centre for Research and Technology Hellas, Greece
Dongbo Min	Advanced Digital Sciences Center, Singapore
Hui-Fuang Ng	Universiti Tunku Abdul Rahman, Malaysia
Cheng Qimin	Huazhong University of Science and Technology, China
Yanyun Qu	Xiamen University, China
Jason Quinlan	University College Cork, Ireland
Miriam Redi	Yahoo, USA
Hong Richang	Hefei University of Technology, China
Stevan Rudinac	University of Amsterdam, The Netherlands
Mukesh Saini	University of Ottawa, Canada
Jie Shao	University of Electronic Science and Technology of China, China
Yu-Wing Tai	KAIST, Republic of Korea
Masayuki Tanaka	Tokyo Institute of Technology, Japan
Chun-Jen Tsai	National Chiao Tung University, Taiwan
Kong-Wah Wan	Institute for Infocomm Research, Singapore
Meng Wang	AKiiRA Media Systems Inc., USA
Yu-Chiang Frank Wang	Academia Sinica, Taiwan
Zhi Wang	Tsinghua University, China
Xiao-Yong Wei	Sichuan University, China
Fei Wu	Zhejiang University, China
Lifang Wu	Beijing University Of Technology, China
Xiao Wu	Southwest Jiaotong University, China

Changsheng Xu   Institute of Automation, Chinese Academy
                                     of Sciences, China
Jizheng Xu   Microsoft Research Asia, China
Yuedong Xu   Fudan University, China
Shuicheng Yan   National University of Singapore, Singapore
Yang Yang   National University of Singapore, Singapore
Mei-Chen Yeh   National Taiwan Normal University, Taiwan
Zheng-Jun Zha   Hefei Institute of Intelligent Machines, Chinese
                                     Academy of Sciences, China
Guangtao Zhai   University of Erlangen Nuremberg, Germany
Yao Zhao   Beijing Jiaotong University, China
Yan-Tao Zheng   Google Inc., USA
Xiangdong Zhou   Fudan University, China
Shiai Zhu   City University of Hong Kong, SAR China
Roger Zimmermann   National University of Singapore, Singapore

Additional Reviewers

Hao, Shijie   Redi, Miriam
Hao, Yanbin   Shao, Xi
Huan, Li   Siyahjani, Farzad
Jingjing, Chen   Song, Ge
Li, Yingbo   Timmerer, Christian
Lo, Li-Yun   Tripathi, Ashish
Mohammed Yakubu, Abukari   Wei, Zhuo
Motiian, Saeid   Wu, Zuxuan
Pan, Tse-Yu   Zhang, Wei
Piccirilli, Marco

Mobile Videos with Contextual Geo-Properties (Keynote Abstract)

Roger Zimmermann

Department of Computer Science
Nantional University of Singapore
Singapore
`rogerz@comp.nus.edu.sg`

Abstract. Multimedia data such audio, images and videos are increasingly combined with other streams of contextual information such as, for example, geo-positions. The widespread availability of smartphones and tablets and the rapid improvement of integrated sensor hardware has enabled the acquisition of high-quality user-generated videos that are annotated with geo-properties. The contextual meta-data (*e.g.*, GPS and the digital compass values) provide a continuous stream of location and viewing direction information and lead to the concept of sensor-rich videos. This contextual information is considerably smaller in size than the visual content and is helpful in effectively and efficiently managing large repositories of videos. This talk presents on overview of our group's work in the areas of geo-referenced, sensor-rich video management. I will describe some of the technologies that we have developed and our experimental experiences. Utilizing sensor meta-data effectively presents many challenges in query processing, especially because real-world sensor data is noisy and sometimes inaccurate. In this talk I will present some of the challenges raised by large-scale geo-referenced applications, some approaches and solutions, and future opportunities.

Table of Contents

Image and Photo 1

Applications

People

Special Session: Image Analysis and Processing under Extra Help

Poster and Demo Session

Region-Based Interactive Ranking Optimization for Person Re-identification

Zheng Wang[1], Ruimin Hu[1,2], Chao Liang[1,2],
Qingming Leng[3], and Kaimin Sun[4]

[1] National Engineering Research Center for Multimedia Software,
School of Computer, Wuhan University, Wuhan, China
[2] Research Institute of Wuhan University in Shenzhen, China
[3] School of Information Science & Technology, Jiujiang University, Jiujiang, China
[4] State Key Laboratory of Information Engineering in Surveying, Mapping, and
Remote Sensing, Wuhan University, Wuhan, China
wangzwhu@whu.edu.cn

Abstract. Person re-identification, aiming to identify images of the same person from various cameras configured in difference places, has attracted plenty of attention in the multimedia community. Previous work mainly focuses on feature presentation and distance measure, and achieves promising results on some standard databases. However, the performance is still not good enough due to appearance changes caused by variations in illuminations, poses, viewpoints and occlusion. This paper addresses the problem through result re-ranking by introducing user feedback. In particular, considering the peculiarity of scarce positive and global similar negative samples in the person re-identification problem, we propose a region-based interactive ranking optimization method, to improve the original query result by labeling locally similar and dissimilar image regions. Experiments conducted on two standard data sets have validated the effectiveness of the proposed method with an average improvement of 10-30% over original basic method. It is proved that the ranking optimization algorithm is both an effective and efficient method to improve the original person re-identification result.

Keywords: Person Re-identification, Ranking Optimization, Human Computer Interaction.

1 Introduction

Person re-identification (a.k.a. person re-id), namely matching the same person across disjoint camera views, represents a valuable task in video surveillance scenarios [2]. Since classical biometric cues, such as face and gait, are usually unreliable or even infeasible in the uncontrolled surveillance environment [3], the appearance of the individual is mainly exploited for person re-id. Generally speaking, person re-id can be regarded as a pedestrian-oriented content-based image retrieval (CBIR) problem [4], i.e. given a query person image taken from one camera, the algorithm is expected to search the images of the same person

W.T. Ooi et al. (Eds.): PCM 2014, LNCS 8879, pp. 1–10, 2014.

captured by other cameras, and generate a ranking list where top results are more likely of the same person to the query image.

Previous research efforts address this problem mainly from two aspects, feature representation and distance measure. The former aims to construct discriminative visual descriptions that are robust and stable among different cameras [2][5][6][7]. The latter focuses on utilizing abundant training samples to learn a proper distance metric, where feature distance of the same person is smaller than that of different persons [3][4][8]. However, due to the influence of various surveillance conditions on illumination, viewpoints and scales, the correct target can hardly rank in top positions of the initial generated ranking list.

Besides feature representation and distance measure, the ranking optimization method is investigated in person re-id to optimize the original results generated by the feature-based or distance-based methods. Recently, few automatic re-ranking methods without any human interaction process have been proposed. Hirzer et al. [10] refined the results with positives obtained from the trajectory of the query person and negatives drawn from the worst matches. Leng et al. [9] utilized the individual and social relationships among images, and proposed bidirectional ranking method. But the improvements in these methods are limited, because an inappropriate optimization direction may be selected with the diversity of the intrinsic relationships among images. Whereas, interactive ranking optimization methods were seldom studied in person re-id. The purpose of this paper is discussing the interactive ranking optimization method for person re-id.

2   Motivation

For interactive ranking optimization method in CBIR, a core idea is that several positives and negatives are selected by user to update the original matching model, and hence generate a more accurate ranking list [11][16]. However, interactive ranking optimization work for face retrieval [17][18] seems improper in person re-id problem, because visual cues are often of low-resolution, ambiguous, and lack of relative details. Traditional methods designed for CBIR also cannot be directly used in person re-id, the reasons are as the following:

Scarce Positives and Global Similar Negatives - In general CBIR, there are usually multiple images of the same object/category in the gallery, providing sufficient positives to describe the query object. But in person re-id, the positive is extremely rare, and in some data sets single image or sequence merely appears once in the gallery. Therefore, the user is usually impossible to pick the genuine positive among huge negatives in the interaction stage. On the other hand, if they can pick the positive, the re-identification process is done, but the method degrades into manual browsing.

Almost all the top exhibited results in the original ranking list are negatives. Intuitively, a user needs to refine their search merely by several sparse negative selections. However, the negatives that hold global similarity to the probe are still hard to be picked. Whereas, a local similar patch may be relatively easy to

Table 1. Proportion (%) of user selecting global similar negatives, local similar negatives and unlabeled in the returned results

		global	local	unlabeled
VIPeR[14]	top 10	4.27	95.61	0.12
	top 20	8.53	91.38	0.09
CUHK02[13]	top 10	3.11	96.74	0.15
	top 20	7.81	92.05	0.14

obtain. We made a preliminary experiment to explain this phenomenon in person re-id. Five volunteers are invited for the feedback behavioral study. They are asked to manually annotate global similar negatives and local similar negatives respectively in the top 10 and the top 20 returned results, which are ranked by a ranking model given a set of random probe images. It is evident from Table 1 that the proportion of global similar negatives is very low, which makes interaction with global similarity negatives a highly difficult mission.

Due to the above two underlying characteristics, that positives and global similar negatives are sparse and hard to select, few works in person re-id which directly utilize positives or regard the global similar negatives as pseudo-positives to achieve the interactive selections [1][12], seem difficult to execute.

However, Table 1 also shows that the proportion of global similar negatives and local similar negatives are extremely imbalanced with the local similar negatives outnumbers the global similar negatives significantly.

Abundant Local Similar Negatives - Motivated by the abundant local similar negatives, if defining the basic unit of relevance feedback as local regions rather than the whole image, we can obtain multiple interactive selections for ranking optimization. Different with conventional relevance feedback methods, this paper proposes a region-based interactive ranking optimization method, based on those locally similar negatives, which are more suitable for human visual perception and judgement.

This paper considers both local neighbor similarity and local cluster similarity among labeled and unlabeled images. Section 4 shows that the method not only improves nearly two times of search efficiency compared to the typical exhaustive search strategy, but also brings about as much as over 20% performance improvement on average over the basic method, even with a poor initial ranking result.

3 Region-Based Interactive Ranking Optimization

For the convenience of following discussion, we consider a probe person image p and a gallery set $G = \{g_i | i = 1, ..., N\}$, where N is the size of the gallery and i is the index of an image. Through computing their similarity scores $S(p, g_i)$ by feature-based or distance-based methods, an initial query result can be obtained. Ranking the similarity scores from high to low, the R most similar gallery images are exhibited to the user for feedback.

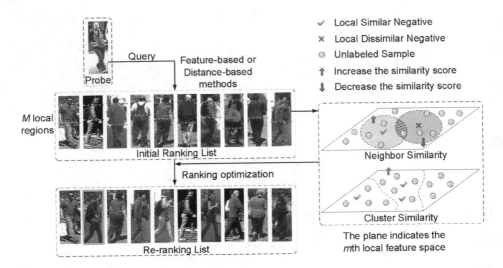

Fig. 1. An overview of the proposed interactive ranking optimization method for person re-identification

Considering the structural consistence of person appearance, where each corresponding region has relatively fixed position and proportion in the human body, we divide the whole person image into M horizontal local regions, and utilize $S^m(p^m, g_i^m)$ to stand for the similarity between mth local region of image p and that of image g_i, where $m \in \{1, ..., M\}$.

In interaction stage, the user picks X pairs of local regions from the exhibited images with their judgements, as showed in Fig.1. And two kinds of negative labels are marked, including local similar negative which means that the region of the labeled sample looks like that of the target, and local dissimilar negative which means the indicated region does not look like that of the target. Let $O^x = \{g_s^{(x,m)}, g_d^{(x,m)}\}$ denote a pair of samples respectively marked similar and dissimilar by the user, where the symbol s and d indicate image indexes, m indicates the similar or dissimilar region of the image, $x \in \{1, ..., X\}$ indicates the xth selection. Here, (x, m) indicates that x and m are corresponding with each other. For each selection O^x, the selected region of the images should be the m region.

Through exploiting these marked samples, we calculate the neighbor similarity score and the cluster similarity score for every gallery image. According to these two scores, we adjust initial similarity score $S(p, g_i)$ to a new similarity score $S_*(p, g_i)$. And then a new ranking list can be acquired.

3.1   Neighbor Similarity

We firstly use pairs of local similar negative and local dissimilar negative to refine the ranking results. Intuitively, an image would be in a greater probability to be the positive, if its corresponding local region is close to that of the indicated local similar negative in the local feature space, meanwhile away from that of the indicated local dissimilar negative. Whereas, it would be in a smaller probability to be the positive, if close to the local dissimilar negative and away from local similar negative.

That is to say, the similarity score should be increased for the former, but decreased for the latter. And we name this score neighbor similarity. As presented in neighbor similarity part of the Fig.1, the samples within the green neighbor region of the local similar negative would get a higher score, within the red neighbor region of the local dissimilar negative would get a lower score, while the cross region would not change. We calculate the neighbor similarity score of the image by equation (1), which is the cumulative sum of each selection O^x.

$$
\begin{aligned}
Score_{ns}(g_i) &= \sum_{x=1}^{X} Score_{ns}^x(g_i) \\
&= \sum_{x=1}^{X} (S^m(g_i^m, g_s^{(x,m)}) - S^m(g_i^m, g_d^{(x,m)})) I_{ns}(g_i, O^x)
\end{aligned}
\tag{1}
$$

Here, $I_{ns}(g_i, O^x)$ is the criteria for the adjustment scope of the neighbor region with the parameter δ, and is discriminated by the equation (2).

$$
I_{ns}(g_i, O^x) = \begin{cases} 1, & |S^m(g_i^m, g_s^{(x,m)}) - S^m(g_i^m, g_d^{(x,m)})| > \delta \\ 0, & others \end{cases}
\tag{2}
$$

3.2   Cluster Similarity

Because our feedback selection depends on the similarity other than the genuine positives, this may bring in the selection discrepancy (Fig 2 left), which leads to unsatisfactory result. In addition, due to the effects of various conditions, the appearance of the target may have some changes beyond the user's imagination, which means that the variation discrepancy (Fig 2 right) may also affect the adjustment results.

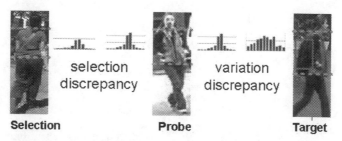

Fig. 2. Illustration for the selection discrepancy and the variation discrepancy. For the left, although the selection image looks like the probe, the feature differences still exist, leading to the selection discrepancy. For the right, a bag exists on the target's shoulder as indicated in the red box, but it cannot be seen in the probe image, which brings the variation error.

In order to smooth the errors of feedback selections, we use cluster similarity to make further adjustment on the similarity score. In a labeled local feature space, if majority of the labeled locally similar negatives are in the same cluster, the cluster would be more proper as the adjustment direction, and the images in this cluster should increase their similarity score. For every m local region, all the images in the local feature space are divided into multiple clusters by K-centers off-line, and the number of clusters is C. As presented in cluster similarity part of the Fig.1, the samples in the same cluster with two different local similar negatives would get a higher score.

For the mth local region which has been labeled by user, we find the cluster c^* by equation (4). We define $g_{c^*}^m$ as the cluster center of cluster c^*.

$$I_k(g_s^{(x,m)}, c) = \begin{cases} 1, & g_s^{(x,m)} \ is \ in \ the \ cluster \ c \\ 0, & else \end{cases} \tag{3}$$

$$c^* = \arg\max_c \sum_{x=1}^{X} I_k(g_s^{(x,m)}, c) \tag{4}$$

Then, we calculate the cluster similarity score of the image by equation (5).

$$Score_{cs}(g_i) = \sum_{m=1}^{M} S^m(g_i^m, g_{c^*}^m) I_{cs}(g_i^m, g_{c^*}^m) \tag{5}$$

$$I_{cs}(g_i^m, g_{c^*}^m) = \begin{cases} 1, & g_i^m \ and \ g_{c^*}^m \ in \ the \ same \ cluster \\ 0, & else \end{cases} \tag{6}$$

Based on the above neighbor similarity and cluster similarity, we adjust the similarity score by the following equation (7). Here, the parameter α tries to achieve a balance between the neighbor similarity score and cluster similarity score.

$$S_*(p, g_i) = S(p, g_i) + \alpha Score_{ns}(g_i) + (1 - \alpha) Score_{cs}(g_i) \tag{7}$$
$$s.t. \quad 0 < \alpha < 1$$

Therefore, every image in the gallery would get a new score. According to these scores, we can get a new ranking list.

4   Experiments

Data sets - The proposed approach is validated on two publicly available datasets, the VIPeR dataset [14], the CUHK02 dataset [13]. These data sets cover a wide range of problems faced in the real world person re-identification applications, e.g. viewpoint, pose, and lighting changes. The widely used VIPeR dataset contains 1,264 outdoor images obtained from two views of 632 persons. Some example images are shown in Fig.3(a). CUHK02 dataset is a larger dataset recently proposed by Wang et al. [13] and contains 971 identities from two disjoint camera views. Some example images are shown in Fig.3(b). The evaluation

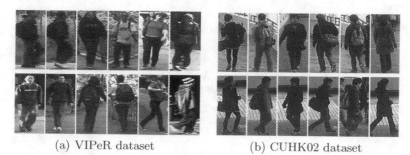

(a) VIPeR dataset   (b) CUHK02 dataset

Fig. 3. Some typical samples of three public datasets. Each column shows two images of the same person from two different cameras.

criteria uses the cumulative match characteristic (CMC) curve [15], which describes the the expectation of finding the true match within the first r ranks.

Evaluating on VIPeR and CUHK02 data set - As other works [2][6] did, we divided the global person image into the head, the torso and the leg parts [2], which meant that $M = 3$. For VIPeR, 316 persons were randomly selected for test, and for CUHK02, 200 persons were randomly selected. The features were described by HSV color feature and LBP texture feature. L2 distance measure was used to generate initial ranking list. We normalized the similarity, neighbor similarity score and cluster similarity score, and initialized parameters $\alpha = 0.5, \delta = 0.2$. In interaction process, 5 users were invited to label several pairs of local similar negative and local dissimilar negative for reducing the subjectivity of individual. Two iterations were carried out in this experiment. And the average CMC of above procedures was adopted in every iteration. The obtained results are shown in Fig.4.

As can be seen, our approach has nearly 15-23% improvement on VIPeR compared with the initial ranking result after feedback once. And after two iterations, our approach can get nearly 19-31% improvement. For CUHK02, our approach has nearly 9-29% improvement compared with the initial ranking result after feedback once. And after two iterations, our approach can get nearly 11-34% improvement.

Comparison with other methods. We constructed an automatic re-ranking method [9] used in person re-id. Based on the idea of the Ali et al. [1] method, we also constructed a relevance feedback method using pseudo-positives and negatives. Comparing with these two methods, the result is shown in Fig.5.

As can be seen, our method outperforms the other two optimization methods over the whole range of ranks. It can prove that interactive ranking optimization may be more efficient than auto re-ranking, and that directly regarding the global similar negatives as the pseudo-positives may be an unsuitable choice, especially in person re-id.

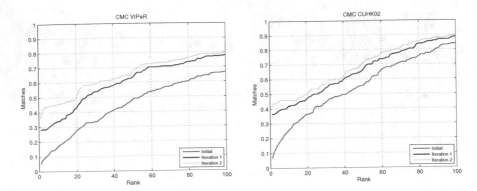

Fig. 4. Our approach on the two data sets

Table 2. Person re-id matching rates(%) at rank 1 on the VIPeR data set

	Method	rank@1
feature-based	SDALF[2]	20
	PartsSC[6]	16
	AIR[5]	17
	SDC[7]	26
distance-based	KISSME[8]	20
	PRDC[3]	16
Our method	L2	6
	L2+RIRO	**28**

We also compared our method with the-state-of-the-art person re-id methods on widely used VIPeR dataset with 316 persons as training/testing set size. For a fair comparison, the results for these methods are directly taken from the original public papers. In these methods, SDALF, PartsSC, AIR and SDC belong to feature-based methods, and PRDC and KISSME are the distance-based methods. Table 2 shows that even with the simplest color feature and L2 distance, our method (RIRO) can get a better performance with once iteration.

Runtime and search time - We test the runtime of the process after feedback selection. The CPU of our computer includes dual core 2.80 GHz and 2 GB RAM. Our region-based ranking optimization method was tested on the VIPeR and CUHK02 data sets. The size of gallery was the same as above tests. This test showed that it cost $20ms$ for VIPeR, and takes $10ms$ for CUHK02. And the search time of our method is reduced by nearly 2 times on average as compared to conventional exhaustive search by basic ranking and human labor. That is from 58.5 ± 27.5 seconds to 28.5 ± 10.0 seconds. These results prove that the proposed method is suitable for real-time-requiring applications, and suggest that it is able to significantly improve the search efficiency.

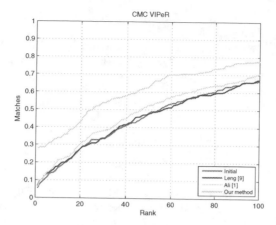

Fig. 5. Performance comparison using CMC curves on VIPeR data set

5 Conclusion

In this work, we present a creatively systematic interactive ranking optimization framework for person re-id, different from recent mainstream methods. The core idea is: concerning the specificity in person re-id work, we only utilize the negatives and introduce local feedback idea. The method is proved to be more effective than the auto re-ranking method and feedback using integral global information. The algorithm is very simple and the effect promotes significantly. These make the proposed region-based interactive ranking optimization method suitable for the practical application.

Acknowledgement. The research was supported by the National Nature Science Foundation of China (61303114, 61231015, 61170023), the Specialized Research Fund for the Doctoral Program of Higher Education (20130141120024), the Technology Research Project of Ministry of Public Security (2014JSYJA016), the Fundamental Research Funds for the Central Universities (2042014kf0250), the China Postdoctoral Science Foundation funded project (2013M530350), the major Science and Technology Innovation Plan of Hubei Province (2013AAA020), the Key Technology R&D Program of Wuhan (2013030409020109), the Guangdong-Hongkong Key Domain Break-through Project of China (2012A090200007), and the Special Project on the Integration of Industry, Education and Research of Guangdong Province (2011B090400601).

References

1. Ali, S., Javed, O., Haering, N., Kanade, T.: Interactive retrieval of targets for wide area surveillance. In: ACM International Conference on Multimedia (ACM MM) (2010)

2. Farenzena, M., Bazzani, L., Perina, A., Murino, V., Cristani, M.: Person re-identification by symmetry-driven accumulation of local features. In: IEEE Conference on Computer Vision and Pattern Recognition (CVPR) (2010)
3. Zheng, W.-S., Gong, S., Xiang, T.: Person re-identification by probabilistic relative distance comparison. In: IEEE Conference on Computer Vision and Pattern Recognition (CVPR) (2011)
4. Li, X., Tao, D., Jin, L., Wang, Y., Yuan, Y.: Person re-identification by regularized smoothing kiss metric learning. IEEE Transactions on Circuits and Systems for Video Technology (TCSVT) (2013)
5. Layne, R., Hospedales, T.M., Gong, S.: Person Re-identification by Attributes. In: British Machine Vision Conference (BMVC) (2012)
6. Kviatkovsky, I., Adam, A., Rivlin, E.: Color invariants for person reidentification. IEEE Transactions on Pattern Analysis and Machine Intelligence (TPAMI) (2013)
7. Rui, Z., Wanli, O., Xiaogang, W.: Unsupervised salience learning for person re-identification. In: IEEE Conference on Computer Vision and Pattern Recognition (CVPR) (2013)
8. Kostinger, M., Hirzer, M., Wohlhart, P., Roth, P., Bischof, H.: Large scale metric learning from equivalence constraints. In: IEEE Conference on Computer Vision and Pattern Recognition (CVPR) (2012)
9. Leng, Q., Hu, R., Liang, C.: Bidirectional ranking for person re-identification. In: IEEE International Conference on Multimedia and Expo (ICME) (2013)
10. Hirzer, M., Beleznai, C., Roth, P.M., Bischof, H.: Person re-identification by descriptive and discriminative classification. In: Heyden, A., Kahl, F. (eds.) SCIA 2011. LNCS, vol. 6688, pp. 91–102. Springer, Heidelberg (2011)
11. Kherfi, M.L., Ziou, D.: Relevance feedback for CBIR: a new approach based on probabilistic feature weighting with positive and negative examples. IEEE Transactions on Image Processing (TIP) (2006)
12. Liu, C., Loy, C.C., Gong, S.: POP: Person re-identification post-rank optimisation. In: IEEE International Conference on Computer Vision (ICCV) (2013)
13. Li, W., Zhao, R., Wang, X.: Human reidentification with transferred metric learning. In: Lee, K.M., Matsushita, Y., Rehg, J.M., Hu, Z. (eds.) ACCV 2012, Part I. LNCS, vol. 7724, pp. 31–44. Springer, Heidelberg (2013)
14. Gray, S.B.D., Tao, H.: Evaluating appearance models for recognition, reacquisition, and tracking. In: IEEE International Workshop on Performance Evaluation of Tracking and Surveillance (PETS) (2007)
15. Wang, X., Doretto, G., Sebastian, T., Rittscher, J., Tu, P.: Shape and appearance context modeling. In: IEEE International Conference on Computer Vision (ICCV) (2007)
16. Parkash, A., Parikh, D.: Attributes for classifier feedback. In: Fitzgibbon, A., Lazebnik, S., Perona, P., Sato, Y., Schmid, C. (eds.) ECCV 2012, Part III. LNCS, vol. 7574, pp. 354–368. Springer, Heidelberg (2012)
17. BNauml, M., Fischer, M., Bernardin, K., Ekenel, H.K., Stiefelhagen, R.: Interactive person-retrieval in tv series and distributed surveillance video. In: ACM International Conference on Multimedia (ACM MM) (2010)
18. Fischer, M., Ekenel, H.K., Stiefelhagen, R.: Interactive person re-identification in tv series. In: CBMI (2010)

Adaptive Tag Selection for Image Annotation

Xixi He[1,2], Xirong Li[1,2,3,*], Gang Yang[1,2], Jieping Xu[1,2], and Qin Jin[1,2]

[1] Key Lab of DEKE, Renmin University of China, 100872 China
[2] Multimedia Computing Lab, Renmin University of China, 100872 China
[3] Shanghai Key Laboratory of Intelligent Information Processing, 200443 China
xirong.li@gmail.com

Abstract. Not all tags are relevant to an image, and the number of relevant tags is image-dependent. Although many methods have been proposed for image auto-annotation, the question of how to determine the number of tags to be selected per image remains open. The main challenge is that for a large tag vocabulary, there is often a lack of ground truth data for acquiring optimal cutoff thresholds per tag. In contrast to previous works that pre-specify the number of tags to be selected, we propose in this paper *adaptive tag selection*. The key insight is to divide the vocabulary into two disjoint subsets, namely a seen set consisting of tags having ground truth available for optimizing their thresholds and a novel set consisting of tags without any ground truth. Such a division allows us to estimate how many tags shall be selected from the novel set according to the tags that have been selected from the seen set. The effectiveness of the proposed method is justified by our participation in the ImageCLEF 2014 image annotation task. On a set of 2,065 test images with ground truth available for 207 tags, the benchmark evaluation shows that compared to the popular top-k strategy which obtains an F-score of 0.122, adaptive tag selection achieves a higher F-score of 0.223. Moreover, by treating the underlying image annotation system as a black box, the new method can be used as an easy plug-in to boost the performance of existing systems.

Keywords: Image annotation, adaptive tag selection, ImageCLEF evaluation.

1 Introduction

Annotating images by computers is crucial for accessing the many unlabeled images at a semantic level. Due to the semantic gap, i.e., the lack of correspondence between visual features extracted from the pictorial content and a user's interpretation of the content, image auto-annotation is challenging. Labeling arbitrary images on the Internet is even more difficult, as a relatively simple concept may exhibit significant diversity in its visual appearance. The imagery of a concept does not limit to realistic photographs, but can also be artificial correspondences such as posters, drawings, and cartoons, as exemplified in

---

* Corresponding author.

W.T. Ooi et al. (Eds.): PCM 2014, LNCS 8879, pp. 11–21, 2014.

car cloudless lake
plant reflection road
sky vehicle water

cloud sky snow sport

building car
cityscape nighttime
painting road vehicle

Fig. 1. Internet images and ground truth tags from the development set of the ImageCLEF 2014 image annotation task [5]. The fact that the number of relevant tags varies over images motivates us to study how to assign a proper number of tags for annotating unlabeled images.

Fig. 1. On the one hand, a large array of tags need to be modeled for depicting the diverse content of Internet image collections, while on the other hand, as not all tags are relevant to a specific image, we need to make binary assignments of the tags to that image.

Quite a few methods have been proposed for image annotation, either by building visual classifiers per tag [1, 2] or by propagating tags from visually similar images [3, 4]. Given a novel image and a tag vocabulary, these methods first compute each tag's relevance score with respect to the given image, and sort the tags in descending order by their scores. The top-k ranked tags are preserved as predicted annotations of the image. The choice of k reflects the trade-off between precision and recall. In previous works, a fixed value of k is used for all images, where $k = 5$ is a common choice [4]. Notice however that the number of relevant tags varies over images. Hence, it is not surprising that such a top-k strategy gives suboptimal results.

A good method for image annotation shall be able to adaptively determine which tags to be selected per image. Since choosing a proper k per image is difficult, one might consider a thresholding strategy that a specific tag is selected if its relevance score is larger than a given threshold. In [1], the thresholds are optimized by maximizing a combined metric of precision and recall, say F-score, on training data. Despite its good performance, optimizing thresholds per tag requires ground-truthed data, which indicates the relevance of an image with respect to a given tag, for all tags in consideration. Consequently, this strategy is inapplicable to novel tags which have no ground truth available.

This paper studies adaptive tag selection for image annotation. In particular, given a ranked list of tags produced by a specific image annotation system for a test image, we aim to answer the question of how to adaptively determine a proper number of tags for annotating the test image. To that end, an adaptive top-k tag selection method is proposed, which beats the standard top-k strategy with ease.

2   Related Work

Image annotation, as an important topic in the multimedia field, has been actively studied. A noticeable effort is to build scalable image annotation systems based on large-scale user-contributed data instead of limited-scale expert-labeled training data [3, 6–8]. While the number of tags that can be modeled is increasing, it remains unclear how to select a proper number of tags to annotate an unlabeled image. This problem is overlooked, because most of the existing works either assess the top-k ranked tags [3, 4] or the entire tag ranking list [6].

To make the number of selected tags adaptive, several works employ a thresholding strategy by keeping tags that have scores larger than specified thresholds. In [9], the authors use the sum of the average and the standard derivation of the scores w.r.t. a tag as its threshold. However, according to our observation, there is a lack of evidence supporting that the score of a relevant tag is indeed larger than this threshold. To find thresholds that are more related to the image annotation performance, the authors in [1] find a global threshold for all tags by cross-validation on ground-truthed data. Due to the diversity in the many tags and their corresponding models, it is unlikely that one threshold is suitable for all tags. On the other hand, obtaining optimal threshold for each tag is difficult as this would require full annotations. Hence, it is worthwhile to study adaptive tag selection given incomplete ground truth, and this has not been well explored in the literature.

3   Adaptive Tag Selection

As aforementioned, for a given tag vocabulary, optimizing the cutoff thresholds per tag is inapplicable to novel tags. Nevertheless, we can safely assume that we have access to a set of training images manually yet incompletely labeled using a subset of the vocabulary. Consequently, depending on whether a tag has a number of ground-truthed images available, the vocabulary can be divided into two disjoint subsets, i.e., the *seen set* consisting of tags with ground truth and the *novel set* consisting of tags without ground truth. We assume tags are uniformly assigned to the two sets, and consequently for a given image, its relevant tags have the same occurrence probability in the seen set and in the novel set. Hence, we can estimate the number of tags to be selected from the novel set according to the number of tags that have been selected from the seen set.

To describe the above idea more formally, we introduce some notation. Let x be an image, t be a tag, and $\mathcal{V}$ be a tag vocabulary. We use $\mathcal{V}_{seen}$ to denote the seen set, and $\mathcal{V}_{novel}$ for the novel set. Let $\mathcal{X}_{train}$ be a set of training images which are manually labeled using $\mathcal{V}_{seen}$ only. In order to select from $\mathcal{V}$ relevant tags to annotate x, we need an image tag relevance function $f(x, t)$ which computes the relevance score of t with respect to x. The popular top-k strategy annotates x by sorting $\mathcal{V}$ in descending order by $f(x, t)$ and selecting the top k ranked tags. In contrast to previous works which designate k in advance, we make k variable by selecting tags from $\mathcal{V}_{novel}$ based on the selection on $\mathcal{V}_{seen}$. Given a specific test image, let $\mathcal{A}$ be the tags that have been selected from $\mathcal{V}_{seen}$. Based on our hypothesis that relevant tags of the test image have the same occurrence probability in $\mathcal{V}_{seen}$ and $\mathcal{V}_{novel}$, we propose to estimate the number of tags to be selected from $\mathcal{V}_{novel}$ as

$$k_{novel} := |\mathcal{V}_{novel}| \cdot \frac{|\mathcal{A}|}{|\mathcal{V}_{seen}|}, \tag{1}$$

where $|\cdot|$ returns the set cardinality. Concerning $\mathcal{A}$, we obtain it by thresholding:

$$\mathcal{A} := \{t \in \mathcal{V}_{seen} | f(x, t) > \tau_t\}, \tag{2}$$

where τ_t is the corresponding threshold found by maximizing the tag's F-score on $\mathcal{X}_{train}$. Since $\mathcal{A}$ is image dependent, the proposed method will select a variable number of tags. The number of selected tags is $|\mathcal{A}| + k_{novel}$. In a rare case where $\mathcal{A}$ is empty, we switch back to the top-k strategy.

Further, as $\mathcal{A}$ is constructed based on the learned thresholds, we consider refining the relevance scores for $\mathcal{V}_{novel}$ by exploiting $\mathcal{A}$ as pseudo labels. Tags that are semantically close to $\mathcal{A}$ shall be strengthened, and in the meanwhile tags from $\mathcal{A}$ that are more reliable shall have more weights. We implement this thought by updating the relevance score of $t \in \mathcal{V}_{novel}$ as

$$f(x, t) \leftarrow w \cdot f(x, t) + (1 - w)\frac{1}{|\mathcal{A}|} \sum_{t' \in \mathcal{A}} sim(t, t') \cdot (\frac{f(x, t')}{\tau_{t'}} - 1), \tag{3}$$

where w is a weighting parameter, and $sim(t, t')$ measures semantic similarity between two tags. In Eq. (3), $f(x, t')$ is divided by $\tau_{t'}$ as an effect of scale normalization. We compute $sim(t, t')$ using the Flickr Context Similarity [10], which is based on the Normalized Google Distance [11], but with tag statistics acquired from Flickr image collections instead of Google indexed web pages.

Notice that the proposed tag selection method treats the underlying image annotation system as a black box. Hence, it can be easily used as a plug-in to boost the performance of existing methods.

4   Image Annotation System

As we aim for modeling many tags, we build an image annotation system with its classifiers trained purely on web data with no need of extra manual labeling. This property makes the system more scalable with respect to the number of tags compared to systems relying on manually labeled data. The main components of the system, namely visual features, training data, and image annotation models, are depicted as follows.

Visual Features. For each image, we extract a bag of visual words using the color descriptor software [12]. A precomputed codebook of size 4,000 is used to quantize densely sampled SIFT descriptors. To improve the spatial discriminativeness of the feature, we further consider 1x1+1x3 spatial pyramids. This results in a visual feature vector of 16,000 dimensions per image.

Training Data Acquisition. We leverage three sources of training data, all of which were acquired with manual annotation for free. The first set is a set of

250K images [13], collected by querying web image search engines. The second set contains one million images with user-click count, released by the MSR Bing [14]. The third set consists of four million user-tagged images from Flickr. As the training sets come from different sources with different (noisy) annotation information, we describe how to select positive training examples for a specific tag t from the individual sets.

For the 250K web images, as they were collected from three web image search engines, namely Google, Yahoo, and Bing, each image x can be described by a triplet $< q, r, s >$, where q represent a query tag, r is the rank of x in the search results of q returned by an specific search engine s. Because a given image might be retrieved by different queries or by the same query but with different search engines, it can be associated with multiple triplets, denoted as $< q_i, r_i, s_i >$, $i = 1, \ldots, l$, where l is the number of triplets. To estimate the relevance of x with respect to t, we propose to compute a search engine based score as

$$relevance_{search}(x,t) = \sum_{i=1}^{l} \delta(q_i, t) \frac{w(s_i)}{\sqrt{r_i}}, \tag{4}$$

where $\delta(q_i, t)$ returns 1 if q_i and t are the same, and 0 otherwise. The variable $w(s_i)$ indicates the weight of a specific search engine, which is empirically set to be 1, 0.5, and 0.5 for Google, Yahoo, and Bing, respectively.

For the user-clicked set, each image is associated with a textual query and the accumulated count of user clicks. A larger click count indicates that the image is more likely to be relevant to the query [14]. We thus match t with queries and use the corresponding click count as the relevance score.

For the Flickr set, we compute tag relevance scores using the semantic field method [8], which is computationally more efficient than visual based approaches [6]. Given an image with its user tags including t, the semantic field method estimates the relevance of t to the image by considering the semantic similarity between t and the other tags. We again use the Flickr Context Similarity to measure the tag-wise similarity.

For the given tag, we obtain its positive training examples from each of the three sets by sorting images in descending order by their relevance scores and preserve the top 1,000 ranked images.

Annotation Models. For each tag we instantiate its $f(x, t)$ by learning two-class SVM classifiers from the three training sets separately. As the training data is overwhelmed by negative examples, we train classifiers by the Negative Bootstrap algorithm [15]. Different from sampling negative examples at random, Negative Bootstrap iteratively selects negative examples which are most misclassified by present classifiers, and thus most relevant to improve classification. Per iteration, the algorithm randomly samples $10 \times 1,000 = 10,000$ examples to form a candidate set. An ensemble of classifiers obtained in the previous iterations are used to classify each candidate example. The top 1,000 most misclassified examples are selected and used together with the 1,000 positives to train a new classifier. For the consideration of efficiency, we use Fast intersection kernel SVMs

(FikSVM) [16]. For each of the three sets, we conduct Negative Bootstrap with 10 iterations, producing in total 3×10 FikSVMs per tag. These FikSVMs are further compressed into a single model such that the annotation time complexity depends only on the feature dimensionality.

We observe that models trained on the three sets are complementary to each other to some extent. We therefore combine the models in a linear late fusion manner. Our previous study shows that weights optimized by coordinate ascent consistently outperforms averaging [17]. So we continue this good practice, and learn the fusion weights by coordinate ascent on $\mathcal{X}_{train}$.

5   Experiments

5.1   Experimental Setup

To verify the effectiveness of the proposed method, we participated in the ImageCLEF 2014 image annotation task [5], a benchmark for developing scalable image annotation systems without using manually labeled training examples. The task asks the participated systems to annotate unlabeled test images using a vocabulary of 207 tags, see the Appendix. There are 2,065 test images manually labeled using the vocabulary. Notice that the ground truth of the test set is unavailable to the participants, so the result reported in this paper are from the official evaluation[1] and extra evaluation provided by the organizers on our request. A development set of 1,000 labeled images are provided for 107 tags, whilst no ground truth is given for the remaining 100 tags. This setting allows us to evaluate the viability of the proposed method.

Baselines. In addition to the common top-5 strategy, we compare with [9], which computes for each tag the average (μ) and the standard deviation (σ) of the scores, and selects the tag having a score above $\mu+\sigma$. Since the development set allows us to find optimal thresholds for each tag in $\mathcal{V}_{seen}$, we also try to reconstruct the thresholds by linear combination of μ and σ with the tag-independent coefficients solved by least square fitting. The threshold of $t \in \mathcal{V}_{novel}$ is estimated by linearly combining μ and σ with the learned coefficients. We denote this strategy as $\mathrm{lsq}(\mu,\sigma)$. Notice that for a fair comparison, all methods are given the same tag rank lists produced by the system described in Section 4.

Performance Metrics. We report mean F-score (mF) and mean Average Precision (mAP) at the image level, which measures the quality of the selected tags and the quality of the entire tag ranking list, respectively.

5.2   Results

As shown in Table 1, the proposed method clearly outperforms the other methods for tag selection. In order to reveal if the gain is mainly contributed by selecting

---

[1] http://imageclef.org/2014/annotation/results. We ignore test images which do not have full ground truth with respect to the 207 tags, so our numbers differ from the original results.

Table 1. Performance of different methods for tag selection. As the methods are given the same tag rankings, they have the same mAP score of 0.151.

Method	mF-score
top-5	0.122
$\mu + \sigma$ [9]	0.127
$\mathrm{lsq}(\mu, \sigma)$	0.108
learned τ for $\mathcal{V}_{seen}$, $\mu + \sigma$ for $\mathcal{V}_{novel}$	0.153
learned τ for $\mathcal{V}_{seen}$, $\mathrm{lsq}(\mu, \sigma)$ for $\mathcal{V}_{novel}$	0.150
proposed method	**0.223**

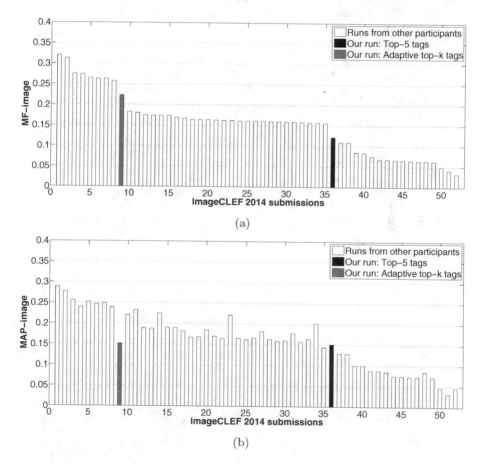

(a)

(b)

Fig. 2. Comparing with the ImageCLEF 2014 submissions. The submissions are sorted in descending order in terms of the mF scores, as shown in Fig. 2(a), with the corresponding mAP scores given in Fi.g 2(b). The inconsistency between the two ranks shows the importance of tag selection for making final annotations.

Table 2. Tagging results produced by our image annotation system. Tags uniquely selected by the adaptive tag selection method are shown in an *italic* font.

Test image	Predicted tags

Good results:

image-id: D1_8UEAb9SNy2krb

Top-5: tiger forest felidae mammal wolf

Adaptive top-k: tiger forest mammal *plant tree wild outdoor leaf mud grass rock branch daytime drink*

image-id: H77a832fe3nqwoHq

Top-5: cat felidae leopard wolf snow

Adaptive top-k: cat felidae leopard wolf snow *dog mammal tiger wild outdoor plant shadow*

Bad results:

image-id: x-tyO58pTX8y0hRZ

Top-5: banana broccoli avocado food pasta

Adaptive top-k: banana avocado *plant tree berry fruit daytime drink rock leaf outdoor cloudless orange bus*

image-id: zuglKl_fHHeRoVW2

Top-5: phone logo bottle guitar pan

Adaptive top-k: phone bottle *hat drink violin indoor knife spoon shadow orange apple person human grape outdoor pear fruit bus strawberry rock leaf*

tags from $\mathcal{V}_{seen}$ using the learned thresholds, for both $\mu + \sigma$ and $\text{lsq}(\mu, \sigma)$ methods, we use the learned thresholds as an alternative to the predicted thresholds for $\mathcal{V}_{seen}$. Though their mF-score increases as shown in the fifth and sixth rows in Table 1, the proposed method maintains the leading position.

Fig. 2 shows the performance of our system in the context of all submissions by the 11 teams participated in the ImageCLEF 2014 task. Even though our submission is ranked 34 out of the 52 submissions in terms of mAP, adaptive tag selection brings us to the 9th position in terms of mF. This result shows the importance of top-k tag selection and the power of the proposed method for top-k tag selection.

For a more intuitive understanding, we present several machine tagging results in Table 2.

6   Conclusions

This paper introduces adaptive tag selection, a new function to enable an image annotation system to label images with a variable number of tags, rather than a fixed number of tags as commonly done in previous works. On the base of our experiments in the ImageCLEF 2014 image annotation task, we offer the following conclusions. Adaptive tag selection is important when performing binary assignments of tags to individual images. Given the same image annotation system, with and without adaptive tag selection make a clear difference in annotation performance measured in terms of F-score. The proposed method is found to be effective for adaptive tag selection.

Acknowledgments. The authors are grateful to the ImageCLEF coordinators especially dr. Mauricio Villegas for helping evaluate our results. This research was supported by the Fundamental Research Funds for the Central Universities and the Research Funds of Renmin University of China (No. 14XNLQ01), NSFC (No. 61303184), SRFDP (No. 20130004120006), BJNSF (No. 4142029), SRF for ROCS, SEM, and Shanghai Key Laboratory of Intelligent Information Processing, China (Grant No. IIPL-2014-002).

Appendix

ImageCLEF 2014 annotation vocabulary. The vocabulary consists of 107 dev tags for which we have access to a ground truth set of 1,000 images, and 100 novel tags with no ground truth available.

The 107 dev tags as $\mathcal{V}_{seen}$: aerial airplane baby beach bicycle bird boat book bottle bridge building bus car cartoon castle cat chair child church cityscape closeup cloud cloudless coast countryside daytime desert diagram dog drink drum elder embroidery female fire firework fish flower fog food footwear forest furniture garden grass guitar harbor hat helicopter highway horse indoor instrument lake lightning logo male monument moon motorcycle mountain newspaper nighttime outdoor overcast painting park person phone plant portrait poster protest rain rainbow reflection river road sand sculpture sea shadow sign silhouette sky smoke snow soil space spectacle sport sun sunset table teenager toy traffic train tree tricycle truck underwater unpaved vehicle violin wagon water

The 100 novel tags as $\mathcal{V}_{novel}$: antelope apple arthropod asparagus avocado banana bear berry blood branch bread broccoli buffalo butterfly camel canidae captive carrot cauliflower cervidae cheese cheetah chimpanzee corn crocodile cucumber donkey egg eggplant elephant equidae felidae flamingo fox fried fruit galaxy giraffe gorilla grape hippopotamus human hunting kangaroo knife koala leaf leopard lettuce lion mammal marsupial meat monkey mud mushroom nebula onion orange ostrich pan pasta pear penguin pig pineapple pinniped pool potato pumpkin rabbit raccoon reptile rhino rice rifle roasted rock rodent sausage soup spider spoon squirrel strawberry submarine tiger tomato trunk tuber turtle vegetable walrus warthog watermelon wild wolf yam zebra zoo

Performance Metrics

F-image. Given a test image x, its relevant tag set R_x, and a predicted tag set P_x, its F-image score is computed as

$$\text{F-image}(x) = \frac{2 * \text{precision}(x) * \text{recall}(x)}{\text{precision}(x) + \text{recall}(x)}, \tag{5}$$

where precision(x) is $|R_x \cap P_x| / |P_x|$, and recall(x) is $|R_x \cap P_x| / |R_x|$. Consequently, MF-image is obtained by averaging F-image scores of all test images.

AP-image. Given a test image x with m tags sorted in descending order by predicted scores, its AP-image score is computed as

$$\text{AP-image}(x) = \frac{1}{|R_x|} \sum_{i=1}^{m} \frac{r_i}{i} \delta(i), \tag{6}$$

where r_i is the number of relevant tags among the top i tags, and $\delta(i)$ is 1 if the i-th tag is in R_x, 0 otherwise. MAP-image is obtained by averaging AP-image scores of all test images.

References

1. van de Sande, K., Snoek, C.: The University of Amsterdam's concept detection system at imageclef 2011. In: CLEF (2011)
2. Li, X., Snoek, C.: Classifying tag relevance with relevant positive and negative examples. In: ACM MM (2013)
3. Wang, X.J., Zhang, L., Ma, W.Y.: Duplicate-search-based image annotation using web-scale data. Proceedings of the IEEE (2012)
4. Ballan, L., Uricchio, T., Seidenari, L., Del Bimbo, A.: A cross-media model for automatic image annotation. In: ICMR (2014)
5. Villegas, M., Paredes, R.: Overview of the ImageCLEF 2014 scalable concept image annotation task. In: CLEF (2014)
6. Li, X., Snoek, C., Worring, M.: Learning social tag relevance by neighbor voting. TMM 11(7), 1310–1322 (2009)
7. Tang, J., Hong, R., Yan, S., Chua, T.S.: Image annotation by knn-sparse graph-based label propagation over noisily-tagged web images. In: TIST (2010)
8. Zhu, S., Jiang, Y.G., Ngo, C.W.: Sampling and ontologically pooling web images for visual concept learning. TMM 14(4), 1068–1078 (2012)
9. Borgne, H.L., Popescu, A., Znaidia, A.: CEA LIST@imageclef 2013: Scalable concept image annotation. In: CLEF Working Notes (2013)
10. Jiang, Y.G., Ngo, C.W., Chang, S.F.: Semantic context transfer across heterogeneous sources for domain adaptive video search. In: ACMMM (2009)
11. Cilibrasi, R., Vitanyi, P.: The Google similarity distance. TKDE 19(3), 370–383 (2004)
12. van de Sande, K., Gevers, T., Snoek, C.: Evaluating color descriptors for object and scene recognition. TPAMI 32, 1582–1596 (2010)
13. Villegas, M., Paredes, R., Thomee, B.: Overview of the imageclef 2013 scalable concept image annotation subtask. In: CLEF (2013)

14. Hua, X.S., Yang, L., Wang, J., Wang, J., Ye, M., Wang, K., Rui, Y., Li, J.: Clickage: Towards bridging semantic and intent gaps via mining click logs of search engines. In: ACMMM (2013)
15. Li, X., Snoek, C., Worring, M., Koelma, D., Smeulders, A.: Bootstrapping visual categorization with relevant negatives. TMM 15(4), 933–945 (2013)
16. Maji, S., Berg, A., Malik, J.: Classification using intersection kernel support vector machines is efficient. In: CVPR (2008)
17. Li, X., Liao, S., Liu, B., Yang, G., Jin, Q., Xu, J., Du, X.: Renmin University of China at ImageCLEF 2013 scalable concept image annotation. In: CLEF Working Notes (2013)

Twitter Food Photo Mining and Analysis
for One Hundred Kinds of Foods

Keiji Yanai and Yoshiyuki Kawano

Department of Informatics, The University of Electro-Communications
1-5-1 Chofugaoka, Chofu-shi, Tokyo 182-8585 Japan
{yanai,kawano-y}@mm.inf.uec.ac.jp

Abstract. So many people post photos as well as short messages to Twitter every minutes from everywhere on the earth. By monitoring the Twitter stream, we can obtain various kinds of images with texts. In this paper, as a case study of Twitter image mining for specific kinds of photos, we describe food photo mining from the Twitter stream. To collect food photos from Twitter, we monitor the Twitter stream to find the tweets containing both food-related keywords and photos, and apply a "foodness" classifier and 100-class food classifiers to them to verify whether they represent foods or not after downloading the corresponding photos. In the paper, we report the experimental results of our food photo mining for the Twitter photo data we have collected for two years and four months. As results, we detected about 470,000 food photos from Twitter. With this data, we made spatio-temporal analysis on food photos.

Keywords: photo tweet, Twitter mining, food photos, food classifier, UEC-FOOD100.

1 Introduction

Twitter is a unique microblog, which is different from conventional social media in terms of its quickness and on-the-spot-ness. Many Twitter's users send messages, which is commonly called "tweets", to Twitter on the spot with mobile phones or smart phones, and some of them send photos and geotags as well as tweets. Most of the photos are sent to Twitter soon after taken. In case of food photos, most of them are taken just before eating on the spot.

In this paper, we focus on food photos embedded in Tweets as a case study on a large-scale Twitter photo analysis. Food is one of frequent topics in Tweets with photos. In fact, we can see many food photos in lunch and dinner time in the Twitter stream.

Then, in this paper, by combining keyword-based search and food image recognition, we mine food photos from the Twitter stream. To collect food photos from Twitter, we monitor the Twitter stream to find the tweets containing both food-related keywords and photos, and apply a "foodness" classifier and 100-class food classifiers to them to verify whether they shows foods or not after downloading the corresponding photos. We used the state-of-the-art Fisher Vector coding with HoG and Color patches for food classifiers which is slightly modified with the rapid food recognition method for mobile environments proposed in [6], and trained them with the UEC-FOOD100 dataset [9][1]

---

[1] http://foodcam.mobi/dataset/

W.T. Ooi et al. (Eds.): PCM 2014, LNCS 8879, pp. 22–32, 2014.
© Springer International Publishing Switzerland 2014

which consists of 100 kinds of foods commonly eaten in Japan. Since we employ the improved method of the real-time mobile recognition, it takes only 0.024 seconds to recognize one image and it achieved about 83% classification rate within the top five candidates.

In the experiments, we report the results of our food photo mining on 100 kinds of foods in the UEC-FOOD100 dataset from the photo tweet log data we have collected for two years and four months. As results, we detected about 470,000 food photos from Twitter with about 99% accuracy. With this data, we have made spatio-temporal analysis on food photos. In addition, we have implemented the real-time food photo detection system from the Twitter stream.

2 Related Works

In this section, we mention about some representative works on Twitter photo mining.

As a representative work on Twitter mining, Sakaki et al. [12] regarded Twitter users as social sensors which monitor and report the current status of the places where the users are. They proposed a system which estimates the location of natural events such as typhoons and earthquakes. They used geotagged tweets to estimate event locations but no photos attached to tweets.

As early works on tweet photos, Yanai proposed World Seer [13] which can visualize geotagged photo tweets on the online map in the real-time way by monitoring the Twitter stream. Nakaji et al. [10] proposed a system to mine representative photos related to the given keyword or term from a large number of geo-tweet photos. They extracted representative photos related to events such as "typhoon" and "New Year's Day", and successfully compared them in terms of the difference on places and time. However, their system needs to be given event keywords or event term by hand. Kaneko et al. [5] extended it by adding event keyword detection to the visual Tweet mining system. As results, they detected many photos related to seasonal events such as festivals and Christmas as well as natural phenomena such as snow and Typhoon including extraordinary beautiful sunset photos taken around Seattle. All of these works focused on only geotagged tweet photos.

On the other hand, Chen et al. [2] treated with photo tweets regardless of geo-information. They analyzed relation between tweet images and messages, and defined the photo tweet which has strong relation between its text message and its photo content as a "visual" tweet. In the paper, they proposed the method which is based on the LDA topic model to classify "visual" and "non-visual" tweets. However, because their method was generic and assumed no specific targets, the classification rate was only 70.5% in spite of two-class classification. Because we focus on only food photos unlike their work, we use specialized object classifiers and have achieved very high classification accuracy. By using the method for real-time mobile food recognition, we can apply it to more than one million and seven hundred thousand tweet images and implement a real-time food Tweet photo detection system.

3 Overview

In this section, we describe an overview of the proposed method to mine food photos from the stored Twitter logs as well as the Twitter stream. We employ the following three-step processing.

(1) We perform keyword-based search with the names of target foods over a set of photo Tweets.

(2) We apply a newly-proposed "foodness" classifier to the tweet photos selected by the keyword-based search for classifying then into either of "food" or "non-food" photo.

(3) We apply individual food classifiers corresponding to the food names. In the experiments, we prepared multi-class discriminative classifiers trained by SVM with the UEC-FOOD100 dataset in the one-vs-rest manner.

The food classifiers employed in the third step is a slight modification of the method for mobile food recognition proposed in [6], while the foodness classifier is newly proposed for removing non-food photos.

4 Detail of the Proposed Method

4.1 Keyword-Based Photo Tweet Selection

In the first step, we select photo tweets by keyword-based search with the names of the target foods. We search tweet message texts for the words of the target food names.

As the target foods, we used 100 kinds of foods in the UEC-FOOD100 dataset in the experiments. Because the UEC-FOOD100 dataset includes common foods in Japan such as ramen noodle, curry, and sushi, we searched only photo tweets the message texts of which are written in Japanese language. We can easily select them by checking the language attribute of each tweet obtained from the Twitter Streaming API.

4.2 Foodness Classifier

We construct a "Foodness" Classifier (FC) for discriminating food images from non-food images. FC evaluates if the given image is a food photo or not. We use FC to remove noise images from the images gathered from the tweet photos selected by the food names.

We construct a FC from the existing multi-class food image dataset. Firstly, we train linear SVMs [4] in the one-vs-rest strategy for each category of the existing multi-class food image dataset. As image features, we adopt HOG patches [3] and color patches in the same way as [6]. Although HOG patches are similar local features to SIFT [8], HoG can be extracted much faster than SIFT. Both descriptors are coded by Improved Fisher Vector (IFV) [11], and they are integrated in the late fusion manner. We perform multi-class image classification in the cross-validation using the trained liner SVMs, and we build a confusion matrix according to the classification results. In the experiments, we used 64 GMMs for IFV coding and two-level spatial pyramid [7], which is much improved from mobile food recognition [6] in terms of the feature dimension.

Table 1. 13 food groups and their member foods for the "foodness" classifier

type of food group	food categories
noodles	udon nooles, dipping noodles, ramen
yellow color	omlet, potage, steamed egg hotchpotch
soup	miso soup, pork miso soup, Japaneses tofu and vegetable chowder
fried	takoyaki, Japaneses-style pancake, fried noodle
deep fried	croquette, sirloin cutlet, fried chicken
salad	green salad, macaroni salad, macaroni salad
bread	sandwiches, raisin bread, roll bread
seafood	sashimi, sashimi bowl, sushi
rice	rice, pilaf, fried rice
fish	grilled salmon, grilled pacific saury, dried fish
boiled	seasoned beef with potatoes
and	simmered ganmodoki
seasoned	seasoned beef with potatoes
sauteed	sauteed vegetables, go-ya chanpuru, kinpira-style sauteed burdock
sauce	stew, curry, stir-fried shrimp in chili sauce

Secondly, we make some category groups based on confusion matrix of multi-class classification results. This is inspired by Bergamo et al.'s work [1]. They grouped a large number of categories into superordinate groups the member categories of which are confusing to each other recursively. In the same way, we perform confusion-matrix-based clustering for all the food categories. We intend to obtain superordinate categories such as meat, sandwiches, noodle and salad automatically. As results, in the experiments, we obtained 13 food groups as shown in Table 1.

To build a "foodness" classifier (FC), we train a linear SVM of each of the superordinate categories. The objective of FC is discriminating a food photo from a non-food photo, which is different from the objective of the third step for discriminating a specific food photo from other kinds of food photos. Therefore, abstracted superordinate categories are desirable to be trained, rather than training of all the food categories directly. The output value of FC is the maximum value of SVM output of all the superordinate food groups.

When training SVMs, we used all the images of the categories under the superordinate category as positive samples. For negative samples, we built a negative food image set in advance by gathering images using the Web image search engines with query keywords which are expected to related to noise images such as "street stall", "kitchen", "dinner party" and "restaurant" and excluding food photos by hand. All the images are represented by Fisher Vector of HoG patches and color patches. SVMs are trained in the late fusion manner with uniform weights.

In the second step, we apply FC for the selected tweet photos and remove non-food photos from the food photo candidates.

4.3   Specific Food Classifiers

In this step, we classify a given photo into one of the prepared food classes.

First, we extract HOG patches and Color patches in a dense grid sampling manner in the same way as the previous step. Then, we apply PCA to all the extracted local features, and encode them into Improved Fisher Vectors. The method to extract features is the same as the previous step including the parameter settings. Next, we evaluate linear classifiers in the one-vs-rest way by calculating dot-product FVs. Finally we output the top-N categories in terms of the descending order of evaluation scores of all the linear classifiers.

In the experiments, we regarded the given tweet photo as a photo of the corresponding food if the food names contained in the tweet messages are ranked in the top five categories by evaluation of 100-kind food classifiers. This is because the top-5 classification rate exceeds 83%, while the top-1 rate is still around 60%.

5   Experimental Results

In this section, we describe the detail of the 100-class food dataset, the results of food photo mining from the Twitter stream, and some analysis on Twitter food data.

5.1   100 Food Categories and Their Classifiers

In the experiments, as target foods, we used 100 foods in the UEC-FOOD100 [9][2], because we employ supervised food photo classification which requires training data to select the target foods in the third step. It contains more than 100 images per category, and all the food item in which are marked with bounding boxes. For training and evaluation, we used only the regions inside the given bounding boxes. The total number of food images in the dataset is 12,905. Figure 1 shows all the category names and their sample photos. As shown in the figure, the dataset consists of common foods in Japan. Then, we restricted tweets from which we mine food photo tweets to only the tweets with Japanese messages, as mentioned in Section 4.1.

In [6], they implemented a mobile food recognition system using the same dataset. Although basically we followed their method for individual food classification in the third step, we extended the parameter setting to improved accuracy. To say it concretely, we doubled the size of GMM for FV encoding from 32 to 64, and added two-level spatial pyramid. As a result, the total feature dimension are raised from 3072 to 35840, which boosted the classification performance evaluated by 5-fold cross-validation as shown in Figure 2. Regarding the processing time, it takes only 0.024 seconds per image to recognize on Core i7-3770K 3.50GHz with multi-threaded implementation optimized for a quad-core CPU.

5.2   Twitter Food Mining

In this subsection, we describe the experimental results on twitter food photo mining. We have been collecting photo tweet logs by monitoring the Twitter stream by using

---

[2] http://foodcam.mobi/dataset/

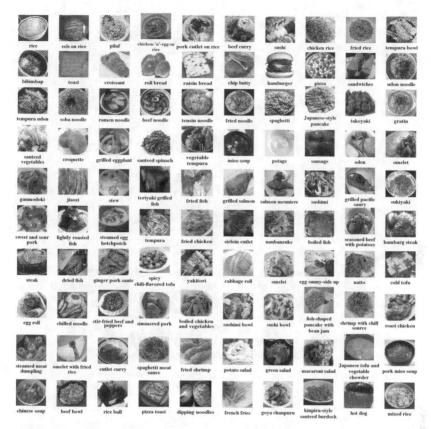

Fig. 1. 100 kinds of foods prepared in the UEC-FOOD100 dataset. See this figure with magnification in the PDF viewer.

Twitter Streaming API. Here, we used 122,328,337 photo tweets with Japanese messages out of 988,884,946 photo tweets over all the world collected from May 2011 to August 2013 for two years and four months.

From these photo tweets, we selected 1,730,441 photo tweets the messages of which include any of the name words of the 100 target foods in the first step of the proposed processing flow. Then, in the second step, we applied a "foodness" classifier (FC) to all the selected images. After applying FC, we applied 100-class one-vs-rest individual food classifiers. As a result, we obtained 470,335 photos which are judged as food photos corresponding to any of the 100 target food categories by our proposed processing pipeline.

For the 470,335 selected photos as food photos, we evaluate the number of selected photos for each category. Table 2 shows the ranking of 100 food categories in terms of the number of mined tweet food photos. The number of "Ramen noodle" and "curry" photos are the most and the second most with the large margin to the third or less ranked food categories, respectively. In fact, "ramen" and "curry" are regarded as the most popular foods in Japan. "Sushi", "dipping noodle (called as Tsukemen in Japanese)"

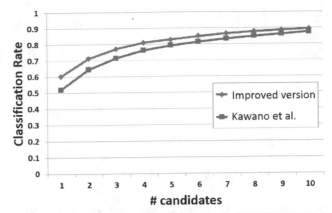

Fig. 2. Comparison on the top-k classification rates with the UEC-FOOD100 dataset evaluated by 5-fold cross validation between [6] and this paper.

Fig. 3. Examples of "omelet" photos. Most of them have drawings drawn by ketchup.

and "omelet with fried rice (called as Ome-rice in Japanese)" are also popular foods in Japan. The results of twitter food image mining reflects food preference of Japanese people. In addition, we found that many of "ome-rice" had drawings or letters drawn with ketchup, as shown in Figure 3. These are estimated to be made at home, while most of "ramen" and "sushi" photos are taken at food restaurants, because there are many ramen noodle and sushi restaurant in Japan. Although "hamburger" and "beef bowl" are also popular in Japan as fast food served at fast-food restaurants such as McDonald and Yoshino-ya, they are ranked at more than twentieth. This is because the foods provided by nation-wide fast-food chain restaurants such as McDonaldo are the same everywhere in the same chain restaurants, and they are not worth posting their photos to Twitter. On the other hand, since there are no monopolistic restaurant chains on ramen noodle and curry in Japan, the foods served at every ramen or curry restaurants have originality and are different from each other.

Next, we evaluated the precision rate of the selected food photos in the each steps regarding the top five foods and two sub-categories of "ramen noodle" and "curry". Table 3 shows the results in case of four types of the combinations of the three kinds of the selection methods, (1) only keywords, (1)+(2) keywords and foodness classifier (FC), (1)+(3) keywords and individual food classifier(IFC), and (1)+(2)+(3) keywords, FC and IFC. Note that this evaluation was done for the 300 random-sampled photos for each cell in the table.

Table 2. The ranking of 100 foods in terms of the number of mined tweet food photos

#	Food	Count	#	Food	Count	#	Food	Count
1	ramen noodle	80021	34	fish-shaped pancake with bean jam	3281	67	dried fish	563
2	curry	59264	35	pork cutlet on rice	3188	68	steamed meat dumpling	561
3	sushi	25898	36	omelet with grilled minced meat	2592	69	french fries	561
4	dipping noodle	22158	37	bibimbap	2368	70	beef ramen noodle	555
5	omelet with fried rice	17520	38	spaghetti	2171	71	sandwiches	551
6	pizza	16921	39	lightly roasted fish	2162	72	cold tofu	517
7	jiaozi	16014	40	seasoned beef with potatoes	2129	73	boiled chicken and vegetables	352
8	Japanese-style pancake	15234	41	natto	2094	74	sirloin cutlet	331
9	steamed rice	14264	42	spaghetti with meat source	1994	75	nanbanzuke	323
10	sashimi	13927	43	steamed egg hotchpotch	1843	76	fried chicken	314
11	hambarg steak	11583	44	egg sunny-side up	1635	77	stir-fried beef and peppers	312
12	beef stake	9503	45	croissant	1579	78	roll bread	288
13	takoyaki	9004	46	udon noodle	1500	79	roast chicken	263
14	fried rice	8383	47	simmered pork	1443	80	macaroni salad	239
15	fried noodle	7905	48	mixed sushi	1371	81	boiled fish	228
16	oden	7453	49	pork miso soup	1229	82	kinpira-style sauteed burdock	225
17	toast	6350	50	ginger-fried pork	1158	83	tempura udon	213
18	cutlet curry	6339	51	potato salad	1150	84	raisins bread	205
19	tempura	5905	52	egg omelet	1146	85	goya chanpuru	198
20	rice ball	5462	53	eels on rice	1071	86	green salad	145
21	gratin	5223	54	egg roll	1058	87	chinese soup	141
22	croquette	4837	55	sweet and sour pork	1049	88	Japanese tofu and vegetable chowder	137
23	stew	4797	56	fried shrimp	1049	89	salmon meuniere	96
24	sashimi bowl	4730	57	sauteed vegetables	1040	90	grilled pacific saury	84
25	chicken-'n'-egg on rice	4513	58	shrimp with chill source	1003	91	chip butty	76
26	tempura bowl	4464	59	cabbage roll	965	92	fried fish	72
27	beef bowl	4285	60	mixed rice	901	93	begitable tempura	71
28	spicy chili-flavored tofu	4081	61	pilaf	891	94	tensin noodle	69
29	yakitori	3829	62	soba noodle	880	95	ganmodoki	34
30	hamburger	3662	63	potage	816	96	grilled salmon	25
31	chilled noodle	3473	64	hot dog	795	97	sauteed spinach	12
32	sukiyaki	3408	65	chicken rice	736	98	teriyaki grilled fish	3
33	miso soup	3295	66	wiener sausage	577	99	grilled eggplant	2
						100	pizza toast	0

Regarding (1), the precision of two sub-categories, "beef ramen noodle" and "cutlet curry", are relatively higher, 94.3% and 92.7%, than "ramen noodle" and "curry". From this results, we can assume that when tweeting detailed food names with photos, the photos probably represent the corresponding foods. After applying both FC and IFC, (1)+(2)+(3), the precision of all the seven foods achieved the best compared to the cases of applying only single methods or only keyword-based search, (1), (1)+(2) and (1)+(3). Except for "sushi", the precision reached 99.0%, which means nearly perfect. This shows the effectiveness of introducing both FC and IFC after keyword-based search. Exceptionally, "sushi" is a difficult food to recognize by object recognition methods, because the appearances of "sushi" varies greatly depending on the kinds of the ingredients on the pieces of hand-rolled rice.

Finally, we describe simple spatio-temporal analysis on Twitter food photos. Figure 4 shows the prevailing-food map where the red marks, the yellow marks and the blue marks represent the areas where "ramen noodle", "curry" and "okonomiyaki" are most popular in terms of the number of food photo tweets, respectively. The left map, the center map, and the right map show the prevailing-food map on all the term (May 2011-Aug. 2013), Dec. 2012 (in winter), and Aug. 2013 (in summer), respectively. From the leftmost map, "ramen noodle" is the most popular over Japan on average through a year. However, compared between the center map and the rightmost map, popularity of "curry" increases in summer, while "ramen noodle" becomes the most popular

Table 3. The number of selected photos and their precision(%) with four different combinations

food category	(1)	(1)+(2)	(1)+(3)	(1)+(2)+(3)
ramen noodle	275652 (72.0%)	200173 (92.7%)	84189 (95.0%)	80021 (99.7%)
beef ramen noodle	861 (94.3%)	811 (99.0%)	558 (99.7%)	555 (99.7%)
curry	224685 (75.0%)	163047 (95.0%)	62824 (97.0%)	59264 (99.3%)
cutlet curry	10443 (92.7%)	9073 (98.0%)	6544 (98.7%)	6339 (99.3%)
sushi	86509 (69.0%)	43536 (86.0%)	48019 (72.3%)	25898 (92.7%)
dipping noodle	33165 (88.7%)	24896 (96.3%)	28846 (93.7%)	22158 (99.0%)
omelet with fried rice	34125 (90.0%)	28887 (96.3%)	18370 (98.0%)	17520 (99.0%)

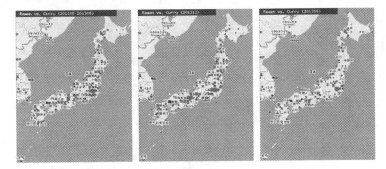

Fig. 4. The prevailing-food map of Japan. See the text.

in winter. Exceptionally, in the area around Hiroshima where the blue marks appear, "okonomiyaki" is always the prevailing food in Twitter food photos, this is partly because Hiroshima has a very popular regional food, "Hiroshima-yaki", which is a variant of "okonomiyaki".

As another temporal analysis on the mined food photos, we examined the time when each food are eaten the most frequently over a day. As results, the most frequent time when "ramen noodle" and "curry" are eaten is between 12pm and 2pm, while the most frequent time of "sushi" and "okonomiyaki" is between 7pm and 9pm. This reflects the difference of the characteristic of the foods. As shown in this subsection, the data we collected through Twitter food photo mining is useful for food habit analysis.

5.3   Real-Time Food Photo Detection System

We implemented a real-time Twitter food photo detection system which continuously detects 100 kinds of food photos from the Twitter stream. We detect the photo tweets including any of 100 kinds of Japanese food names about ten times per minute at most. Because the time to download a thumbnail image is about 2 or 3 seconds and the processing time for food recognition for each image is less than 0.1 seconds, we can process all the pipeline on a single machine in the real-time way. The very fast food recognition method which was originally designed for a mobile application made it possible.

As shown in Figure 5, the detected food photos are shown on the map if they have geotags or geo-related words such as place names in their Tweet messages, and on the right side the photos are displayed as the results by online k-means clustering. This system can be accessible via `http://mm.cs.uec.ac.jp/tw/`.

Fig. 5. Detected food photos are displayed on the map when geo-information are available

6   Conclusions

In this paper, we described food photo mining from the Twitter stream as a case study of specific categories of tweet photo mining. To do that, we proposed the three-step processing consisting of keyword-based selection with food category names, classification of food or non-food photos, and visual verification of the correspondence between the extracted food words and the food category of the tweet photo. In addition, we showed the collected data was useful for various kinds of analysis of foods.

Currently we always keep running the real-time food photo detection system and collecting new food photos. For example, we are collecting about 20,000 "ramen noodle" and 15,000 "curry" photos per month. When the number of both food images exceeds one million, we will release them as a large-scale food dataset for research purpose, which we expect enables fine-grained food recognition research.

References

1. Bergamo, A., Torresani, L.: Meta-class features for large-scale object categorization on a budget. In: Proc. of IEEE Computer Vision and Pattern Recognition (2012)
2. Chen, T., Lu, D., Kan, M.Y., Cui, P.: Understanding and classifying image tweets. In: Proc. of ACM International Conference Multimedia, pp. 781–784 (2013)

3. Dalal, N., Triggs, B.: Histograms of oriented gradients for human detection. In: Proc. of IEEE Computer Vision and Pattern Recognition (2005)
4. Fan, R.E., Chang, K.W., Hsieh, C.J., Wang, X.R., Lin, C.J.: LIBLINEAR: A library for large linear classification. The Journal of Machine Learning Research 9, 1871–1874 (2008)
5. Kaneko, T., Yanai, K.: Visual event mining from geo-tweet photos. In: Proc. of IEEE ICME Workshop on Social Multimedia Research (SMMR) (2013)
6. Kawano, Y., Yanai, K.: Rapid mobile object recognition using fisher vector. In: Proc. of Asian Conference on Pattern Recognition (2013)
7. Lazebnik, S., Schmid, C., Ponce, J.: Beyond bags of features: Spatial pyramid matching for recognizing natural scene categories. In: Proc. of IEEE Computer Vision and Pattern Recognition (2006)
8. Lowe, D.G.: Distinctive image features from scale-invariant keypoints. International Journal of Computer Vision 60(2), 91–110 (2004)
9. Matsuda, Y., Yanai, K.: Multiple-food recognition considering co-occurrence employing manifold ranking. In: Proc. of IAPR International Conference on Pattern Recognition (2012)
10. Nakaji, Y., Yanai, K.: Visualization of real world events with geotagged tweet photos. In: Proc. of IEEE ICME Workshop on Social Media Computing (SMC) (2012)
11. Perronnin, F., Sánchez, J., Mensink, T.: Improving the fisher kernel for large-scale image classification. In: Daniilidis, K., Maragos, P., Paragios, N. (eds.) ECCV 2010, Part IV. LNCS, vol. 6314, pp. 143–156. Springer, Heidelberg (2010)
12. Sakaki, T., Okazaki, M., Matsuo, Y.: Earthquake shakes twitter users: real-time event detection by social sensors. In: Proc. of the International World Wide Web Conference, pp. 851–860 (2010)
13. Yanai, K.: World seer: a realtime geo-tweet photo mapping system. In: Proc. of ACM International Conference on Multimedia Retrieval (2012)

Improving Color Constancy with Internet Photo Collections

Shuai Fang[1], Chuanpei Zhou[1], Yang Cao[2], and Zhengjun Zha[3]

[1] Hefei University of Technology
[2] University of Science and Technology of China
[3] Hefei Institute of Intelligent Machines, Chinese Academy of Sciences

Abstract. Color constancy is the ability to measure colors of objects independent of the color of the light source. A well-known color constancy method makes use of the specular edge to estimate the illuminant. However, separating specular edge from input image is under-constrained and existing methods require user assistance or handle only simple scenes. This paper presents an iterative weighted Specular-Edge color constancy scheme that leverages large database of images gathered from the web. Given an input image, we execute an efficient visual search to find the closest visual context from the database and use the visual context as an image-specific prior. This prior is then used to correct the chromaticity of the input image before illumination estimation. Thus, introducing the prior can provide a good initial guess for the successive iteration. In the next, a specular-map guided filter is used to improve the precision of specular edge weighting. Consequently, it can increase the accuracy of estimating the illuminant. To the end, we evaluate our scheme on standard databases and the results show that the proposed scheme outperforms state of the art color constancy methods.

Keywords: Visual context, color constancy, specular map, guilded filter.

1 Introduction

A computational approach to recover the actual color of objects is referred to as color constancy [1]. For machine vision, it is of importance for many multimedia applications, such as color based object recognition, image retrieval, image classification, and object tracking [1]. One of the core issues of color constancy is to build up an optimal illumination estimation model.

Approaches to illumination estimation can be divided into two groups depending on the estimation model they use [2]. For the first group, some predefined estimation models is used based on certain assumption on the illuminant. Typical approaches include Grey World (GW)[3], White-patch[4], Shades of Grey (SoG)[5], Grey Edge (GE)[6]. For the second group of approaches, the estimation model needs to be trained by using a subset of training data. Typical approaches include Color-by-Correlation (C-by-C)[7], Spatio-Spectral statistics-based method (Spatio-Spectral)[8] Net-works-based method(NN)[9], Support Vector Regression-base method

W.T. Ooi et al. (Eds.): PCM 2014, LNCS 8879, pp. 33–43, 2014.

(SVR)[10]. Both types of approaches have their respective advantages and limitations. The first group of approaches is much simple and the additional computational complexity is low. However, a limitation of these approaches is that determining the appropriate parameters for a specific image is under constraint. Many approaches rely on simple heuristics to constrain the parameters, but these heuristics can fail on many scenes. The second group of approaches has better universality. This, however, the training process increases the computational complexity. Moreover, if the color distribution of input image is not covered by the sampling distribution of training data, the model needs to be retrained using additional training data.

One highly novel color constancy methods is Photometric Edge Weighting proposed by A. Gijsenij [11]. This method is based on the observation that the specular edges have a significant influence on the performance of the illuminant estimation. However, its specular edges detection algorithms still contain other type of edges (shadow edge & material edge). One feasible solution is to use the specular-map to improve the setting of specular edge weight value.. And so far automatic techniques are limited to assume that incident light is pure-white, while the images captured in the real world require to be color corrected.

Image collections aggregate many images containing the similar visual context. Inspired by existing work on coherent intrinsic image [12], image collection can provide a rich source of information to compute color constancy. In this work, we exploit this information as a meaningful and image-specific prior, and then use the prior to correct the chromaticity of the input image. Furthermore, we apply the rectified image as an initial guess for the iteration process of the separation of specular-map. Moreover, an iterative optimization framework is further proposed to improve the accuracy of the separation of specular-map from input image.

2   Color Constancy

In this section, we will firstly introduce the image representation in dichromatic model, the traditional edge-based color constancy methods, and the separation method of specular and diffuse reflection. These methods provide a base of the work of this paper.

2.1   Dichromatic Reflection Model

The dichromatic reflection model states that reflection of object surfaces can be modeled as a linear combination of diffuse and specular reflections [13].

$$I_c(\mathbf{x}) = w_d(\mathbf{x}) \int_{\Omega} S_d(\lambda, \mathbf{x}) E(\lambda) q_c(\lambda) \, d\lambda + w_s(\mathbf{x}) \int_{\Omega} S_s(\mathbf{x}) E(\lambda) q_c(\lambda) \, d\lambda \qquad (1)$$

Where c means image channel $\{r, g, b\}$, $w_d(\mathbf{x}), w_s(\mathbf{x})$ are the geometrical parameters for diffuse and specular reflection, respectively. $S_d(\lambda, \mathbf{x})$ is the diffuse spectral reflectance function and $S_s(\mathbf{x})$ is the specular spectral reflectance function. $E(\lambda)$ is the illumination spectral function and λ is wave length. Here we assume that the illumination is constant throughout the whole scene, so $E(\lambda)$ is independent of the image coordinate position x. $q_c(\lambda)$ is a camera sensor function. Tan[1] defines image chromaticity $\sigma_c(x)$, diffuse chromaticity $\Lambda_c(x)$, and specular chromaticity (or illumination chromaticity) Γ_c as follows:

$$\sigma_c(x) = I_c(x) \Big/ \sum_i I_i(x), \quad \Lambda_c(x) = I_c^D(x) \Big/ \sum_i I_i^D(x), \quad \Gamma_c = I_c^S \Big/ \sum_i I_i^S \tag{2}$$

where $I_c^D(x) = \int_\Omega S_d(\lambda, x) E(\lambda) q_c(\lambda) d\lambda$ and $I_c^S = \int_\Omega E(\lambda) q_c(\lambda) d\lambda$.

For simplicity, equation (1) can be rewritten as follows:

$$I_c(x) = m_d(x)\Lambda_c(x) + m_s(x)\Gamma_c \tag{3}$$

where $m_d(x) = w_d(x) \sum_i I_i^D(x)$ represents the diffuse component[23], and $m_s(x) = w_s(x) S_s(x)(x) \sum_i I_i^S$ is the specular component[23].

2.2   Edge Based Color Constancy

Grey edge method[6] is one of state-of-the-art color constancy methods. This method is based on gray edge assumption that the average of the reflectance differences in a scene is achromatic. Grey edge can be represented by general Minkowski-norm:

$$\mathbf{e}^{n,p,\sigma} = \left(\int |\partial^n f_{c,\sigma}(x) / \partial x^n|^p \right)^{1/p} = k\mathbf{e}_c \tag{4}$$

where n is the order of the derivative and p is the Minkowski norm. k is the constant parameter. $f_{c,\sigma}(x)$ is defined as the convolution of the image with the derivative of a Gaussian filter with scale parameter σ [14]. Many different algorithms[15]can be generated with the change of parameters n, p, σ. For $n=0$, $p=1$, $\sigma=0$, the Eq. (4) is equal to the Grey-World assumption; for $n=0$, $p=\infty$, $\sigma=0$, it is equal to color constancy by White-patch; for $n=1$, it is equal to the 1stGrey-Edge assumption; for $n=2$, it becomes2ndGrey-Edge; and for $n>2$, it becomes high-order Grey-Edge method.

2.3   Separating Specular and Diffuse Reflection

Traditional strategies to separate specular and diffuse component require explicit color segmentation, which is impossible for complex scene. For such a reason, Tan et al. [16]propose a separating algorithm based solely on colors and chromaticity information, without any segmentation information. A specular free image can be obtained by intensity logarithmic differentiation.

2.3.1   Specular-to-Diffuse Mechanism

Specular-to-diffuse mechanism bases its techniques on chromaticity and intensity value of specular and diffuse pixels. The maximum chromaticity and maximum diffuse chromaticity are defined as follows:

$$c^{max}(x) = \max(I_r(x), I_g(x), I_b(x)) \bigg/ \sum I_i(x), \quad \Lambda^{max} = \max(\Lambda_r, \Lambda_g, \Lambda_b) \quad (5)$$

The total diffuse intensity of specular pixels is computed using:

$$\sum I_i^{diff} = I^{max}(3c^{max} - 1) \bigg/ c^{max}(3\Lambda^{max} - 1) \quad (6)$$

where $I^{max} = \max(I_r, I_g, I_b)$.

Specular-free image is a pseudo-code of diffuse components. It is geometrically identical to diffuse component of the input image. The difference of both is in their surface colors. Formally, the specular-free image can be described as:

$$\hat{I}_c(x) = m_d(x)\hat{\Lambda}_c(x) \quad (7)$$

where the arbitrary scalar value of maximum diffuse chromaticity ($\hat{\Lambda}$) is obtained from the smallest maximum chromaticity of the input image.

2.3.2   Intensity Logarithmic Differentiation

According to Eq. (3), given only one colored pixel, it is an ill posed problem to determine whether it is diffuse or specular pixel. Therefore, Tan et al. uses intensity logarithmic differentiation of two adjacent pixels instead. By computing the logarithmic differentiation between input image(according to Eq. (3)) and specular-free image (according to Eq. (7)), the two neighboring pixels can be determined as follows.

$$\Delta(x) = \mathrm{dlog}(I_c(x)) - \mathrm{dlog}(\hat{I}_c(x))$$

$$\Delta(x) = \begin{cases} = 0 : diffuse \\ \neq 0 : \text{specular or boundary} \end{cases} \quad (8)$$

3   Overview

The overview of our proposed approach is illustrated in Fig. 1. There are three main modules, searching visual context, chromaticity correction and illumination estimation. Given an input image, our approach firstly queries an image database to retrieve the k closest images to the input image using visual search. The results from the search are defined as the visual context for the input. From the visual context, we estimate the parameters of chromaticity correction to remove the distortion in the input image. In the next, the corrected image is used as the initial guess for illumination estimation. In the step of illumination estimation, the specular edge is firstly estimated based on STD separation and then a specular-map guided filter is applied to improve the weighting of specular edge. Finally, after the illumination is estimated, a modified diagonal transformation [15] is applied to obtain the color-corrected images.The above processes are iteratively continued until the expected color correction results are achieved.

The three modules are integrated in a cognitive loop. For each image, the chromaticity correction module receives initial visual context and the corrected image from the other two modules and feeds back the refined chromaticity rectified result. Thus, the modules exchange information that helps compensate for their individual disadvantages and improves overall system performance. The pipeline of our algorithm is detailed in the following sections.

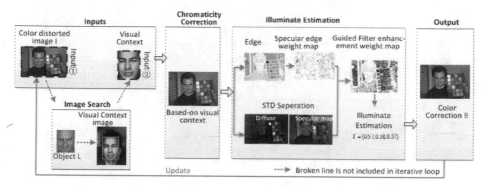

Fig. 1. The flowchart of our algorithm

4   Weighted Grey-Edge Color Constancy Based on Visual Context

4.1   Visual Context

For the object with obvious color feature in image, we search similar images taken under approximately standard illumination. The image is used as a prior visual context to improve the illumination estimation. In order to obtain appropriate visual context, the Visual Query Suggestion(VQS)[17] image search scheme is used to find the prior visual context. It provides user a better user interface by simultaneously providing keyword and image suggestion. Because of our special needs, the following steps on the base of VSQ scheme are done:

(1) The establishment of database: We established a database of 10000 images downloaded from Flickr and the images include some objects with regular color, such as faces of white, black and yellow race, cars with various type, and flowers with various color.

(2) Spatial feature: Camera angle has a great influence on color of object in image. The object's position   in image reflects to some extend the azimuth of the camera and the object. Thereforce, spatial feature is included in the search of visual context. In order to avoid the impact of image resolution and size, the spatial feature is

normalized, i.e. $(x/M, y/N)$ .Here (x, y) is pixel coordinate of object, and (M, N) is the size of image.

(3) Input and Output: Inputs include the manually-selected saliency image block (object), image size and the normalized position of the object, and use it as image suggestion for VQS scheme. Outputs are similar objects (image block) and scores representing the similar degree. In order to avoid searching accidental error, we choose the mean of top k (approximately 20) similar blocks as our visual context.

4.2   Specular-Edge Weighting Based on Specular-Map Guided Filter

Edge-based color constancy algorithms have poor adaptability to images with relatively little texture. For these images, grey edge algorithms perform even worse than those pixel-based color constancy methods [15]. According to [18], the image edge can be classified into three types: Material edge, Specular Edge and Shadow edge. At the same time, Arjan[11]et al. points out that specular edge can provide objective evidence of conformance to estimate illumination. Inspired by the two works, we propose a novel color constancy method that uses specular-map guided filter to improve the specular edge weighting schemes.

Firstly, we focus on the first order derivative information, and the image can be expressed as follows:

$$\mathbf{I}'(x) = m_d(x)\mathbf{\Lambda}'(x) + m'_d(x)\mathbf{\Lambda}(x) + \mathbf{m}'_s(x)\mathbf{\Gamma} \tag{9}$$

Here the superscript represents spatial derivative. Since we assume a constant light illuminates for the whole scene, the position of light source is independent of image coordinate location x. we know that image derivative is composed of three vectors($\mathbf{\Lambda}'(x), \mathbf{\Lambda}(x), \mathbf{\Gamma}$), which have a close relationship with Material edge, Shadow edge and specular edge. Specular edge direction is equal to the illumination direction $\mathbf{\Gamma}$, when we project image derivative to specular edge direction and form specular variant.

$$\mathbf{O}' = (\mathbf{I}' . \mathbf{\Gamma})\mathbf{\Gamma} \tag{10}$$

Assuming that the image taken under standard illumination, we can get $\mathbf{\Gamma} = (1/\sqrt{3}, 1/\sqrt{3}, 1/\sqrt{3})$. If the specular variant only contains sepcular edge, the specular edge weight is defined as:

$$W_{spe}(x) = |\mathbf{O}'(x)| / \|\mathbf{I}'(x)\| \tag{11}$$

In fact, the direction vectors $\mathbf{\Lambda}'(x), \mathbf{\Lambda}(x)$ are not perpendicular to specular edge direction $\mathbf{\Gamma}$. Therefore, the projections onto specular variant still contain other edges (shadow edge &material edge). This results in the error of illumiant estimation.

In order to overcome the problem, the specular component (described as a specular-map) is calculated according to [16].Then the specular-map I^{spe} is used as a guide image to improve the weighting of specular-edge by the guided filter[19]. The kernel function of guided filter can be explicitly expressed by:

$$w_{ij}(I^{spe}) = \frac{1}{|\omega|^2} \sum_{k:(i,j)\in\omega_k} (1 + \frac{(I_i^{spe} - \mu_k)(I_j^{spe} - \mu_k)}{\sigma_k^2 + \varepsilon}) \qquad (12)$$

where ω_k is k-th windows, $|\omega|$ is the number in ω_k, μ_k is the mean in ω_k, δ_k^2 is the variance in ω_k, and ε is the smoothing factor. Then the new specular edge weighting can be obtained as follows.

$$W_{spe}^{GF}(x) = W_{spe}(x) \cdot w(x) \qquad (13)$$

Therefore, combining with Eq. (4), the new color constancy model can be obtained as follows.

$$(\int |(w_{specu}^{GF}(x))^k f'(x)|^p)^{\frac{1}{p}} = k\mathbf{e}_c \qquad (14)$$

4.3   Chromaticity Correction

Corrected image is used to recaculate the specular edge weights and reseparate the specular-map the in the iterative framework. With respect to Arjan[11]method, it use corrected image to recalculate highlight edge, the problem is that the loop probably doesn't converge, which result in the more iterations, the worse image recovery. Because of the specially chosen of visual context, the mean chromaticity of corrected image and visual context should be approximately close ,if our iterative illumination toward to the true value. The following calculation is carried out as Eq. (15) to keep chromaticity uniformity of the corrected image and visual context, in order to ensure that the iteration can rapidly converge to near-optimal solutions.

$$\Gamma_c = \begin{cases} \Gamma_c * \sigma_c^B / \sigma_c & if\ |\sigma_c^B - \sigma_c| > \varepsilon \\ \Gamma_c & else \end{cases} \qquad (15)$$

where $|\sigma_c^B - \sigma_c|$ is the absolute value of mean chromaticity difference between the block on corrected imag B and visual context.

5   Experiments

5.1   Performance Measure

For all images in the data set, the true light source $\hat{e}_c$ is known as a priori. To measure how close the estimated illuminant $\hat{e}_l$ resembles the true value, the angular error ℓ is used [20]

$$\ell = \cos^{-1}(\hat{e}_l . \hat{e}_c) \qquad (16)$$

The smaller the error ℓ is, the closer the result of estimation is to the true light source.

5.2   Comparing Precision with Popular Color Constancy Algorithm

We used the same error metric and ground truth to evaluate our scheme on Color-checker[21] database and SFU Grey-ball[22] database, and compared with state of the art methods including grey world [3], white path [4],grey edge [6], and Arjan method[11]. Images (T) of each database were manually separated into two classes: simple scene images (S) and complex scene images (C). Table1 shows the errors of the algorithms on the Color-checker database, which has 223 simple images and 345 complex images. Table2 shows the errors of the algorithms on the SFU Grey-ball database, which has more than 11,000 images extracted from 2 hours video and includs 2275 simple images and 9071 complex images. Some example results are shown in Fig. 2.the lower left subscript denotes the angle error.

Inputs   GreyWorld   Whitepatch   1stgreyedge   2ndgreyedge   Arjan[11]   Ours

Fig. 2. Some results of color constancy

Table 1. The errors of various methods(p=5, $\sigma = 2$) onColor-Checker database

Method	S set			C set			T set		
	Mean	Median	Max	Mean	Median	Max	Mean	Median	Max
Grey World	7.8	7.9	46.0	10.3	8.9	31.6	9.8	7.4	46.0
White Patch	7.5	7.4	36.3	8.4	9.3	29.5	8.2	6.1	36.3
1st Grey Edge	8.5	9.4	36.7	6.8	7.2	30.7	7.1	5.6	36.7
2st Grey Edge	8.8	9.2	37.3	6.9	6.8	26.9	7.3	5.5	37.3
Arjan [11]	7.6	8.3	31.3	6.5	7.5	42.1	6.7	5.0	42.1
proposed	5.7	6.2	28.6	5.6	5.4	45.6	5.6	5.1	45.6

According to Table 1 and Table.2,we can find that the mean absolute error of the proposed scheme is smaller than other method. It means that the the overall precision of the scheme is higher than others. For other methods, some better for simple scene and some better for complex scene. However, our scheme can obtain better results for both simple scene and complex scene. Therefore, the proposed scheme has not only higher precision but also more outstanding universality. In a word, the introduction of guide-filter and visual context prior make expecting and better result.

Table 2. The errors of various methods(p=5, $\sigma = 2$) on SFU Grey-balldatabase

method	S set			C set			T set		
	Mean	Median	Max	Mean	Median	Max	Mean	Median	Max
Grey World	11.4	12.1	55.2	13.4	8.9	37.8	13.0	10.6	55.2
White Patch	10.1	10.6	38.9	12.8	9.0	34.3	12.3	10.0	38.9
1st Grey Edge	12.5	9.8	35.3	10.2	8.2	48.2	10.7	9.1	48.2
2st Grey Edge	12.1	10.3	46.5	10.5	7.3	30.5	10.8	9.2	46.5
Arjan [11]	11.5	9.5	50.2	10.1	7.3	44.3	10.4	9.0	50.2
Our proposed	7.0	6.5	33.2	6.9	6.3	43.4	6.9	6.7	43.4

5.3   Comparing Iteration Number with Arjan's Method

The iteration number of our scheme and the method in [11] on Color-checker database and SFU Grey-ball database are recorded and shown in Fig.3. The proposed scheme can generally complete in 4-5 iteration and converge faster than the method in [11]. The maximum number of iterations is set to 10. In a word, the visual context prior to chromaticity correction in every loop makes our scheme converge fast and get ideal result in almost occasion.

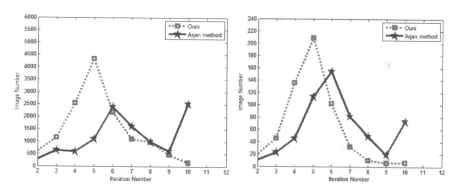

Fig. 3. Statistics for the number of iteration   Left: SFU grey-ball Right: Color Checker

6   Conclusion

In this paper, we introduced a method to obtain color constancy from photo collections. Such collections contained similar visual context to input image and were leveraged to correct the chromaticity of the input image to the standard white illuminant. Therefore, the corrected images were used as the initial guess of illumination estimation. Moreover, we proposed a novel filtering method for specular edge weighting that is under the guidance of specular map. To increase the accuracy of the separation of specular map, an iterative optimal framework was presented. We presented results on several benchmark datasets and have computed the color constancy for more than

11000 images. The results shows that use of visual context from photo collection can improve the estimated illuminant significantly, and that we can produce better color correction results to different lighting conditions. This correction has the potential to benefit to color based object recognition and tracking tasks that assume coherent color when matching features from multiple views of a scene.

References

[1] Tan, R., Nishino, K., Ikeuchi, K.: Color constancy through inverse-intensity chromaticityspace. J. Opt. Soc. Am. A 21(3), 321–334 (2004)
[2] Xiong, W.: Separating Illumination from Reflectance in Color Imagery (2007)
[3] Buchsbaum, G.: A spatial processor model for object colour perception. Journal of the Franklin Institute 310(1), 1–26 (1980)
[4] Land, E.H.: The retinex theory of color vision. Scientific American 237(6), 108–128 (1977)
[5] Finlayson, G.D., Trezzi, E.: Shades of gray and colour constancy. In: Proceeding of IS&T/SID 12th Color Imaging Conference (CIC), pp. 37–41 (2004)
[6] Weijer, J.V., Gevers, T., Gijsenij, A.: Edge-Based Color Constancy. IEEE Trans. on Image Processing 16(9), 2207–2214 (2007)
[7] Finlayson, G., Hordley, S., Hubel, P.: Color by correlation: A simple,unifying framework forcolor constancy. IEEE TPAMI 23(11), 1209–1221 (2001)
[8] Rezagholizadeh, M., Clark, J.J.: Edge-Based and Efficient Chromaticity Spatio-spectral Models for Color Constancy. In: Computer and Robot Vision, pp.188–195 (2013)
[9] Cardei, V., Funt, B., Barnard, K.: Estimating the Scene Illumination Chromaticity Using aNeural Network. J. Opt. Soc. Am. A 19(12), 2374–2386 (2002)
[10] Xiong, W., Funt, B.: Estimating Illumination Chromaticity via Support Vector Regression. Journal of Imaging Science and Technology 50(4), 341–348 (2006)
[11] Gevers, T., Gijsenij, A., van de Weijer, J.: Improving Color Constancy by Photometric Edge Weighting. IEEE TPAMI 34(5), 918–929 (2012)
[12] Dale, K., Johnson, M.K., Sunkavalli, K., Matusik, W., Pfister, H.: Image Restoration using Online Photo Collections. In: ICCV, pp. 2217–2224 (2009)
[13] Shafer, S.: Using color to separate reflection components. Color Res. Appl. 10, 210–218 (1985)
[14] Freeman, W.T., Adelson, E.H.: The design anduse of steerable filters. IEEE TPAMI 13(9), 891–906 (1991)
[15] Gijsenij, A., Gevers, T.: Color constancy using natural image statisticsand scene semantics. IEEE TPAMI 33(4), 687–698 (2011)
[16] Tan, R.T., Ikeuchi, K.: Separating ReflectionComponents of Textured Surfaces Using a Single Image. IEEE TPAMI 27(2), 178–193 (2005)
[17] Zha, Z.J., Yang, L., Mei, T., Wang, M., Wang, Z.: Visual query suggestion. ACM Multimedia, 15–24 (2009)
[18] van de Weijer, J., Gevers, T., Geusebroek, J.: Edge and corner detection by photometricquasi-invariants. IEEE TPAMI 27(4), 625–630 (2005)
[19] He, K., Sun, J., Tang, X.: Guided image filtering. In: Daniilidis, K., Maragos, P., Paragios, N. (eds.) ECCV 2010, Part I. LNCS, vol. 6311, pp. 1–14. Springer, Heidelberg (2010)

[20] Lu, R., Gijsenij, A., Gevers, T., Nedovic, V., Xu, D.: Color constancyusing 3D scene geometry. In: Proc. 12th ICCV, pp. 1749–1756 (September/October 2009)
[21] Shi, L., Funt, B.: Re-processed Version of the Gehler Color Constancy Dataset of 568 Images
[22] Ciurea, F., Funt, B.: A Large Image Database for Color Constancy Research. In: Proceedings of the Imaging Science and Technology Eleventh Color Imaging Conference, Scottsdale, pp. 160–164 (November 2003)
[23] Tan, R.T., Nishino, K., Ikeuchi, K.: Separating reflection components based on chromaticity and noise analysis. IEEE Trans. PAMI 26, 1373–1379 (2004)

A Background Modeling Scheme Based on High Efficiency Motion Classification for Surveillance Video Coding

Pei Liao[1], Xiaofeng Huang[2], Huizhu Jia[2,*], Kaijin Wei[2], Binbin Cai[2], Guoqing Xiang[1], and Don Xie[2]

[1] SECE of Shenzhen Graduate School, Peking University, 518055, Shenzhen, China
[2] National Engineering Laboratory for Video Technology, Peking University, Beijing, China
{pliao,xfhuang,hzjia,kjwei,gqxiang,xdxie}@jdl.ac.cn

Abstract. Recently, high-efficiency video coding becomes more and more demanded as the explosive requirements of network bandwidth and storage space for surveillance video applications. In this paper, we propose a background modeling scheme based on high efficiency motion classification. Firstly, pixels at each location are classified into three motion states, namely the static, the gentle motion and the severe motion states, according to the motion vectors of the corresponding current block and neighboring blocks. Then based on the classification and pixel differential value, the segmentation is performed for the co-located pixels in the training frames, and the mean pixel value of each segment can then be calculated. Finally, the background modeling frame can be obtained by an optimized weighted average of the segmented mean pixel values. Experimental results show that our proposed scheme achieves an average PSNR gain of 0.65dB than the AVS surveillance baseline video encoder, and it gets the best performance among several high efficiency background modeling methods in fast motion and large foreground sequences.

Keywords: surveillance video coding, motion classification, background modeling, weighted average.

1 Introduction

Surveillance video systems are widely used for safety and communication applications recently. The huge required network bandwidth and the increasing demands of storage space are two key challenges in its applications. The compression efficiency of existing video coding standards, like H.264 [1] and AVS [2], is usually not high enough because they are basically designed for general video applications. Consequently, it is necessary to take the specifics of the surveillance video, like static background, into account for high-efficiency surveillance video coding.

In most surveillance applications, cameras are usually fixed at a certain location and direction to capture the scene for a long time. And the background in these frames

---

* Corresponding author.

W.T. Ooi et al. (Eds.): PCM 2014, LNCS 8879, pp. 44–53, 2014.

is always the same except the noise generated by the camera or the slow change of the environment. Typically, a framework of background modeling based video encoder is used which is shown as Fig. 1. In this framework, a good background modeling frame which is long-term referenced by surveillance video sequences will improve the coding efficiency dramatically.

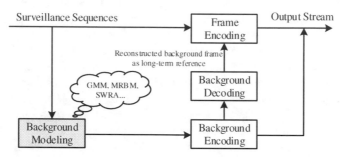

Fig. 1. Framework of background modeling based video encoder

Recently, many effective background modeling methods have been proposed, which can be classified into two categories named parametric and non-parametric methods. In parametric methods, an adaptive Gaussian mixture model (GMM) [3] is used for background subtraction and object detection, with better Gaussian mixture quality compared to earlier GMM methods. However, the huge bandwidth requirement limits its application in hardware realization. In non-parametric methods, an effective background generation method named most reliable background mode (MRBM) [4] is used, which generates clear background and is robust to noise and camera shaking. In paper [5], a segment-and-weight based running average (SWRA) method is proposed to alleviate the computational complexity of background modeling, and achieves comparable coding performance as GMM-5 method.

Actually, the background modeling methods in papers [3-4] are designed for object detection or background subtraction, while the main objective of background modeling in Fig. 1 is to save encoded bits for background frame and provide better long-term reference for surveillance video sequences. Although the algorithm in work [5] is oriented for video coding, besides the pixel level information, the global level information, such as motion vectors, can be adaptively used to improve surveillance video coding performance. In this paper, a motion classification based background modeling scheme is proposed. This scheme firstly classifies pixels at each location into three motion states, namely the static, the gentle motion and the severe motion states. Secondly, the segmentation of the co-located pixels in the training frames is performed based on both its motion state and pixel differential value, and the mean pixel value of each segment can be calculated. Finally, the background modeling frame can be obtained by an optimized weighted average of the segmented mean pixel values.

The rest of the paper is organized as follows. Section 2 in detail describes the proposed method. Section 3 presents the experimental results. And conclusion of this paper is made in section 4.

2   Proposed Method

Background modeling increasingly plays an important role in surveillance video coding, which aims at segmenting foreground and background in object-oriented video encoder [6] and providing better long-term prediction efficiency in background-prediction-based video encoder as shown in Fig. 1. In this section, a motion classification based background modeling scheme is presented for background-prediction-based video encoder.

The overall framework of the proposed scheme is shown in Fig. 2. The motion classification is based on the motion vectors of the corresponding current block and the neighboring blocks. For the (x,y) pixels in the training frames, the segmentation is based on both its motion state and pixel differential value (SAD). The mean pixel value avg_i and the segment length L_i can be calculated for the i_{th} segment, respectively. The weight w_s_i for i_{th} segment is proportional to the L_i, and a weighted average of the previous segments and the current segment is calculated for background modeling pixel value BGV. The background modeling pixel value BGV is updated at the end of each segment. When BGV value equals to 0 at the last of the training frames, value 128 is directly assigned to BGV. The algorithm of our proposed scheme is shown in Fig. 3, which is a detailed description of Fig. 2.

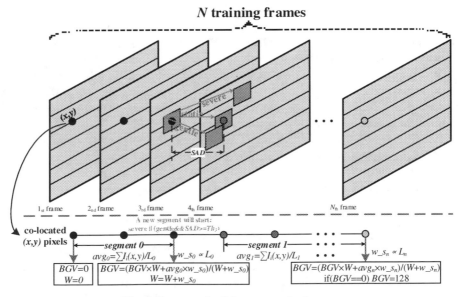

Fig. 2. Framework of the proposed scheme

(1) Initialization: For each (x,y) pixel location, the background modeling pixel value BGV and the total previous segment weight W are initialized to 0. The first segment length L and its mean pixel value avg are also initialized to 0.

(2) Motion classification: There are three motion states defined in our proposed method, namely the static, the gentle motion and the severe motion states. Each pixel

Input:
  $\mathbf{T} = \{I_i=f_i(x,y)|i=1\sim N\}$, where $f_i(x,y)$ is i_{th} frame in the N training frames.
Background Modeling Procedure:
For each pixel location (x,y), **Begin**
  $\{BGV, W, L, avg\}=0$;   (1)
  While $i=1\sim N$, **Begin**
    if$((abs(MV_x)+abs(MV_y))_c<Th_1$ && $\forall(abs(MV_x)+abs(MV_y))_n<Th_1)$ (2)
      cs = static;
    else if$(abs(MV_x)_c<$Th$_2$ && $abs(MV_y)_c<$Th$_2)$
      cs = gentle motion;
    else
      cs = severe motion;
    if$(($static $\parallel$ (gentle motion && $abs(I_i-I_{i+1})<$Th$_3$)) && $i<N$) **Begin**   (3)
      $avg = (avg\times L+I_i)/(L+1)$;
      $L = L+1$;
    End
    else **Begin** /*a new segment*/
      $w\_s \propto L$
      $BGV = (BGV\times W+ avg\times w\_s)/(W+w\_s)$;
      if$(i==N$ && $BGV==0)$
        $BGV = 128$;
      $W = W+w\_s$;
      $L = avg = 0$;
    End
  End
End
Output the background frame.   (4)

Fig. 3. Algorithm of the proposed scheme

in the training frames will be classified into these three states based on the motion vectors of the corresponding current block $((MV_x,MV_y)_c)$ and eight neighboring blocks $((MV_x,MV_y)_n)$ as shown in Fig. 4(a). The motion vector is searched by using the $(i+1)_{th}$ frame as reference for the current i_{th} frame as in Fig. 2. For simplicity, the motion vectors derived from motion estimation module in encoder can be directly used instead, which will decrease the complexity greatly. The static state is assigned when the addition of absolute motion vector values of current block and eight neighboring blocks are smaller than a threshold, simultaneously. In order to integrate our proposed scheme into existing hardware architecture [7] easily, only four neighboring searched blocks instead of total eight neighboring blocks are used for motion classification as shown in Fig. 4(b). The threshold Th1 is set to 1 for the tolerance of noise and small camera shaking.

In order to segment the co-located pixels in the training frames accurately, two motion states are distinguished in the proposed method as shown in Fig. 2. The partition of the motion states is based on the actual motion vector value of the current block, when its absolute values of horizontal and vertical motion vector are both less than a threshold Th_2, the gentle motion is assigned, and otherwise the severe motion is assigned. The principle of the motion state partition is shown in Fig. 5. When the motion vector exceeds the block size (b_s), as MV_b in Fig. 5, there will be no overlap between the reference block and the current block. Accordingly, the current block is named as a "*scene change block*" and is set to the severe motion state. Otherwise, for the MV_a case as shown in Fig. 5, the current block is set to the gentle motion state. Thus, the threshold Th_2 is equal to the block size b_s.

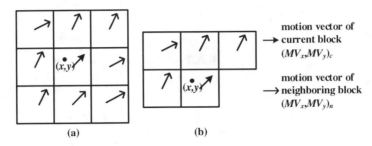

Fig. 4. (a) Motion vectors of current block and eight neighboring blocks for motion classification. (b) Motion vectors of current block and four neighboring blocks for motion classification.

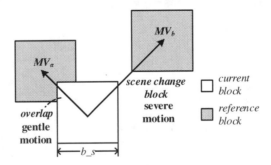

Fig. 5. Principle of motion state partition

(3) Segmentation and parameter calculation: When the current pixel is in a severe motion state or the pixel differential value is larger than the third threshold Th_3 in gentle motion state as in Fig. 2, a new segment will start. Otherwise, the segment is continued and its mean pixel value *avg* and segment length L are calculated, separately. The threshold value Th_3 is dynamically adjusted as Fig. 3 in [8], except that the $Diff(m,n)$ is the difference between $I_i(m,n)$ and $I_{i+1}(m,n)$.

The weight of the segment w_s is proportional to the segment length L. In order to evaluate the proportion accurately, six piecewise functions as shown in Fig. 6 are listed for testing. The test platform is illustrated as in Part 3.1. When the segment length L is smaller than a threshold Th_0 as in Fig. 6, the segment weight w_s is assigned to 0 in order to eliminate false segments. Typically, the value of Th_0 is set to $N/20$. For the segment length L larger than the Th_0, the six piecewise functions are listed based on the fact that the larger segment weight w_s is assigned with larger segment length L. As shown in Table 1, compared to the AVS surveillance baseline video encoder, the quadratic function achieves the best performance among these six piecewise functions, which achieves an average PSNR gain of 0.645dB and an average bitrate decrease of 21.24%. In the piecewise linear functions as shown in Fig. 6(a), the linear function achieves the lowest performance (0.617dB PSNR gain in average), and the piecewise1 function achieves the highest performance (0.630dB PSNR gain in average). In the piecewise power functions as shown in Fig. 6(b), the performance is decreasing as the order of the power function increased. Table 1 illustrates that the quadratic function accords with the proportional relationship between the segment weight w_s and the segment length L. This conclusion will be useful for other background modeling methods.

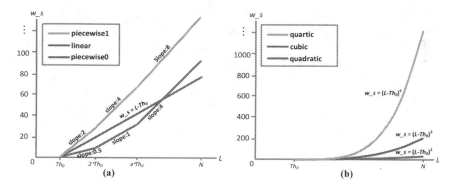

Fig. 6. The curves of six piecewise functions. (a) Piecewise linear functions. (b) Piecewise power functions.

Table 1. Performance comparisons of the six piecewise functions (VS SM-2)

Format	Sequence	piecewise0		linear		piecewise1		quadratic		cubic		quartic	
		PSNR (ΔdB)	Bitrate (Δ%)	PSNR (ΔdB)	Bitrate (Δ%)	PSNR (ΔdB)	Bitrate (Δ%)	PSNR (ΔdB)	Bitrate (Δ%)	PSNR (ΔdB)	Bitrate (Δ%)	PSNR (ΔdB)	Bitrate (Δ%)
CIF	Crossroad	0.533	-13.89	0.526	-13.71	0.550	-14.29	0.573	-14.86	0.550	-14.31	0.497	-13.05
	Overbridge	0.575	-17.91	0.567	-17.75	0.585	-18.24	0.596	-18.65	0.557	-17.46	0.496	-15.74
	Snowroad	0.736	-25.07	0.734	-25.04	0.740	-25.22	0.729	-24.78	0.702	-23.95	0.690	-23.56
	Snowway	0.467	-17.13	0.461	-16.97	0.477	-17.57	0.511	-18.88	0.485	-18.03	0.406	-15.17
SD	Bank	0.814	-31.77	0.818	-31.78	0.828	-32.13	0.846	-32.85	0.822	-31.92	0.792	-30.60
	Crossroad	0.547	-16.35	0.547	-16.35	0.558	-16.65	0.567	-17.05	0.541	-16.20	0.493	-14.89
	Office	0.403	-16.43	0.399	-16.31	0.412	-16.74	0.421	-17.05	0.403	-16.43	0.371	-15.36
	Overbridge	0.882	-24.86	0.880	-24.83	0.899	-25.31	0.918	-25.80	0.886	-24.97	0.832	-23.40
Avg.		0.620	-20.43	0.617	-20.34	0.630	-20.77	**0.645**	**-21.24**	0.618	-20.40	0.57	-18.97

Instead of buffering each segment weight such as w_s_0, w_s_1..., W is used to represent the total weights of previous segments in order to save memory as shown in Fig. 7. The background modeling pixel value BGV is updated at the end of each segment, by weighted average of total previous segments and current segment w_s_c.

The *BGV* value may be still 0 when it runs into the last training frame N. This is because the pixel is always in the severe motion state or its segment length L is too short, all of which indicate that the pixel is always in foreground. Pixel value 128 is assigned to that pixel as in [6], which will improve the compression efficiency.

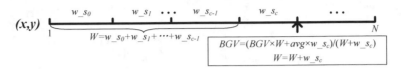

Fig. 7. The calculation of *BGV* and *W*

(4) Output: After the background modeling process at the last training frame, the background picture can be generated and output.

From the above analysis, the proposed motion classification based background modeling method mainly depends on the motion vector, pixel differential value and piecewise weighting function. Besides, the proposed method updates the model value of each pixel at the end of a segment, and background frame is updated at the end of training frames. As a result, the proposed method satisfies the demand of periodical background updating and real-time requirement of hardware realization. In section 3, the performance of the proposed method is evaluated.

3   Experimental Results

3.1   Experiment Setup

In this section, AVS surveillance baseline video encoder (SM-2) is used to evaluate the efficiency of the proposed method. Besides our proposed method, other three background modeling methods, the SWRA method in [5], the GMM using 5 models for each pixel in [3], and the MRBR method in [4], are implemented and embedded into SM-2 encoder for comparison. Corresponding Encoders are named SM-MC, SM-SW, SM-GMM5 and SM-MR. The SM parameters configuration is listed in Table 2. In the table, profile ID 44 corresponds to the surveillance video coding.

Table 2. Parameters configuration.

Parameter	Vaule	Parmeter	Vaule	Parmeter	Vaule
Profile ID	44	Search Range	32	RD Optimization	Used
Rate Control	Disable	Reference Num.	2	Frame Structure	IPPP
Entropy Coding	CABAC	Motion Est.	UMHexagonS	QP of P frame	24,31,37,44
Intra period	25	Use Mode	ALL	Frame rate	25

For fair comparison, background frames in the encoders should be updated and encoded simultaneously. A sequence structure dividing input frames into super group of pictures (S-GOP) is used for the surveillance video encoders as shown in Fig. 8.

The background frame Bg_1 which is long-term referenced by sequences in S-GOP$_1$ is modeled by TrainSet$_0$, and encoded at the first frame of S-GOP$_1$. The same structure will be utilized for S-GOP$_n$, where Bg_n is modeled by TrainSet$_{n-1}$ in S-GOP$_{n-1}$. In our experiments, the number of training frames is set to 120 and the size of an S-GOP is set to 600. Besides, to simplify the bit-allocation of the background frame, the quantization parameter for background frame is equal to that of P frames minus 5, and only intra predictions are utilized.

Eight surveillance sequences captured by static camera are used in our experiments. For the data set, the first 900 frames of SD surveillance sequences of *Crossroad*, *Overbridge*, *Office*, *Bank* and CIF sequences of *Crossroad*, *Overbridge*, *Snowroad*, *Snowgate* [9] are used to evaluate the five encoders. The content of these sequences are shown in Fig. 9. In these sequences, *Crossroad* (CIF), *Overbridge* (CIF&SD) and *Office* (SD) have relatively large foreground regions and fast motion.

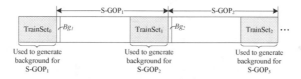

Fig. 8. Sequence structure for background modeling

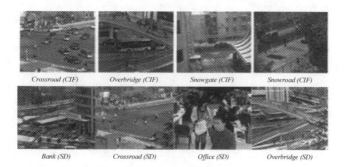

Fig. 9. Test sequences and their content

3.2   Results

In this part, the encoding performance of the SM-MC, SM-GMM5, SM-SW, SM-MR and the SM-2 encoders are compared. Results show that the proposed motion classification based background modeling scheme encoder (SM-MC) achieves better performance than the other three background modeling based video encoders, especially in fast motion and large foreground sequences.

Compared to the SM-GMM5/SM-SW/SM-MR/SM-2 encoders, the SM-MC encoder achieves both PSNR gain and bitrate reduction. The SM-MC achieves an average gain of 0.02dB/0.06dB/0.12dB/0.65dB than the SM-GMM5/SM-SW/SM-MR/SM-2 encoders. For CIF sequences, our method achieves an average PSNR gain of 0.03dB/0.05dB/0.13dB gain than the SM-GMM5/SM-SW/SM-MR encoders. In these four CIF sequences, the SM-MC encoder shows the highest gain for Crossroad and Overbridge sequences, and SM-GMM5 shows the best performance for Snowroad and Snowway sequences. And an average PSNR gain of 0.02dB/0.07dB/0.12dB is achieved than the SM-GMM5/SM-SW/SM-MR encoders for SD sequences. In SD sequences, the SM-MC encoder shows the best performance for Office and Overbridge sequences, and SM-GMM5 encoder gets the highest gain for Bank and Crossroad sequences. These all imply that our proposed method shows the highest performance in large foreground and fast motion sequences, and the existing GMM-5 method shows the highest gain in small foreground and slow motion sequences. The SM-M2 encoder shows the worst performance in these five encoders, and the second and the third worst are SM-MR and SM-SW encoders, respectively. An example rate-distortion curve of Office (SD) sequence is shown in Fig. 10.

Table 3. Performance Comparisons

Format	Sequence	SM-MC VS SM-GMM5		SM-MC VS SM-SW		SM-MC VS SM-MR		SM-MC VS SM-2	
		PSNR (ΔdB)	Bitrate (Δ%)	PSNR (ΔdB)	Bitrate (Δ%)	PSNR (ΔdB)	Bitrate (Δ%)	PSNR (ΔdB)	Bitrate (Δ%)
CIF	Crossroad	0.08	-1.99	0.08	-2.05	0.21	-5.19	0.57	-14.86
	Overbridge	0.04	-1.01	0.06	-1.88	0.16	-4.35	0.60	-18.65
	Snowroad	0	0.26	0.02	-0.64	0.04	-1.22	0.73	-24.78
	Snowway	-0.02	0.93	0.03	-0.79	0.09	-3.27	0.51	-18.88
CIF Avg.		0.03	-0.45	0.05	-1.34	0.13	-3.51	0.60	-19.29
SD	Bank	0	0.11	0.06	-2.15	0.07	-2.94	0.85	-32.85
	Crossroad	-0.01	0.25	0	-0.06	0.09	-2.44	0.57	-17.05
	Office	0.05	-2.01	0.16	-4.79	0.15	-4.75	0.42	-17.05
	Overbridge	0.02	-0.6	0.06	-1.46	0.15	-4.58	0.92	-25.80
SD Avg.		0.02	-0.56	0.07	-2.12	0.12	-3.68	0.69	-23.19
Avg.		**0.02**	**-0.51**	**0.06**	**-1.46**	**0.12**	**-3.59**	**0.65**	**-21.24**

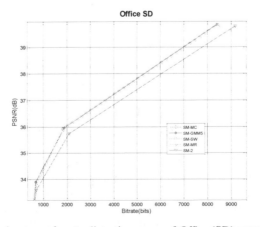

Fig. 10. An example rate-distortion curve of *Office* (SD) sequence

4   Conclusion

In this paper, a high-efficiency motion classification based background modeling scheme is proposed. Firstly, pixels at each location are classified into three motion states, namely the static, the gentle motion and the severe motion states. Then based on the classification and pixel differential value, the segmentation is performed for the co-located pixels in the training frames, and the mean pixel value of each segment can be calculated. Finally, the background modeling frame can be obtained by an optimized weighted average of the segment mean pixel values. In order to resolve the invalid background modeling in fast motion sequences in existing methods, our method takes the block level motion status into consideration for more accurate background modeling. Experimental results show that our proposed scheme achieves an average PSNR gain of 0.65dB than the AVS surveillance baseline video encoder, and gets the best performance among several high efficiency background modeling methods in fast motion and large foreground sequences.

Acknowledgements. This work is partially supported by grants from the Chinese National Natural Science Foundation under contract No.61171139, and National High Technology Research and Development Program of China (863 Program) under contract No.2012AA011703.

References

1. Wiegand, T., Sullivan, G., Bjøntegaard, G., Luthra, A.: Overview of the H.264/AVC Video Coding Standard. IEEE Transactions on Circuits and Systems for Video Technology 13(7), 560–576 (2003)
2. Gao, W., Ma, S., Zhang, L., Su, L., Zhao, D.: AVS Video Coding Standard. Intelligent Multimedia Communication: Techniques and Applications, 125–166 (2010)
3. Haque, M., Murshed, M., Paul, M.: Improved Gaussian mixtures for robust object detection by adaptive multi-background generation. In: IEEE International Conference on Pattern Recognition, pp. 1–4 (2008)
4. Liu, Y., Yao, H., Gao, W., Chen, X., Zhao, D.: Nonparametric background generation. In: IEEE International Conference on Pattern Recognition, vol. 4, pp. 916–919 (2006)
5. Zhang, X., Tian, Y., Huang, T., Gao, W.: Low-Complexity and High-Efficiency Background Modeling for Surveillance Video Coding. In: IEEE International Conference on Visual Communications and Image Processing, pp. 1–6 (2012)
6. Zhang, S., Wei, K., Jia, H., Xie, X., Gao, W.: An efficient foreground-based surveillance video coding scheme in low bit-rate compression. In: IEEE International Conference on Visual Communications and Image Processing, pp. 1–6 (2012)
7. Yang, W., Yin, H., Gao, W., Qi, H., Xie, X.: Multi-stage motion vector prediction schedule strategy for AVS HD encoder. In: IEEE International Conference on Consumer Electronics, pp. 339–340 (2010)
8. Zhang, X., Tian, Y., Liang, L., Huang, T., Gao, W.: Macro-Block-Level Selective Background Difference Coding for Surveillance Video. In: IEEE International Conference on Multimedia and Expo, pp. 1067–1072 (2012)
9. ftp://124.207.250.92/public/seqs/video/ (accessed by AVS member)

An Adaptive Perceptual Quantization Algorithm Based on Block-Level JND for Video Coding

Guoqing Xiang[1,2], Xiaodong Xie[2,*], Huizhu Jia[2], Xiaofeng Huang[2], Jie Liu[2],
Kaijin Wei[2], Yuanchao Bai[2], Pei Liao[1,2], and Wen Gao[2]

[1] SECE of Shenzhen Graduate School, Peking University, 518055, Shenzhen, China
[2] Engineering Lab for Video Technology School of EECS, Peking University, Beijing, China
{gqxiang,hzjia,xfhuang,liuzimin,xdxie}@jdl.ac.cn

Abstract. It has been widely demonstrated that integrating efficient perceptual measures into traditional video coding framework can improve subjective coding performance significantly. In this paper, we propose a novel block-level JND (just-noticeable-distortion) model, which has not only adjusted pixel-level JND thresholds with more block characteristics, but also integrated them into a block-level model. And the model has been applied for adaptive perceptual quantization for video coding. Experimental results show that our model can save bit rates up to 24.5% on average with negligible degradation of the perceptual quality.

Keywords: JND model, masking effect, adaptive perceptual quantization, visual quality, video coding performance.

1 Introduction

Traditional hybrid video coding aims to remove spatial and temporal statistical redundancies for signal compression. However, most of these methods often neglect perceptual features for better subjective video coding. Considering that human eyes are the ultimate receivers, it is worthwhile to dedicate perceptually friendly coding researches to further remove perceptual redundancies and improve subjective quality. The just noticeable distortion (JND) threshold, i.e. the distortion that observers just begin to notice, is one of the popular perceptual methods used for such applications.

There have been abundant research efforts to develop rational JND models and apply them into video coding. The existing JND models can be classified into pixel domain and transform domain, respectively. In the pixel domain, most JND models use luminance adaptation and texture masking to compute pixel-level JND [1]. X.K.Yang et al. extended the JND model with a Nonlinear Additively Masking Model (NAMM) by integrating the luminance and texture masking together [2]. Except the luminance and texture that affect the perceived distortion, some other important factors have been studied, such as Chen et al. introduced a famous FJND model [3] integrated with retina foveation model to account for the relationship between visibility and eccentricity. Most of the transform domain JND methods are modeling in DCT domain, and the

---

* Corresponding author.

W.T. Ooi et al. (Eds.): PCM 2014, LNCS 8879, pp. 54–63, 2014.
© Springer International Publishing Switzerland 2014

subband JND features higher accuracy with the consideration of channel's interactions. Among the popular DCT domain JND models, researchers focus on luminance adaptation, the spatial CSF (contrast sensitivity function) and temporal CSF effects [4] to get useful spatial-temporal JND models. However, Jia's model [4] only refers the magnitude of motion contribution to final spatial-temporal JND threshold, but the directionality of motion is neglected. Wei et al. furthered the model by introducing a gamma correction to compensate the original luminance adaptation effect and a novel temporal modulation factor to integrate temporal properties [5].

The existing JND methods have been widely used in hybrid video coding to enhance coding performance. In [7], a low-complexity perceptual rate distortion model has been introduced to replace the Lagrange RD cost model and demonstrated inter mode decision performance improvement. Chen et al. [3] optimized the quantization parameter for each MB (macro-block) based on its FJND information. The Lagrange multiplier in the rate-distortion optimization is adapted to ensure that the MB noticeable distortion is minimized. Some researchers have utilized JND models to improve the compression rates mainly by residues [6] [8] or DCT coefficients [9-10] filtering based on a hard or soft JND threshold.

From the JND models and application methods in video coding mentioned before, we can find that most the popular JND models have some characteristics as follows. First, although these models have considered adjacent characteristics of a pixel or a subband during the modeling procedure, they are just applied by separate pixel-level filtering [6, 8, 9, 10], and these methods do not consider that traditional video coding is on the basis of block units, which prefers more on smooth compression. And pixel-level filtering may introduce much artifact distortion fluctuation in a block. Second, the computational complexity is too large for most JND models, such as using the canny operator in [5]. At last, none of them pay attention to the fact that the quantization parameter decided by traditional mode decision has not considered perceptual properties, namely, the QP can be perceptually adjusted to remove more visual redundancies or enhance perceptual quality. In order to resolve these problems, an adaptive perceptually quantization method for video coding based on a block-level JND model is proposed in this paper. It not only integrates the pixel-level JND into the block-level JND, but also uses perceptual quantization to improve compression efficiency.

The reminder of this paper is organized as follows. In Section 2, the main structure of the proposed block-level JND model is introduced. Based on our block-level JND model, an adaptive perceptual quantization method is integrated into video coding framework in Section 3. In Section 4, the experimental results are shown and discussed. The Section 5 draws the conclusions of our work.

2   Proposed Block-Level JND Model

The proposed block-level JND model is computed by two steps, computing pixel-level JND and then integrating it into block-level JND. The pixel-level JND in [5] can be expressed for every 4×4 block in an image as the integration of a spatial JND value JND_s and a temporal modulation factor F_T,

$$JND_T\left(n,i,j\right) = JND_S\left(n,i,j\right) \cdot F_T\left(n,i,j\right) \tag{1}$$

where n is the index of a 4×4 block in the image, i and j are the DCT coefficients indices. Then the block-level JND can be computed as,

$$JND_{blcok}(k) = \alpha \cdot lnD_{block}(k) \tag{2}$$

where k denotes the kth 8×8 block in a macro-block, JND_{block} is the JND threshold of a 8×8 block, α is an empirical control parameter. D_{block} means the HVS perceptual sensitivity of a block computed by the integration of its pixel-level JND_T and the block energy, which will be detailed later.

2.1   The Pixel-Level JND Threshold

The spatial JND threshold is the product of luminance adaptation factor, the contrast masking factor and the frequency property of DCT sub-band, it can be calculated as,

$$JND_S(n,i,j) = T_{basic}(n,i,j) \cdot F_{lum}(n) \cdot F_{contrast}(n,i,j) \tag{3}$$

where T_{basic} is the base JND threshold generated by the CSF effect, F_{lum} is the luminance adaptation and $F_{contrast}$ denotes the contrast masking effect.

First, the basic threshold is considered. The HVS is sensitive to spatial frequencies, and the spatial frequency of the (i, j)th subband in the nth DCT block is related to block dimension N, which can be computed as in [5],

$$\omega_{i,j} = \frac{1}{2N}\sqrt{(i/\theta_x)^2 + (j/\theta_y)^2} \tag{4}$$

where θ_x, θ_y denotes the horizontal and vertical visual angle respectively, and they are the same as,

$$\theta_x = \theta_y = 2 \cdot \arctan\left(1/\left(2 \cdot R_{vd} \cdot P_{ch}\right)\right) \tag{5}$$

where R_{vd} stands for the ratio of viewing distance [5] to picture height P_{ch}. Then the basic threshold for DCT subband can be calculated as,

$$T_1(n,i,j) = s \cdot \frac{1}{\phi_i \phi_j} \cdot \frac{\exp(c\omega_{ij})/(a+b\omega_{ij})}{r+(1-r)\cdot\cos^2\varphi_{ij}} \tag{6}$$

where s accounts for the spatial summation effect with an empirical value 0.25, parameter r is set to 0.6, and the normalization factors ϕ_i or ϕ_j are expressed as

$$\phi_m = \begin{cases} \sqrt{1/N}, m = 0 \\ \sqrt{2/N}, m > 0 \end{cases} \tag{7}$$

and more parameters can be found in [5]. And because of the difference between 4×4 block and 8×8 block, for ϕ_m in (7), the $\phi_i\phi_j$ will make the basic perceptual distortion fail by half, so we set the basic threshold of a 4×4 block as follows to ensure perceptual distortion consistency with [5]

$$T_2 = 2 \cdot T_1 \tag{8}$$

In order to take more block-level luminance characteristics into account [12], we adjust each basic threshold as

$$T_{basic} = T_2 \cdot \left(\frac{C(0,0,n)_{4\times4}}{C(0,0)_{8\times8}} \right)^{\tau} \qquad (9)$$

where $C(0,0,n)_{4\times4}$ and $C(0,0)_{8\times8}$ is the nth 4×4 block DC coefficient in a 8×8 block and the 8×8 block DC coefficient respectively, the parameter τ here equals to 0.649.

The luminance adaptation for HVS mainly depends on average luminance intensity value $\bar{I}$ of each 4×4 block, and it is described as in [5]

$$F_{lum}(n) = \begin{cases} (60 - \bar{I})/150 + 1, & \bar{I} \leq 60 \\ 1, & 60 < \bar{I} < 170 \\ (\bar{I} - 170)/425 + 1, & others \end{cases} \qquad (10)$$

It is easy to know that in smooth and edge areas the distortion can be more easily recognized than texture areas with high energy, so we should compute the contrast masking factor according to different type of blocks, namely Plane, Edge and Texture. However, the Canny operator may be too complicated in [5] for block classification, here we replaced it with a famous and useful block classification method in DCT domain [13] used for JPEG encoder. According to [13], the 4×4 block can be divided into four indicative areas as show in Fig.1, where DC, L, M, and H denotes the absolute sums of DCT coefficients in different areas respectively. By easily computing the $(L+M)/H$, L/M, $M+H$ and comparing them with experimental threshold μ_1, μ_2, α_1, α_2, β_1, β_2, γ etc., we can decide the block type as Plane, Edge and Texture quickly. Generally, the larger values of L/M and $(L+M)/H$ for a block means higher possibility to be Edge area, and usually smaller $M+H$ indicates Plane block. More details and the comparison procedure can be found in [13].

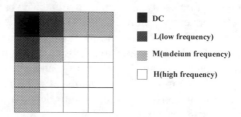

Fig. 1. Classification indicators for a 4×4 DCT block

Therefore through employing the elevation factor in [5],

$$\psi(n,i,j) = \begin{cases} 1, & \text{for Plane and Edge block} \\ 2.25 & \text{for } (i^2 + j^2) \leq 4 \text{ in Texture block} \\ 1.25 & \text{for } (i^2 + j^2) > 4 \text{ in Texture block} \end{cases} \qquad (11)$$

we can compute $F_{contrast}$ as follows,

$$F_{contrast} = \begin{cases} \psi, & \text{for } \left(\text{i}^2 + \text{j}^2\right) \leq 4 \text{ in Plane and Edge block} \\ \psi \cdot \min\left(4, \max\left(1, \left(\dfrac{C(n,i,j)}{T_{basic}(n,i,j) \cdot F_{1um}(n)}\right)^{0.36}\right)\right), & \text{others} \end{cases} \tag{12}$$

where $C(n,i,j)$ denotes the (i,j)th DCT coefficient in the nth block.

In order to consider the temporal effect, we need to evaluate the temporal modulation factor F_T. Many works have demonstrated that there are no independent characteristics between the spatial and temporal frequencies, so we should take spatial factor into account for the temporal modulation factor. In [5], F_T is derived as

$$F_T = \begin{cases} 1 & f_s < 5cpd \ \& \ f_t < 10Hz \\ 1.07^{(f_t - 10)} & f_s < 5cpd \ \& \ f_t \geq 10Hz \\ 1.07^{f_t} & f_s \geq 5cpd \end{cases} \tag{13}$$

where f_s accounts for spatial frequency discussed above and f_t denotes the temporal frequency, which have relationships with motion vectors acquired by motion estimation, current frame rate, and eyes move velocity, etc. Detailed calculation procedure has been described in [5].

Finally, the pixel-level JND threshold in classic 4×4 DCT domain can be obtained as (1), which is the basis of block-level JND model.

2.2   The Block-Level JND Threshold

The block-level JND threshold is proposed based on the following facts. Firstly, observers are more easily attracted to a block or an area than a pixel in an image. Among most natural scenes, the distortion beyond a block-level perceptual distortion threshold in a block will be more easily noticed than a pixel-level difference out of its threshold. Secondly, observers can be more easily attracted to high frequency content and is more sensitive to the distortion of low frequency areas, such as edges and noise in plane area, respectively. And human eyes have less interest in the medium frequency areas which contain much information and energy, and become less sensitive to their distortion. Therefore, it is rational to take subband pixel-level JND of a block and its energy distribution characteristics together into account to find a reasonable JND threshold for each block in a picture.

Similar to [11], a block-level JND for image is proposed, it has considered the energy distribution characteristics in a block and the difference of block types in JND modeling, and it can be expressed as,

$$D_{block}(k) = \sum_{n=0}^{M} \sum_{i=0}^{N} \sum_{j=0}^{N} JND_T(n,i,j) \cdot |C(n,i,j)|^2 \tag{14}$$

where M is the number of sub-block divided in a block, here its value is 4. The larger the D_{block}, the less sensitivity to distortion of the block for HVS, i.e. the more redundancies can be removed for better compression without much visual difference. And from the expression of D_{block}, we can find that in the very low and very high frequency sub-bands, such as very simple plane area with near-zero JND_T and very complex edge areas with very low energy respectively, their values tend to be smaller than in medium frequency sub-bands with larger JND_T and energy such as modest texture areas. As a consequence, there will be less artifact distortion fluctuation in a block than pixel-level JND and we can avoid introducing too much artifact distortion in low frequency areas and protect more details in high frequency areas. At the end, the block-level JND can be computed as (2), which will be incorporated in perceptual video coding.

3   Adaptive Perceptual Quantization for Video Coding

In the traditional hybrid video coding standards, an offset of quantization parameter QP, namely ΔQP, will be used in Differential Pulse Code Modulation (DPCM) and transmitted in coded bit stream, and the QP used for residual DCT coefficient quantization or inverse quantization can be expressed as,

$$QP = QP_0 + \Delta QP \tag{15}$$

where QP_0 is original quantization parameter of current macro-block, and it will be used for uniform quantization in a macro-block. However, the QP used in the best mode coding does not explore perceptual characteristics very well. According to [14], the quantization error should be limited to

$$\mid e_{QP} \mid = \mid C - C_{rec} \mid \le JND_{block} \tag{16}$$

where C_{rec} stands for the reconstructed DCT coefficient. Taking the maximum unnoticeable distortion into account, a perceptual quantization step should be limited to the block-level noticeable distortion, so we can get

$$QP_{step} = 2 \cdot JND_{block} \tag{17}$$

where QP_{step} is a uniform quantizer step applied to each DCT coefficient C. Then we can combine the procedure in [14] with the proposed block-level JND model in Section 2 and get the perceptual ΔQP_{JND} as,

$$\Delta QP_{JND} = Ceil(K \cdot \log_2 JND_{block}) \tag{18}$$

where $Ceil(x)$ denotes the closest integer not more than the argument, K means the relationship between QP and QP_{step} and it varies from different video coding standard. At the end, the perceptually adaptive quantization parameter QP_{JND} is computed as

$$QP_{JND} = QP_0 + \Delta QP_{JND} \tag{19}$$

In order to comply with traditional macro-block video coding standard, we should average all the mentioned QP_{JND} of each 8×8 block in a macro-block for uniform quantization as,

$$QP_{JND\_MB} = \frac{1}{B} \sum_{k=0}^{B} QP_{JND}(k) \tag{20}$$

where B is the number of blocks in a coding macro-block. The QP_{JND_MB} will be used for the best mode coding to remove more perceptual unnoticeable redundancies and integrate it into video coding procedure as shown in Fig. 2. The quantization offset ΔQP_{JND_MB} values' mapping will be transmitted to final bit stream as follows

$$\Delta QP_{JND\_MB} = QP_{JND\_MB} - QP_0 \tag{21}$$

4   Experimental Results

In order to evaluate the performance of our proposed block-level JND scheme, the integration procedure is implemented on AVS Jizhun profile. The GOP length is 15 with structure IBBPBBP….The frame rate is 30 frames per second, the motion estimation is carried out at a quarter pixel resolution with search range of 16 and the RDO is enabled. The test sequences are 4:2:0 YUV format covering CIF, 720p and 1080p resolutions. We have compared the subjective quality and bit rate compression performance of video encoded by the proposed method with Yang's model [2], which is a famous and useful JND model [10]. The chosen subjective distortion measure is the Multiple Scale-Structural Similarity (MS-SSIM) [15] calculated on the luminance frames and averaged for the whole sequence. The Table 1 shows our experimental results and the Fig.3 (a) to Fig.3 (b) demonstrate subjective performance improvement directly. And we can make some discussions as follows.

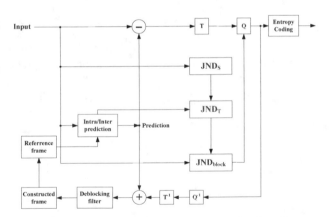

Fig. 2. Perceptual adaptive quantization video coding diagram

According to the Table 1, we can conclude that our block-level JND model shows less negligible subjective loss and lower MS-SSIM reduction than Yang's with the value of -0.3265% and -0.5654%, respectively. It means that the perceptually processed sequences almost have the same visual quality as original reference encoded sequences and are better than Yang's. We can see that the proposed block-level JND model yields an average 24.5% bit rate savings compared to Yang's 18.8% bit rate reduction. Meanwhile, Yang's PSNR loss is almost twice than our model, which means that because of considering block-level JND characteristics rather than directly using separate JND filtering, our model can shape perceptual noise more uniformly. The Fig.3 (a) and (b) show the integral performance by bit rate versus MS-SSIM exampled by sequence "Optics" and sequence "Life". We can easily find that our model has a better subjective performance, which is in accordance with the result of Table 1.

Table 1. Performance of proposed block-level JND model

Sequences	QP	Proposed model			Yang's model		
		ΔBitrate (%)	ΔPSNR (dB)	ΔMS-SSIM (%)	ΔBitrate (%)	ΔPSNR (dB)	ΔMS-SSIM (%)
Football (CIF)	16	-12.07%	-1.18	-0.1347%	-5.00%	-2.37	-0.2565%
	20	-13.55%	-1.32	-0.2181%	-3.97%	-1.68	-0.2490%
	24	-13.38%	-1.24	-0.3127%	-2.83%	-1.01	-0.2231%
	28	-12.59%	-1.08	-0.3835%	-1.96%	-0.58	-0.1832%
Foreman (CIF)	16	-33.44%	-2.16	-0.2154%	-17.08%	-3.20	-0.2388%
	20	-35.32%	-2.12	-0.2565%	-14.89%	-2.27	-0.2046%
	24	-32.60%	-1.72	-0.2912%	-11.70%	-1.35	-0.1624%
	28	-25.20%	-1.19	-0.2659%	-7.51%	-0.71	-0.1153%
Optis (720p)	16	-24.29%	-1.89	-0.4112%	-32.73%	-5.01	-1.1559%
	20	-36.70%	-2.34	-0.8442%	-35.60%	-3.98	-1.2634%
	24	-35.31%	-1.66	-0.7978%	-29.96%	-2.48	-1.1315%
	28	-26.24%	-1.15	-0.7721%	-20.96%	-1.57	-1.1011%
Sheriff (720p)	16	-22.89%	-1.68	-0.2139%	-33.53%	-5.99	-0.9714%
	20	-30.63%	-2.06	-0.4333%	-35.97%	-4.77	-0.9858%
	24	-33.69%	-1.85	-0.5478%	-34.16%	-3.25	-0.8771%
	28	-28.64%	-1.30	-0.4905%	-26.16%	-1.96	-0.7071%
Life (1080p)	16	-24.42%	-1.43	-0.0725%	-22.62%	-5.72	-0.3959%
	20	-27.86%	-1.54	-0.1194%	-22.86%	-4.50	-0.3994%
	24	-29.10%	-1.35	-0.1417%	-21.28%	-3.07	-0.3692%
	28	-25.95%	-1.00	-0.1609%	-16.92%	-1.93	-0.3128%
Tennis (1080p)	16	-14.94%	-1.02	-0.1425%	-16.07%	-4.22	-0.6586%
	20	-21.75%	-1.06	-0.3198%	-18.15%	-3.22	-0.7470%
	24	-15.17%	-0.58	-0.1433%	-11.44%	-2.05	-0.4653%
	28	-12.26%	-0.51	-0.1470%	-8.85%	-1.53	-0.3960%
Average		-24.50%	-1.43	-0.3265%	-18.84%	-2.85	-0.5654%

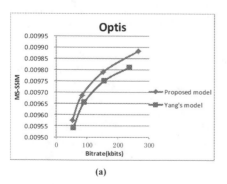

(a)

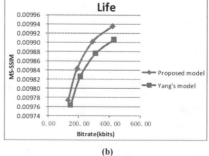

(b)

Fig. 3. Subjective performance comparison (a) sequence "Optics" (b) sequence "Life"

5   Conclusion

In this paper, we have proposed a novel block-level JND model and integrated it with video coding scheme to generate the perceptually adaptive quantization. The proposed algorithm is fully compatible with current mainstream video coding standard and also can be applied for H.264/AVC and HEVC coding framework. The experimental results show average 24.5% bit rate saving with negligible impact on the sequence perceptual quality. In the future, we plan to explore more accurate JND model and integrate its applications into hybrid video coding to further enhance the coding performance.

Acknowledgements. This work is partially supported by grants from the Chinese National Natural Science Foundation under contract No.61171139 and National High Technology Research and Development Program of China (863 Program) under contract No.2012AA011703.

References

1. Chou, C.H., Li, Y.C.: A perceptually tuned subband image coder based on the measure of Just-Noticeable-Distortion Profile. IEEE Transaction on Circuits and Systems for Video Technology 5(6), 467–476 (1995)
2. Yang, X.K., Lin, W.S., Lu, Z., Ong, E.P., Yao, S.S.: Just-noticeable-distortion profile with nonlinear additivity model for perceptual masking in color images. In: IEEE International Conference on Acoustics, Speech, and Signal Processing, pp. 609–612 (2003)
3. Zhenzhong, C., Guillemot, C.: Perceptually-Friendly H.264/AVC Video Coding Based on Foveated Just-Noticeable-Distortion Model. IEEE Transactions on Circuits and Systems for Video Technology 20(6), 806–819 (2010)
4. Yuting, J., Weisi, L., Kassim, A.A.: Estimating Just-Noticeable Distortion for Video. IEEE Transactions on Circuits and Systems for Video Technology 16(7), 820–829 (2006)
5. Zhenyu, W., Ngan, K.N.: Spatio-Temporal Just Noticeable Distortion Profile for Grey Scale Image/Video in DCT Domain. IEEE Transactions on Circuits and Systems for Video Technology 19(3), 337–346 (2009)
6. Hao, C., et al.: Temporal color Just Noticeable Distortion model and its application for video coding. In: IEEE International Conference on Multimedia and Expo (ICME), Suntec City (2010)
7. Huan, W., Xueming, Q., Guizhong, L.: Inter mode decision based on Just Noticeable Difference profile. In: 2010 17th IEEE International Conference on Image Processing (ICIP), Hong Kong (2010)
8. Chun-Man, M., King, N.N.: Enhancing compression rate by just-noticeable distortion model for H.264/AVC. In: IEEE International Symposium on Circuits and Systems, ISCAS 2009, Taipei (2009)
9. Luo, Z., et al.: H.264/Advanced Video Control Perceptual Optimization Coding Based on JND-Directed Coefficient Suppression. IEEE Transactions on Circuits and Systems for Video Technology 23(6), 935–948 (2013)
10. Qi, C., Li, S.: AVS encoding optimization with perceptual just noticeable distortion model. In: 2013 9th International Conference on Information, Communications and Signal Processing (ICICS), Tainan (2013)

11. Wilson, T.A., Rogers, S.K., Myers Jr., L.R.: Perceptual-based hyper spectral image fusion using multiresolution analysis. Optical Engineering 34(11), 3154–3164 (1995)
12. Watson, A.B.: Visually optimal DCT quantization matrices for individual images. In: Data Compression Conference, DCC 19, UT, Snowbird (1993)
13. Tong, H.H.Y., Venetsanopoulos, A.N.: A perceptual model for JPEG applications based on block classification, texture masking, and luminance masking. In: Proceedings of the 1998 International Conference on Image Processing, ICIP 1998, Chicago, IL (1998)
14. Naccari, M., Mrak, M.: Intensity dependent spatial quantization with application in HEVC. In: 2013 IEEE International Conference on Multimedia and Expo (ICME), San Jose, CA (2013)
15. Seshadrinathan, K., Soundararajan, R., Bovik, A.C., Cormack, L.K.: Study of subjective and objective quality assessment of video. IEEE Trans. on Image Proc. 19(6), 1427–1441 (2010)

Improved Prediction Estimation Based H.264 to HEVC Intra Transcoding

Daxin Guang, Pin Tao, Sichao Song, Lixin Feng, Jiangtao Wen, and Shiqiang Yang

Department of Computer Science and Technology, Pervasive Lab of TNList,
Tsinghua University, Beijing, China

Abstract. High Efficiency Video Coding (HEVC) achieves significant coding efficiency improvement at a cost of much higher computation complexity. In this paper, we propose a prediction estimation based H.264 to HEVC intra transcoder. In HEVC intra coding, the most complexity comes from coding unit size decision. The proposed method judges the coding unit size not only by using the intra prediction mode, prediction direction and other information from H.264 decoder, but also by trying the prediction on decoding picture with the most dominant prediction directions of H.264. Both of this information is fed into the classifier. Then the Support Vector Machine (SVM) classifiers were trained and apply to different division level to improve the speed of coding unit size decision accuracy loss as small as possible. Experiment shows that about 1.42 times speed up over the HEVC HM 10.0 reference software at about 0.069dB rate distortion performance loss.

Keywords: HEVC, Video transcoding, Machine learning.

1   Introduction

H.264/AVC standard is the most widely adopted video coding standard which is widely used in internet video broadcasting, video on demand and various storage based video data compressing. In 2013, ITU-T and ISO/IEC MPEG released the next generation of video coding standard named High Efficiency Video Coding(HEVC)[1]. HEVC technology almost double the coding efficiency over H.264/AVC under the rate-distortion performance matrix which means it can save about 50% communication bandwidth or storage volume than the traditional H.264 technology. Nevertheless, HEVC has extremely high computation complexity than H.264 especially in intra encoding. The complexity of HEVC intra encoding mainly comes from trying to traverse all possibility from the largest Coding Unit(CU) size, 64x64, to the smallest one, 8x8, recursively. In order to reduce the encoding complexity by taking into account the industry widely used H.264 encoder and a huge amount of H.264 bit-streams are existed, several transcoding technologies were proposed in recent years. The H.264 to HEVC transcoder could transcode H.264/AVC bit-streams into HEVC bit-streams faster than what the decoder-encoder cascade could do with some encoding quality loss.

W.T. Ooi et al. (Eds.): PCM 2014, LNCS 8879, pp. 64–73, 2014.

Video transcoders have been widely studied by many researchers before HEVC were emerged, e.g. transcoding from H.263 to H.264, MPEG-2 to H.264, etc.[2]. Normally the transcoder will take full advantage of all information of decoder and try to predict various decision mode and parameters of target encoder accurately. In the previous research works, MPEG-1, MPEG-2, H.264 and other ancestor of HEVC technologies use 16x16 size image block as their macroblock in the codec. But one of the most important changing of HEVC is the largest picture block size was enlarged into 64x64, and coding unit will try each level division from 64x64 to 8x8. The block size of H.264 and HEVC are different which may bring some uncertainty of adopting traditional transcoding technology into the-state-of-art transcoder directly.

Zhang et al. [3] proposed a system by modifying the power spectrum based rate-distortion optimization (PS-RDO) model and reducing the candidate settings for CU quadtree structures and PU partitions on intra blocks. This system achieves a speed up of close to 3 times over HEVC HM 4.0 reference software with little R-D performance loss for Intra picture. Shanableh, T. et al.[4] proposed a MPEG-2 to HEVC video transcoder which used the content-based machine learning solution to predict the depth of the HEVC coding units but mainly on inter pictures. About 3 times speed up will be achieved by this method and about 0.1dB loss. ZongYi Chen et. al.[5] re-used the coding information such as motion vector, prediction mode and residual coefficient extracted from the H.264 decoder which can save about 40% encoding time. Wei Jiang et. al.[6] proposed a transcoding method based on region feature analysis. They segmented each picture blocks into three sizes in units of coding tree unit (CTU) based on the correlation between image coding complexities and coding bits of the H.264 source stream. Zhang et. al.[7] proposed a transcoder incorporates SVM classifier based macroblock(MB) partition mode decision and a fast prediction mode decision, but in a reverse transcoding mode, from HEVC to H.264. Tong Shen et. al.[8] reduced the intra mode candidates by calculating the probability condition for QP, H.264 decoding information and reference blocks. Others research work of H.264 to HEVC transcoding, such as [9], mainly focused on inter frame transcoding. For all the above research works, the transcoder for intra frames mainly used the H.264 decoding information, such as intra prediction mode, direction and coding bits, etc. Thinking about the decision of CU division in HEVC should compare the coding cost of splitting the CU or not, we believe the H.264 encoding information is not enough for transcoder to do the decision of the CU division. The motivation of our proposal comes from the transcoder should take into account the both side information of splitting the CU or not. Some information can comes from the H.264 decoder, and we also try to do the prediction on decoding image to get more information for CU splitting decision. Combine the H.264 decoding information and prediction estimation result maybe help to improve the intra transcoding accuracy.

In this paper, we propose a prediction estimation based H.264/AVC to HEVC intra transcoding framework. The remaining of this paper is organized as the following. Section 2 describes the architecture of the proposed intra transcoder. Section 3 gives the detailed implementation of the intra transcoder. Section 4 presents the experiment results on HEVC test sequences. Finally, Section 5 summarizes the work of our paper.

2   Intra Transcoding Framework

The conceptually straightforward cascaded architecture is adopted in this paper as shown in Figure 1. Firstly, we decode the inputting pre-encoded H.264 bit-stream using a H.264 decoder, generating reconstructed pictures and decoding information contained in the H.264 bit-stream. The following HEVC encoder takes the reconstructed pictures and encodes it, outputting the HEVC stream. During the encoding process, the H.264 decoding information is utilized to accelerate the intra RDO process.

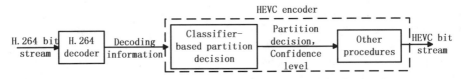

Fig. 1. The architecture of proposed transcoder

It is known that the RDO is a time-consuming process. The information from the H.264 decoding stream is used to help make decision on whether to split the current CU or not during the HEVC encoding. The pre-trained classifier is used to make decision according to the H.264 decoding information and prediction estimation information on decoded picture.

3   Detailed Implementation

3.1   Classifier-Based Partition Decision

In order to improve the encoding quality, several new features are added in HEVC encoding standard. One of the most important features is that the size of the Largest Coding Unit (LCU) is 64x64 which is much larger than the 16x16 size of MB in H.264. A quad-tree structure of CU segmentation is used to calculate the RD cost of the CU in each size level, and then HEVC encoder chooses the best partition for the whole LCU. This process costs a lot of time as the candidates of the CU partitions are drastically increased compared to H.264 intra coding scheme. In order to make the process faster, we treat the CU partition mode decision as a classification problem. Some important information is retrieved from the H.264 decoding stream, such as the mode of MBs, the prediction directions of the sub-blocks and the decoded picture pixels. Furthermore, not only the above information from H.264 bit-stream is used, we also do the prediction with some most dominant prediction directions on the decoding CU image data. This additional information is seemed as the important supplement to describe the both side characters of the CU. The information from H.264 bit-stream is used to describe the character of division of current CU, and the information of prediction on decoding CU image data is used to describe the character of non-division of current CU. Along with other information, e.g. QP value, both of the above characters will be used to compose the feature vector of current CU which will be sent to the pre-trained classifier. The output of the classifier will be a partition decision and a confidence level of the given decision. Figure 2 illustrates this process.

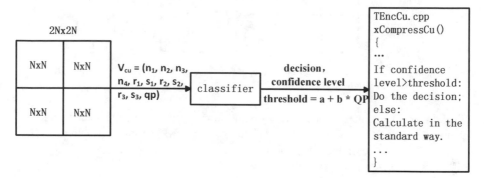

Fig. 2. The process of giving a decision by the classifier

To accelerate the speed of encoding at the expense of minor coding quality loss, a SVM based 2-class classifier is adopted. The input vectors of the classifier are described as following.

In HEVC, there are four types of CU size, from 64x64 to 8x8. For the CU size on 64x64, 32x32 and 16x16, HEVC encoder should recursively search the best decision about splitting the CU or not. The decision is made based on the RD cost in HEVC encoder. On one side, it counts the cost of splitting the CU. On the other side, it counts the cost of not splitting the CU. Inspired by this idea, we propose a vector listed below to describe both side features of a CU.

$$V_{cu} = <n_1, n_2, n_3, n_4, r_1, s_1, r_2, s_2, r_3, s_3, QP>$$

n_i denotes the bits number of four sub-CU with the 1/4 CU size of current CU . We calculate the total bits of each minor square of these four sub-CUs. To some extent, the bit counts of four sub-CUs can describe the coding cost or character of splitting the current CU. The information of bits can be retrieved from the H.264 decoder. As MB is the elementary encoding unit in H.264, if the sub-CU size is no less than MB size, the bit counts is the sum of corresponding bits of MB, otherwise the bit counts can be acquired from the sub-MB information or just the average fraction of MB. Components n_i represents the character of current CU in splitting side.

r_i denotes the average prediction residual of most dominant three directions. In H.264, each MB or sub-MB block will have a prediction direction. Therefore, the most three dominant prediction directions come from the H.264 decoder according to their area proportion in current CU range. We try to do the prediction using the top three dominant directions and derive the prediction average residuals, which are showed in figure 3. Furthermore, s_i denotes the variance of prediction residual of three most dominant directions. Components r_i and s_i represent the character of current CU in non-splitting side.

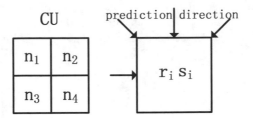

Fig. 3. Input vector for SVM classifier, left denotes the bits number of four minor squares in CU, right denotes the prediction on whole CU

Moreover, QP is included as an additional element in input vector because the CU division decision is strongly correlated with QP value. If the QP is smaller, the CU will more likely to be divided into sub-CUs.

3.2 Confidence Level of Classifier

When the feature vector is generated, it will be sent to the pre-trained two-class classifier. The classifier can give the decision result of whether to split the current CU without calculating the RD-Cost. Certainly, the decision given by the classifier is not always correct. Another label called confidence level is also given by the classifier. The confidence level is a decimal value ranged from 0 to 1, the bigger the value, the credibility is stronger of the splitting decision. When the confidence level is high enough, our proposed transcoder will use the decision to speed up the HEVC intra encoding process. On the contrary, if the confidence level is not high enough, the transcoder had better not adopt the decision provided by the classifier and should calculate the RD-cost to find the best division for the current CU. In order to distinguish between the reliable decision and the uncertain decision, we should designate a threshold. If the confidence level of the decision is larger than or equal to the threshold, we take the result as a reliable one and the CU splitting decision based on RDO is skipped. Then the transcoder is speeded up. Otherwise, the classifier decision will be thought as uncertain. In this case, the encoder will encode the current CU according to the standard HEVC process.

Although the value of the confidence level is from 0 to 1, the confidence level with a value larger than 0.5 will show a reliable decision. Consequently, the value of threshold had better be smaller than 1 and greater than 0.5. Besides, the experiment shows that when the QP is smaller, the picture will be encoded more carefully. In this case, it's more possible for the CU to be split so it is relatively easier for the classifier to make the prediction decision and the confidence level at that time will be somewhat higher. As a result, the value of threshold should be in proportion to the value of QP. Finally, with some experiments, the threshold is formulized as a linear function of QP as denoted by formula (1).

$$threshold = a + b * QP \tag{1}$$

b is the coefficient greater than zero, which can guarantee that the threshold is in proportion to the value of QP. The bigger the QP and b the threshold is higher, and the CU tends to be studied more carefully by the transcoder because the wrong decision will issue the greater loss. The values of a and b could make sure that the value of threshold is smaller than 1 and greater than 0.5.

4   Experimental Results

4.1   Training the Classifier

The Liblinear tool are selected as the SVM two-class classifier because the calculation complexity of training the data is much lower and it costs less time. Besides, the classification performance of Liblinear is almost similar to non-linear classifiers on large scale of data.

Different size YUV sequences from SD to HD are used to check the performance of the proposed transcoder. They are BasketballDrive in 1920x1080, BasketballDrill in 832x480 and BQSquare in 416x240. The H.264 bit-streams of these sequences are encoded and decoded by the H.264 JM reference software in order to get the demanded information of the MBs. Then these decoding sequences are encoded by HEVC standard encoder, HM reference software, so that the best partition choices of HEVC encoder are logged as anchor for future training and comparison. After the information all of H.264 and HEVC are collected, the data file for training will be generated.

In H.264, MB(size 16x16) is the fundamental encoding unit. However, in HEVC, the CU size ranges from 64x64 to 8x8. We unified a 4x4 meta unit to record the information from the H.264 decoder and HEVC encoder. The following two figures show that the process of generating the information of H.264 and HEVC in 4x4 meta units.

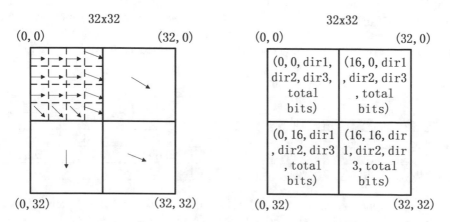

Fig. 4. Generating the information in 4x4 meta unit from the information of the MB in H.264

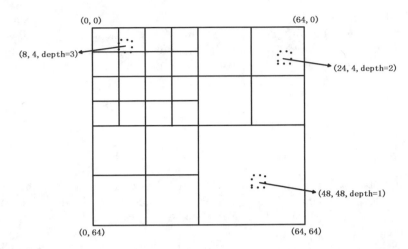

Fig. 5. Generating the information in 4x4 meta unit from the information of the CU in HEVC

Then three data files, containing data of 64x64, 32x32, 16x16 CU, are built from the unified meta data and latter act as the training data and test data of the SVM classifier. For each YUV sequence, the second half of these sequences is fed into the classifier for training. The first half of these sequences is used as the testing data. The feature vector of current CU in them is input to the SVM classifier. The classification result accompany with the confidence level will be given. The proposed transcoder will make decision on splitting, non-splitting or do RDO according to the classification result and its confidence level.

4.2   Evaluation of the Proposed Method

The H.264/AVC bit-stream is generated with JM 16.2 with QP 22, 27, 32, 37. The proposed transcoding scheme is implemented in HM 10.0 and liblinear1.93 which is used as the SVM classifier. To inspect the impact of coefficient in formula (1) on transcoding performance, several coefficient sets are tested as shown in Table 1.

Table 1. Several coefficient sets for threshold

a	b	threshold	QP=22	QP=27	QP=32	QP=37
0.6	0.007	value 1	0.754	0.789	0.824	0.859
0.5	0.009	value 2	0.698	0.743	0.788	0.833
0.4	0.011	value 3	0.642	0.697	0.752	0.807
0.466	0.002	value 4	0.51	0.52	0.53	0.54

With different values of threshold, different BD-PSNR and BD-Bitrate results of the proposed transcoder were derived on three picture size sequence as shown in Tables 2-5.

Table 2. Results of proposed methods with value 1 in Table 1

Sequence	QP	Speed-up	BD-PSNR(dB)	BD-Bitrate
BQSquare (416x240)	22	1.14	-0.000	-0.01%
	27	1.14		
	32	1.10		
	37	1.11		
BasketballDrill (832x480)	22	1.25	-0.011	+0.23%
	27	1.15		
	32	1.08		
	37	1.03		
BasketballDrive (1920x1080)	22	1.10	-0.005	+0.20%
	27	1.03		
	32	1.22		
	37	1.21		
Average		1.13	-0.005	+0.15%

Table 3. Results of proposed methods with value 2 in Table 1

Sequence	QP	Speed-up	BD-PSNR(dB)	BD-Bitrate
BQSquare (416x240)	22	1.15	-0.001	+0.01%
	27	1.14		
	32	1.12		
	37	1.09		
BasketballDrill (832x480)	22	1.59	-0.014	+0.29%
	27	1.42		
	32	1.32		
	37	1.27		
BasketballDrive (1920x1080)	22	1.39	-0.010	+0.35%
	27	1.28		
	32	1.23		
	37	1.20		
Average		1.23	-0.008	+0.22%

Table 4. Results of proposed method with value 3 in Table 1

Sequence	QP	Speed-up	BD-PSNR(dB)	BD-Bitrate
BQSquare (416x240)	22	1.15	-0.000	+0.01%
	27	1.16		
	32	1.11		
	37	1.10		
BasketballDrill (832x480)	22	1.63	-0.015	+0.30%
	27	1.46		
	32	1.35		
	37	1.28		
BasketballDrive (1920x1080)	22	1.42	-0.015	+0.54%
	27	1.30		
	32	1.22		
	37	1.20		
Average		1.28	-0.010	+0.283%

Table 5. Results of proposed methods with value 4 in Table 1

Sequence	QP	Speed-up	BD-PSNR(dB)	BD-Bitrate
BQSquare (416x240)	22	1.16	-0.007	+0.07%
	27	1.16		
	32	1.16		
	37	1.14		
BasketballDrill (832x480)	22	1.71	-0.079	+1.68%
	27	1.69		
	32	1.57		
	37	1.45		
BasketballDrive (1920x1080)	22	1.77	-0.122	+4.81%
	27	1.49		
	32	1.41		
	37	1.33		
Average		1.42	-0.069	+2.19%

According to the experiment results, the proposed transcoder with prediction esti-mation method which uses both information from sub-CUs and un-split CU can achieve the satisfied performance. The encoding time will be reduced to half or one third of the standard HEVC encoder on intra picture with about 0.069dB coding efficiency loss. And we also can see that when the threshold is set reasonably lower, greater speedup can be achieved.

But the speedup mainly comes from the QP, the coefficient in formula (1) mainly affect the coding efficiency. The smaller the value of b, the coding efficiency is better. Finally, we notice that the speedup of proposed transcoder tend to be greater for large-size videos with acceptable loss on BD-PSNR.

5   Conclusion and Future Work

In this paper, we proposed a prediction estimation based H.264/AVC to HEVC intra transcoding method. By using both the character information of sub-CU and unsplit CU, the transcoder can achieve one times or higher speedup with the small coding efficiency loss. The impact of confidence level of threshold for classification result was also explored. Comparing with the HEVC HM-10.0 reference software, about 1.42 speeding up was achieved at a cost of about 0.069dB loss on BD-PSNR.

Acknowledgement. This work was supported by National Significant Science and Technology Projects of China under Grant No. 2013zx01039-001-002-003, and MSRA project under Grant No. FY14-RES-SPONSOR-111.

References

1. Sullivan, G.J., Ohm, J.R., Han, W.J., Wiegand, T.: Overview of the high efficiency video coding (hevc) standard. IEEE Transactions on Circuits and Systems for Video Technology (2012)
2. Ahmad, I., Wei, X., Sun, Y., Zhang, Y.-Q.: Video Transcoding: An Overview of Various Techniques and Research Issues. IEEE Transactions on Multimedia, 7(5) (October 2005)
3. Zhang, D., Li, B., Xu, J., Li, H.: Fast transcoding from h. 264 avc to high efficiency video coding. In: 2012 IEEE International Conference on Multimedia and Expo (ICME), pp. 651–656 (2012)
4. Shanableh, T., Peixoto, E., Izquierdo, E.: MPEG-2 to HEVC video transcoding with content-based modeling. IEEE Transactions on Circuits and Systems for Video Technology 23(7) (2013)
5. Chen, Z.-Y., Tseng, C.-T., Chang, P.-C.: Fast Inter Prediction for H.264 to HEVC Transcoding. In: 3rd International Conference on Multimedia Technology, ICMT 2013 (2013)
6. Jiang, W., Chen, Y., Tian, X.: Fast transcoding from H.264 to HEVC based on region feature analysis. Multimed Tools Appl., 13 (September 2013)
7. Zhang, J., Dai, F., Zhang, Y., Yan, C.: Efficient HEVC to H.264/AVC Transcoding with Fast Intra Mode Decision. In: Li, S., El Saddik, A., Wang, M., Mei, T., Sebe, N., Yan, S., Hong, R., Gurrin, C. (eds.) MMM 2013, Part I. LNCS, vol. 7732, pp. 295–306. Springer, Heidelberg (2013)
8. Shen, T., Lu, Y., Wen, Z., Zou, L., Chen, Y., Wen, J.: Ultra Fast H.264/AVC to HEVC Transcoder. In: Data Compression Conference (2013)
9. Jiang, W., Chen, Y.W.: Low-complexity transcoding from H.264 to HEVC based on motion vector clustering. Electronics Letters 49(19), 1224–1226 (2013)

Unified VLSI Architecture of Motion Vector and Boundary Strength Parameter Decoder for 8K UHDTV HEVC Decoder

Shihao Wang, Dajiang Zhou, Jianbin Zhou,
Takeshi Yoshimura, and Satoshi Goto

Graduate School of Information, Production and Systems, Waseda University, Japan
wshh1216@moegi.waseda.jp

Abstract. This paper presents a VLSI architecture design of unified motion vector (MV) and boundary strength (BS) parameter decoder (PDec) for 8K UHDTV HEVC decoder. PDec in HEVC is deemed as a highly algorithm-irregular module, which is also challenged by high throughput requirement for UHDTV. To solve these problems, four schemes are proposed. Firstly, the work unifies MV and BS parameter decoders to share on-chip memory and simplify the control logic. Secondly, we propose the CU-adaptive pipeline scheme to efficiently reduce the implementation complexity. Thirdly, on-chip memory is organized to meet the high throughput requirement for spatial neighboring fetching. Finally, optimizations on irregular MV algorithm are adopted for 43.2k area reduction. In 90nm process, our design costs 93.3k logic gates with 23.0kB line buffer. The proposed architecture can support 7680x4320@60fps real-time decoding at 249MHz in the worst case.

Keywords: UHDTV, HEVC, parameter decoder, motion vector, boundary strength, real-time decoding.

1 Introduction

Nowadays Ultra High Definition Television (UHDTV) has been a hot topic because of the better visual experience it provides. Up to 8K UHDTV video format has been regarded as the high-end application in the near future. At the same time, the new High Efficiency Video Coding (HEVC) video coding standard has been standardized in Jan. 2013. As a successor to H.264, HEVC can achieve double video compression rate while guaranteeing the equivalent quality [1]. Therefore, video decoder supporting both 8K UHDTV and HEVC is required by the market. However, the VLSI implementation for real-time decoder is challenged since high video resolution directly increases the burden on system throughput because of the huge data volume. Assuming the clock frequency is 250MHz, for 7680x4320@60fps, at least 8 pixel/cycle throughput have to be achieved. On the other hand, compared to H.264/AVC, HEVC introduces new coding tools to achieve better compression performance, such as Advanced Motion Vector Prediction (AMVP) mode, merge mode and flexible quad-tree CU structure [2].

W.T. Ooi et al. (Eds.): PCM 2014, LNCS 8879, pp. 74–83, 2014.

These new algorithms are regarded as the most algorithm-irregular parts in the decoder, which pose problems for high-performance VLSI implementation.

In this paper unified MV and BS parameter decoder is proposed. Figure. 1 illustrates the unified PDec in the whole decoder. The design for unified PDec is feasible and reasonable by following reasons. 1) Original syntaxes from CABAD are too obscure to be directly used by motion compensation (MC) and deblocking filter (DBF) so that pre-process is needed. Data and control flow in pre-process is characterized an irregularity because these processes are highly input-dependent and evolve with the algorithm itself. 2) The pre-process to decoder syntaxes into parameters has less dependency with core calculation in MC and DBF. Further, BS calculation relies on the result of current block's MV, which inspires us to build unified PDec architecture to share internal memory and simplify control logic without much extra overhead. Previously, several parameter decoder approaches for H.264/AVC are reported such as [3–5]. The throughput in worst case is 260 cycles/MB, which is not enough for 8K UHDTV. In Zhou et al.'s work [6], a joint parameter decoder for 4K@60fps UHDTV application is proposed for MV, BS and intra prediction mode. Considering this H.264/AVC approach is not applicable for HEVC and the technique can't support high throughput requirement for 8K video, we can't directly inherit [6] for HEVC solution.

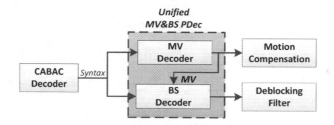

Fig. 1. Unified MV&BS Parameter decoder

In this paper a novel VLSI PDec architecture is proposed with four contributions.

1) Unified MV and BS parameter decoder is proposed to achieve memory sharing and simplify the control logic.

2) CU-adaptive pipeline is proposed to support all the new features adopted by HEVC as well as to efficiently reduce the implementation complexity.

3) On-chip memory is organized by maintaining a line buffer and a set of left-top registers to support data accessing for spatial neighboring blocks.

4) Optimization schemes are designed for implementing the irregular algorithm to reduce 43.2k logic gates.

In total, our proposed parameter decoder architecture can support decoding the 7680x4320@60fps videos at 249MHz clock frequency in worst case.

The rest of the paper is organized as followings. We first give an overview for PDec in HEVC in Sect. 2. Section 3 shows PDec pipeline design in detail. In Sect. 4, on-chip memory organization related to PDec is discussed. Implementation results are shown in Sect. 5. Finally, we conclude the whole paper in Sect. 6.

2   Overview of Parameter Decoder in HEVC

As is shown in Fig. 1, the proposed unified parameter decoder contains two major parts, motion vector calculation and boundary strength calculation. Unlike the work in [5, 6], our proposed PDec excludes intra prediction mode. In HEVC standard, the result of intra prediction is used as a feedback by CABAC for decoding following transform information. Hence, in the view of system level, we capsule the intra prediction mode inside CABAC to avoid long feedback path, instead of PDec. A brief discussion on MV and BS is given below.

2.1   Motion Vector Calculation

MV calculation is in charge of decoding syntax elements into motion parameters which can be directly used by MC. Motion parameters contain locality in both spatial and temporal regions. In HEVC, irregular coding algorithm is employed to eliminate such kinds of redundancy for compression efficiency.

MV calculation in HEVC contains two modes, Advanced Motion Vector Prediction (AMVP) mode and merge prediction mode. Both of them require prediction parameters of five spatial neighboring blocks and 2 temporal co-located blocks as input, as depicted in Fig. 2. Besides, each of the two prediction modes owns its unique algorithm so resource sharing between them is limited. In AMVP, all seven reference blocks are categorized into three regions, A, B and Col regions. Each region produces one candidate so that a list of at most three candidates can be constructed. After removing the identical ones in the list, the final MV will be chosen by a syntax mvp_lx_flag. In merge mode, motion parameter decision starts with constructing merge candidate list. Firstly, valid blocks are pushed into the list in the order of B0, A0, A1, B1, B2 and Col. Then the combined candidates will be assembled and added into list with the valid candidates in the first step. Finally, zero candidates with different reference frame are produced if the list isn't full. After the list is constructed, merge MV result is chosen by the merge_idx from the list. Note that in each mode, the scaling operations will be processed when there is a difference between reference frame of current block and that of reference block.

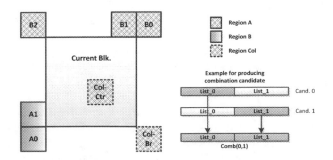

Fig. 2. Candidate Selection in MV Calculation

Table 1. Cycles Allocation for Different CU cases

CU Size	Process Cycles for Each Stage (Cycles for CU contains 2 PUs are shown in brackets)			Max. Cycles for each CU
	Stage 1. Memory Read	Stage 2. MV Calculation	Stage 3. BS Cal. and Mem. Wr.	
8x8	8(8)	4(8)	3(4)	8(8)
16x16	10(12)	4(8)	7(10)	10(12)
32x32	18(20)	4(8)	15(22)	18(22)
64x64	34(36)	4(8)	31(46)	34(46)

2.2 Boundary Strength Calculation

Boundary strength is used in deblocking filter for filter selection. Current block's parameters and parameters of top and left neighboring blocks should be fetched. The neighboring data is compared with current block's to calculate boundary strength. BS value is only calculated at edges of prediction unit and transforms unit which locate at 8x8 block grid.

3 The Unified CU-adaptive Pipeline

3.1 Pipeline Analysis

We propose the unified CU-adaptive pipeline strategy for balancing between performance and design complexity. We define the pipeline granularity is CU instead of fixed size in previous works like 6 for high throughput requirement of UHDTV. HEVC supports maximum 64x64 block size. If fixed 4x4 pipeline granularity is used, large redundant calculation directly degrades the performance (for example, one 64x64 prediction unit needs up to 256 4x4 pipeline blocks, even if pixels in 64x64 block share one motion vector).

Calculation for motion vector can be summarized as following steps. Firstly, memory is read for reference data fetching. Then in the second step, the correspondent algorithm is executed to get the results. Finally, the results of current block needs to be written back to the memory because the following blocks may use them as references. BS calculation follows the similar process. We notice that reference data of BS has overlaps with that of MV. Meanwhile, BS calculation needs the result of MV, so BS calculation should be scheduled after the correspondent process for MV. By incorporating the two processes together, we define the unified architecture as consisting of three main stages: 1) memory reading for reference data fetching, 2) MV calculation for current CU and 3) BS calculation and MV writing back. Table 1 shows the necessary process cycles for each stage and for all kinds of CU sizes and partition modes.

Pipelined architecture is necessary to support 7680x4320@60fps application. We define the worst case is that the whole frame is coded as 8x8 coding unit.

In such situation the cycles for each stage is 8, 8 and 4, as shown in right-most column of Table 1. Non-pipelined sequential implementation needs 20 cycles for one 8x8 block, which is quite slower than the required 8 pixels/cycle. Thus, pipeline structure should be employed for 8K UHDTV application by considering the throughput requirement and the implementation complexity.

3.2   Pipeline Implementation

As we know that a pipeline's throughput is mainly decided by the slowest stage, hence we decelerate the faster stage so that the speed is the same as the slowest one for each kinds of CU. It is achieved by adding NOP cycles in the trailing of fast stages 3(a). Meanwhile, the fixed delay between each two stages (4 cycles for stage 1&2 and 6 cycles for stage 2&3) is designed to achieve seamless joint for processing any two continuous CUs. The proposed pipeline schedule is shown in Fig. 3(b). In detail, the first stage is in charge of fetching neighboring data from line buffer, left-top register and temporal SRAM. The second stage is to calculate MV parameters. This stage is not the bottleneck for the speed but is the core part for irregular algorithm. The last stage is to calculate BS as well as to write MV and BS into memory. The process cycles of this stage also vary with CU's characteristics.

Forward by-passing is utilized for potential memory conflict hazard as in Fig. 4. Because of pipeline, we allocate memory reading and writing in stage 1 and stage 3 correspondingly. Thus two potential memory conflict hazards exist. Firstly, the same memory address may be read and written in one cycle simultaneously. The second is write-after-read hazard caused by delayed writing operation. Forward by-passing schemes can resolve such problems. We have fixed the delays between stages so that stage 3 is 10 cycles delayed than stage 1. Meanwhile, the minimum process cycles are 6 cycles when CU is 8x8. Therefore, when current CU is on its third stage, at most following two CU go through

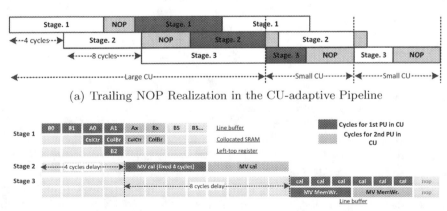

(a) Trailing NOP Realization in the CU-adaptive Pipeline

(b) Cycle Resource Allocation in the CU-adaptive Pipeline

Fig. 3. Analysis on Designing the CU-adaptive Pipeline

its first stage, in which memory access conflict may happen during these 10 cycles. Thus we use registers to store the previous two CUs' result. Whenever current CU wants to access contents of previous two CU, we directly fetch them from registers and bypass the access for line buffer. The forward by-pass scheme efficiently deals with the potential hazards for the line buffer.

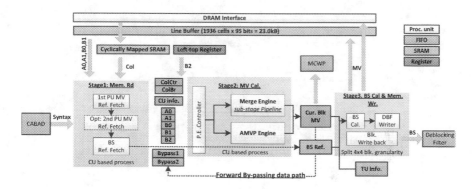

Fig. 4. Blockdiagram for the CU-adaptive Pipeline

The framework of the unified parameter decoder is shown in Fig. 4. As a result, no matter what the current and next CU's partition mode and size are, they can be processed continuously without pipeline pause to guarantee high pipeline performance.

3.3   Optimization on Stage 2

Stage 2 is in charge of Algorithm-irregular MV calculation. We exploit the potential regularity inside and do hardware optimization for area-efficient implementation. Firstly, scaling operation is optimized. Scaling operation contains one division and two multiplications, leading to inefficient area and timing cost, which is around 5ns and 10k gates, correspondingly. For area cost, we notice that though both region A and B contain scaling operation, only one of the operation can occur for each prediction unit, thus we extract scaling operator to save one scale operator. Similar area optimization is used in temporal candidates' scaling as only one of ColCtr and ColBr can be used. For timing issue, a four sub-stage pipeline inside stage 2 (MV calculation stage) is designed to allocate one divider and two multipliers in each stage. Thus 5ns critical path is fragmented so that we can achieve 400MHz timing constrains.

Secondly, merge mode process engine is optimized. Candidate list's construction consists of several steps as is depicted in Fig. 5(a). As steps are data dependent to each other, combined candidates and zero candidates need to be produced sequentially, leading to long data path. Meanwhile, large area cost is needed for MUX and candidate producer's implementation. The performance is

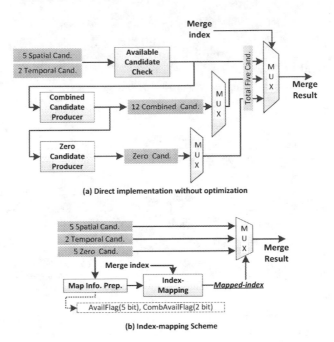

Fig. 5. Index-mapping Scheme for Merge Mode Process

degraded under this structure. Thus we propose Index-mapping scheme to de-couple the dependency as shown in Fig. 5(b). The Index-mapping scheme only receives merge index and available flags of each reference blocks to generate the absolute index of final result. So that we skip the construction of candidate list and directly get merge result based on the decoded index. As the input for merge index decoder is only ten bits, the implementation is quite smaller and simpler than the un-optimized structure.

4   The Proposed On-chip Memory Organization

In this section, the organizations of on-chip memory are discussed. We mainly focus on the memory organization for spatial neighboring's storage because the numbers of temporal reference blocks is much fewer, which is fixed two for each coding unit.

Data accessing is considered as the bottleneck for achieving high throughput requirement of 8K UHDTV application. As is mentioned in section 2, five neighboring blocks are needed for MV process and all left and top neighboring blocks for BS calculation. All the data should be fetched within the scheduled cycles listed in Table 1. In order to meet the throughput, line buffer is maintained as shown in Fig. 6. Considering the ability to distinguish different blocks, the proposed line buffer's cell is set equal to minimum TU size (4x4). Meanwhile, to simplify the control logic, single line buffer is employed to consist of not only

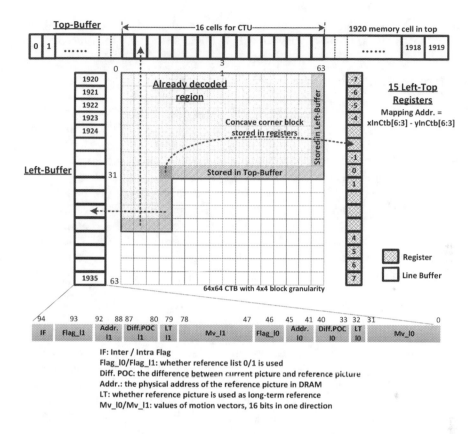

Fig. 6. Line Buffer and Left-top Register Organization

the storage of the top row (Top-Buffer) but also the left blocks of current CTU (Left-Buffer). Under this memory organization, the neighboring block A0, A1 are stored in Left-Buffer while B0, B1 are in Top-Buffer correspondingly. For each prediction unit, these four blocks can be read out from line buffer in four cycles. If one CU is divided into two PUs, at least two blocks (A0, B1 or A1, B0) can be reused so extra two cycles is needed. Further, A1 and B1 can be reused for BS calculation to save extra two cycles.

In addition, fifteen registers are maintained to store the top-left B2 blocks. B2 is not stored in line buffer like others. The motivation comes from two aspects. One is that the proposed replacing strategy for line buffer can't store the blocks at the concave corner in the decoded regions. So the top-left registers help implement the defection of line buffer. On the other hand, we use register instead of SRAM for B2 storage. The reason is that B2 will be read and refreshed frequently inside CTU, implementation by register will eliminate the potential memory accessing conflict problem (read and write simultaneously).

In conclusion, we use line buffer and left-top registers to store the spatial neighboring data. Left-top registers can be read simultaneously while line buffer is accessed, so that four cycles can guarantee the fetching for five spatial neigh-boring blocks and the rest of the schedule can be used for BS reference fetching.

5   Implementation Result

The unified PDec is implemented on RTL level design using Verilog HDL and further synthesized by Synopsys Design Compiler with TSMC 90nm process. In detail, the maximum clock frequency of our proposal can achieve 400MHz. The area under 300HMz is 93.3k. The on-chip line buffer's size is 23.0kB. Our pro-posed architecture is adequately capable for real-time decoding 7680x4320@60fps for the worst case (defined in 3.1) even if worst case is too strict to be happened. We also simulate our proposal on HEVC test sequence and the result shows the average process speed is 17.8pixel/cycle, which is able to finish decoding highest profile 6.3 7680x4320@120fps applications at only around 111.8MHz on average.

Table 2 shows the comparison with other related works. Compared to existing works, ours is the only one that supports HEVC standard. Though in [6] intra prediction mode calculation is included, it affects area and timing cost little since intra algorithm is quite simple compared to that of MV, especially in HEVC. On the other hand, line buffer size is much larger than others for three reasons. 1) HEVC support PU's edge equals to 4 at least, so storage for 4x4 block granularity is needed; 2) MV parameters in HEVC is more than that in H.264; 3) 8K UHDTV's frame width is twice larger than 4K's, leads to double size of line buffer. Finally, as total area is related to the throughput, we define the normalized gate number in the table for a fair comparison. The normalized results show that our proposed unified architecture has around twice efficiency on area cost, even if supported HEVC's complexity is more than that of H.264/AVC.

Table 2. Comparison with state-of-the-art works

	Tao, [5]	Zhou, [6]	This Work
Standard	H.264/AVC&AVS	H.264/AVC	H.265/HEVC
Function	MV/BS/Intra	MV/BS/Intra	MV/BS
Worst-case Throughput (Pixel/cycle)	0.73	4.0	4.0
Avg. Throughput (Pixel/cycle)	1.6	4.0	17.8
SRAM Size (Line buffer)	4.8k(1080p)	3.6k(2160p)	23.0k(4320p)
Logic Gate	63.0k	37.2k	93.3k
Norm. Logic Gate[a]	101.27	7.47	4.69
Max. Resolution	1920x1080@30fps 84MHz	3840x2160@60fps 124MHz	7680x4320@60fps 249MHz
Technology	65nm	90nm	90nm

[a] As the supported maximum resolution varies, we define normalized logic gate equals to logic gate divided by processed pixels per second (x10^{-5}).

6   Conclusion

This paper proposes unified parameter decoder architecture for UHDTV. The design can accomplish the algorithm of MV and BS calculation in new HEVC standard. In particular, CU-based pipeline strategy is proposed to simplify control logic. Moreover, memory organization is designed for spatial neighboring storage to guarantee enough bandwidth. Finally, optimization on irregular algorithm is adopted for 43.2k logic gates reduction. In total, the proposed unified parameter decoder supports 7680x4320@60 fps real-time video decoding at 249MHz in worst case.

Acknowledgments. This work is supported by Waseda University Graduate Program for Embodiment Informatics (FY2013-FY2019).

References

1. Sullivan, G.J., Ohm, J., Han, W., Wiegand, T.: Overview of the High Efficiency Video Coding (HEVC) Standard. IEEE Transactions on Circuits and Systems for Video Technology 22(12), 1649–1668 (2012)
2. JCT-VC: High Efficiency Video Coding (HEVC) Test Model 13 (HM 13) Encoder Description. JCTVC-O1002 (November 2013)
3. Yoo, K., Lee, J., Sohn, K.: VLSI architecture design of motion vector processor for H.264/AVC. In: 15th IEEE International Conference on Image Processing, pp. 1412–1415. IEEE Press, New York (2008)
4. Yin, H., Zhang, D., Wang, X., Xiao, Z.: An efficient MV prediction VLSI architecture for H.264 video decoder. In: 1st International Conference on Audio, Language and Image Processing, pp. 423–428. IEEE Press, New York (2008)
5. Tao, Y., He, G., He, W., Wang, Q., Ma, J., Mao, Z.: Effective multi-standard macroblock prediction VLSI design for reconfigurable multimedia systems. In: IEEE International Symposium on Circuits and Systems (ISCAS), pp. 1487–1490. IEEE Press, New York (2011)
6. Zhou, J., Zhou, D., He, X., Goto, S.: A bandwidth optimized, 64 cycles/MB joint parameter decoder architecture for ultra high definition H.264/AVC applications. IEICE Trans. on Fundamentals E93-A(8), 1425–1433 (2010)

Using Label Propagation to Get Confidence Map for Segmentation

Haoran Li, Hongxun Yao, and Xiaoshuai Sun

School of Computer Science and Technology,
Harbin Institute of Technology, Harbin, China
h.yao@hit.edu.cn

Abstract. We propose a novel algorithm to segment objects from the existed segmentation results of the co-segmentation algorithms[1]. Previous co-segmentation algorithms work well when the main regions of the images contain only the target objects; however, their performances degenerate significantly when multi-category objects appear in the images. In contrast, our method adopts mask transformation from multiple images and discriminatively enhancement from multiple object categories, which can effectively ensure a good performance in both scenarios. We propose to use sift-flow[2] between pre-segmented source images and target image, and transform the source images' segmentation mask to fit the target testing image by the flow vectors. Then we use all the transformed masks to vote the testing image mask and get the initial segmentation results. We also propose to use the ratio between the target category and the other categories to eliminate the side effects from other objects that might appeared in the initial segmentation. We conduct our experiment on internet images collected by Rubinstein .etc[1]. We also do additional experiment to study the multi-object conjunction cases. Our algorithm is effective in computation complexity and able to achieve a better performance than the state-of-the-art algorithm.

1 Introduction

Previous papers considered the task of pixel segmentation of similarly looking objects in a number of images as image co-segmentation[3], where no prior information is available. The recent algorithm proposed by Rubinstein[1] is using saliency map[4] as a prior information based on a hypothesis that the subject object locates in salient region.

Recent research indicates that the task of multi-object semantic segmentation on images with several common objects can be regarded as a combination of two sub-problems, i.e. label propagation[5] and belief voting [6]. Inspired by the previous findings, we argue that patches sampled from two semantic similar images probably share similar visual contents as well as the same semantic labels, and the more similar between the two images, the more similar between the corresponding semantic masks. The above argument forms the main hypothesis of this paper, in which we divide the semantic co-segmentation task

W.T. Ooi et al. (Eds.): PCM 2014, LNCS 8879, pp. 84–92, 2014.

into two sub-stages: 1) label propagation via SIFT-Flow and 2) label validation by probabilistic belief voting. Given a concept, we adopt co-segmentation[1] to automatically obtain a small set of pre-segmented images. Then, at the label propagation stage, we transform the pre-segmented images' masks to fit the target image and get the initial semantic segmentation result using SIFT-Flow[2] based warping algorithm. Finally, we adopt probabilistic voting strategy to obtain the confident segmentation mask by counting the positive responses from all transformed masks.

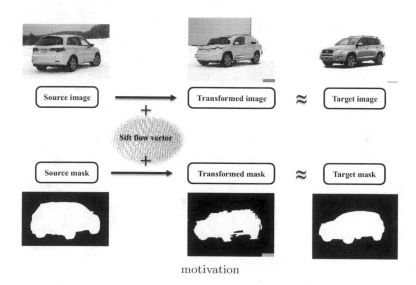

motivation

Fig. 1. We warp the source image to fit target image by sift-flow, so it is practicable to warp the the segmented mask of the source image to get the mask of target image

Our experiments are conducted on the internet images gathered by Rubinstein etc. We randomly select 100 images from each class along with their segmentation masks as the basic concept model, and then select 200 images randomly as testing examples.

2 Related Work

Rubinstein et al. [1] adopted saliency maps to detect the subject region and eliminate the distracting contents in the images. However, when multiple objects are presented in the image, the object belonging to the target concept might not be highlighted or even covered by the saliency map because the other objects might be much more salient within that image context. Thus, in such cases, the Rubinstein's algorithm could probably fail to locate and segment the target semantic region correctly. In this paper, we adopt Rubinstein's algorithm to get a group of rough segmentations as a noisy concept model, and then apply it to propagate semantic labels to new images. We demonstrate that the potential error and noises introduced by Rubinstein's algorithm can be effectively handled by cross-model discriminative enhancement.

Recent works of label propagation can be divided into three types. 1) propagate labels of a image globally, which is weak tagging technique in traditional image annotation research. 2) propagate labels of superpixels extracted from images, which stronger and much more complex than 1). Some algorithms use such kind of label propagation to detect and segment the objects, and use MRF to smooth the segmentation mask. 3) propagate the label of each pixel in the image, which is the strongest label propagation technique. The co-segmentation proposed by Rubinstein[1] is based on this label propagation. Our algorithm is also based on 3), in which we treat each segmentation mask as the label of each pixel in the pre-segmented image.

3   The Model

Our algorithm is based on the pre-segmentation results of Rubinstein et al. [1] which contains a small group of images and the images' segmented binary mask. A concept mode includes two components. One is images $\mathbf{I} = \{I^1, I^2...I^n\} = \{\{I_1^1, I_2^1.....I_i^1\}, \{I_1^2, I_2^2.....I_i^2\}, ...\}$, n=3 and i=100 . $\mathbf{I}$ was constructed by randomly sampling the Internet images collected by Rubinstein [1]. The other component is the semantic binary masks $\mathbf{B} = \{B^1, B^2...B^n\} = \{\{B_1^1, B_2^1.....B_i^1\}, \{B_1^2, B_2^2.....B_i^2\}, ...\}$, n=3 and i=100 , which can be obtained by Rubinstein's algorithm.

3.1   Framework

As showed in Figure 2, we divide our algorithm into 2 stages. First, We calculate the flow vector by using SIFT-Flow[2] between two images to transform the masks to get the target image's mask. Then, we combine all the transformed masks into one confidence map for the target concept. Finally, we discuss how to use cross-concept analysis to further refine the confidence maps. A good confidence map have a strongly effect on accurate object segmentation. Based on the final confidence map, we can use several algorithms like grab cut, graph cut, or even using threshold to get the binary segmentation mask. We show in the discussion section that our framework has a very flexible structure and can also be easily extended to more complex scenarios such as multi-object co-segmentation.

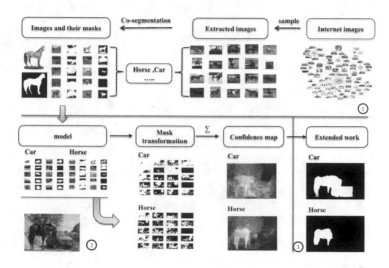

Fig. 2. Stage 1,using sift-flow vector to get the transformed mask; Stage 2, do statistic in transformed masks for confidence map;Stage 3, extended work

3.2   Mask Transformation

Previous algorithm use sparsity correspondence and sift[7] feature to match visual contents between images. Here, we use SIFT-Flow to match the pathes between images. The objective function of SIFT-Flow is:

$$\mathbf{E} = \sum_{p \in X} (min(| S_i(p) - S_j(p + W(p)) |, t)$$
$$+ \eta(| u(p) | + | v(p) |)$$
$$+ \sum_{(q,p) \in \varepsilon} (min(\alpha | u(p) - u(q) |, d)$$
$$+ min(\alpha | v(p) - v(q) |, d))) \tag{1}$$

S_i and S_j are SIFT feature maps of the testing image and the pre-segmented images. p denotes a point in the testing image, $p \in S_i$. $w(p)$ is the flow vector at point p.

Our algorithm can be regarded as a weakly supervised model based on pre-segmented images and the corresponding binary masks. After constructing dense correspondence between the images, we can take the flow vectors as the directions along which the known labels are propagated. $\mathbf{TB} = \{TB^1, TB^2...TB^n\} = \{\{TB_1^1, TB_2^1.....TB_i^1\}, \{TB_1^2, TB_2^2.....TB_i^2\}, ...\}.$

$$\mathbf{TB}_i^n = B_i^n(p + W_i^n(p)) \tag{2}$$

According the hypothesis presented before, we will get several sets of transformed masks which all probably contain the regions of objects that belong to the corresponding class in the testing image.

3.3   Confidence Map

For each class, we choose 100 images from Rubinstein's Internet image collection randomly, which means the images and masks of our pre-defined concept model are very diverse and noisy. Rubinstein et al. found that the images in common class probably have the common objects[1]. Here, we count all the transformed mask to dismiss the distracters and highlight the common semantic regions.

We consider each transformed mask TB_i^n as a vote for all pixels in the testing image on object class n. Therefore, the confidence map $\mathbf{CM} = \{CM^1, CM^2, \dots CM^n \dots\}$ is the statistical result of voting.

$$\mathbf{CM}^n = \sum_{i \leq size(Cn)} (TB_i^n) \tag{3}$$

where, Cn is the the nth object class and $size(Cn)$ is the number of the images in the object class.

After this stage, we get all class-specific confidence maps of the testing images. The confidence map, as the result of our algorithm, is the mid-level representation of semantic segmentation. A simple way to segment the object is to use a threshold such as expectation to binaryzation. As a considerable way to segment the target object using confidence map, we found that the grabcut[8] is a good choice to get binary segmented mask.

4   Experiment

The experiments are conducted on Rubinstein's Internet Dataset. We implemented Rubinstein's approach using the default parameters [1] as the main baseline.

4.1   Concept Localization

Similar with Rubinstein [1], we test our approach on three general concepts including horse, car and air plane. In this test, we select 200 images per concept as testing data and select 100 images to train the initial concept model.

ROC Test - We set threshold from 0 to 1 and stepwise is $\frac{1}{64}$ to get the ROC curve from the confidence maps. The results are shown in Figure 3. The y-axis is true positive ratio

$$tpr = \frac{Area(mask \cap Gtruth)}{Area(mask \cap Gtruth) + Area(\overline{mask} \cap Gtruth)}$$

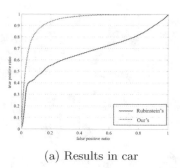 

(a) Results in car   (b) Result in horse

Fig. 3. Results in one-object segmentation. The performance of our algorithm in the one-object dataset is better.

and the x-axis is the false positive ratio

$$fpr = \frac{Area(mask \cap \overline{Gtruth})}{Area(mask \cap Gtruth) + Area(\overline{mask} \cap Gtruth)}$$

.

As shown in Figure 3.a, our algorithm get better performance than Rubinstein's approach. Rubinstein et al. use saliency map and intra image, which is a strong bias on their energy function and will cause potential overfitting of the segmentation region. In contrast, our algorithm is based on label propagation, it can lead to satisfied results when most of the pre-segmented masks are plausible. As shown in Figure 3.b, the results of "horse" for both approach are worse than those of "car". We analysis the horse images and draw a conclusion that the Internet images of "horse" is much more diverse than the other two classes. The images might contain only the head of a horse, the body of a horse, or even a group of horses. In some of the challenging cases, the horse are confused with the background and thus hard to be found even by human observers. Figure 4 shows some examples of the confidence maps for "horse" and "car".

4.2   Extended Experiment

we use the testing images combined with two objects car and horse. we get two ROC curve in both car class and horse class as the same evaluating method used in one object cut.

As shown in the Figure 5, when using Rubinstein's algorithm to segment images with more than one object, the saliency map will misses some of or even the whole part of the target objects and the distraction of the other objects will probably lead to ambiguous segmentations. Our algorithm is based on probabilistic voting, which tend to highlight the regions with the target object rather than those with the other objects as long as most of the pre-segmented masks are correct. Therefore, the segmented result of our algorithm is stably better than Rubinstein's.

Fig. 4. Confidence map in horse and car. The second column is Rubinstein's, the third column is ours. Our confidence map is better by human observation.

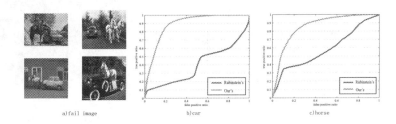

Fig. 5. Results in multi-object segmentation. The performance of our algorithm in the multi-object dataset is better.

5   Discussion

By introducing cross-concept analysis, we can use the confidence map of other concepts to eliminate the distractions from the other salient objects in the image and thus provide a purer mask for the target object. Figure 6.a shows a typical failure case in Rubinstein's framework, where a car and horse are presented in the same image and both objects are quite salient. The ROC curve shown in Figure 6.b is the result of the confidence map for horse dismissing the other concepts' effect. The results demonstrate the effectiveness and potential of cross-concept analysis for post noise removal.

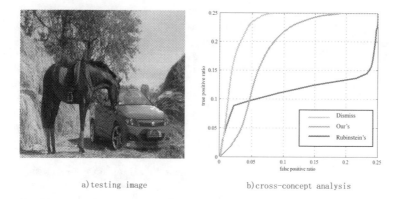

a) testing image   b) cross-concept analysis

Fig. 6. After dismissing the other conceptions' effect, the performance has improved

6   Conclusion

In contrast to previous algorithms, we bring in a novel way to use a small group of roughly segmented Internet images to model general concepts and localize the corresponding object in raw images. The paper has two major contributions: 1) propose a simple yet stable way to model and segment concepts; and 2) provide a novel cross-model approach to ensure the segmentation performance under conceptual distractions. Transferring the label propagation by transforming the segmented mask using sift-flow vector is a effective way to get target object labeled mask, and using belief voting by treating the transformed mask as a voting for all pixels in the image is a effective and stable way to get clean confidence map. We will do further research on how to use other objects' confidence map to refine the target map in large scale scenarios.

Acknowledgement. This work was supported in part by the National Science Foundation of China No. 61472103, and Key Program Grant of National Science Foundation of China No. 61133003.

References

1. Rubinstein, M., Joulin, A., Kopf, J., Liu, C.: Unsupervised joint object discovery and segmentation in internet images. In: 2013 IEEE Conference on Computer Vision and Pattern Recognition (CVPR), pp. 1939–1946. IEEE (2013)
2. Liu, C., Yuen, J., Torralba, A.: Sift flow: Dense correspondence across scenes and its applications. IEEE Transactions on Pattern Analysis and Machine Intelligence 33(5), 978–994 (2011)
3. Vicente, S., Rother, C., Kolmogorov, V.: Object cosegmentation. In: 2011 IEEE Conference on Computer Vision and Pattern Recognition (CVPR), pp. 2217–2224 (2011)

4. Cheng, M.-M., Zhang, G.-X., Mitra, N.J., Huang, X., Hu, S.-M.: Global contrast based salient region detection. In: 2011 IEEE Conference on Computer Vision and Pattern Recognition (CVPR), pp. 409–416. IEEE (2011)
5. Rubinstein, M., Liu, C., Freeman, W.T.: Annotation propagation in large image databases via dense image correspondence. In: Fitzgibbon, A., Lazebnik, S., Perona, P., Sato, Y., Schmid, C. (eds.) ECCV 2012, Part III. LNCS, vol. 7574, pp. 85–99. Springer, Heidelberg (2012)
6. Yang, Y., Liang, Q., Niu, L., Zhang, Q.: Belief propagation stereo matching algorithm using ground control points. In: Fifth International Conference on Graphic and Image Processing. International Society for Optics and Photonics, pp. 90690W–90690W (2014)
7. Lowe, D.G.: Distinctive image features from scale-invariant keypoints. International Journal of Computer Vision 60(2), 91–110 (2004)
8. Rother, C., Kolmogorov, V., Blake, A.: Grabcut: Interactive foreground extraction using iterated graph cuts. ACM Transactions on Graphics (TOG) 23, 309–314 (2004)

Image Region Labeling by Exploring Contextual Information of Visual Spatial and Semantic Concepts

Kai He[1], Wen Chan[1], Guangtang Zhu[1], Lan Lin[2], and Xiangdong Zhou[1]

[1] School of Computer Science, Fudan University, China
{12210240019,11110240007,12210240052,xdzhou}@fudan.edu.cn
[2] School of Electronics and Information Engineering, Tongji University, China
linlan@tongji.edu.cn

Abstract. Region Labeling is to automatically assign semantic labels to the corresponding image regions. Most of the previous works focus on exploiting low level visual features, particularly visual spatial contextual information, to address the problem. However, very few work explore high level semantic information of the whole image to deal with the problem. In this paper, we propose a new region labeling approach by integrating both visual spatial and semantic contextual information into a unified model. In our method, region labeling is regarded as a multi-class classification problem. For each semantic concept, we train a Conditional Random Field (CRF) model respectively. It consists of both the region grid sub-graph and the co-occurred semantic label sub-graph. In our model, the integration of the two kinds of contextual information brings reinforcement effect on the improvement of region labeling. The experiments are conducted on two commonly used benchmark datasets and the experimental results show that our method achieves the best performance compared with the strong baselines and the state-of-the-art methods.

Keywords: Image Region Labeling, Region Graph, Concept Graph, Conditional Random Field.

1 Introduction

Image region labeling (tagging) has attracted significant attentions in broad research and industry communities. Region labeling is usually formulated as a multi-class classification problem, whose goal is to automatically select a class label from a predefined vocabulary and assign it to the corresponding image region (patch). Since predicting semantic label by only exploiting the visual feature of local image region or patch cannot achieve robust labeling results, many researchers take the visual spatial contextual information into consideration. Under the assumption that adjacent regions have a higher likelihood of belonging to the same label, as illustrated in Figure 1(a), the idea of "label smoothing" has attracted much research attentions [13, 14, 17].

W.T. Ooi et al. (Eds.): PCM 2014, LNCS 8879, pp. 93–102, 2014.

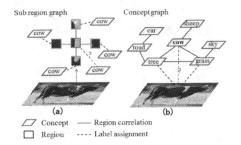

Fig. 1. (a) A Sub Region Graph in an image. (b) An example to show the usage of a concept graph.

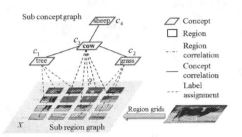

Fig. 2. An example to show the framework of our model for jointly spatial context and semantic context modeling

Despite of focusing on visual features, recent work on utilizing high level semantic information shows promising results in improving the image region labeling performance. Zheng *et al.* [21] learn a hierarchical semantic dictionary for image region labeling. Different from their work of exploiting tag taxonomy which is made manually, we propose to integrate the semantic contextual information into the image region labeling automatically. The semantic context refers to semantic concepts which co-occur frequently in the labels of images. For instance, in Figure 1(b), "cow" and "grass" are two frequently co-occurring concepts. Intuitively, knowing the image labeled with "grass" indicates that the label "cow" will have higher probability of occurring in some regions of the same image. Both semantics and visual features have been exploited in some previous works for whole image annotation [5, 7, 16, 11, 18–20], which is regarded as a multi-label classification problem. To the best of our knowledge, we are the first one to explore the contextual information of visual spatial and semantic concepts for image region labeling. We propose to integrate the traditional visual spatial "label smoothing" and the high level semantic co-occurrence constraints into a CRF model by using a hybrid graph, which consists of the region grid sub-graph and the semantic co-occurrence sub-graph. The main idea is illustrated in Figure 2. In the Figure, the concept "cow" has a sub concept graph which is learned from the data. When judging the correlation between the concept "cow" and the grid in the middle of the green regions, if one or some neighbouring concepts such as "tree" and "grass" have already been labeled in the image, the probability of classifying the region to "cow" will be increased. Meanwhile, if most of the four neighbouring regions have the same positive label of concept "cow", the probability will also be increased. Consequently, exploiting these two kinds of contextual information in our model inference form a reinforcement power to improve the region labeling performance significantly.

In summary, our contributions in this work are two fold: (1) we explore the high level semantic contextual information for image region labeling and propose a unified model for both visual spatial and semantic context modeling; and (2) we conduct extensive experiments on two widely used benchmark datasets: the MSRC-v2 [14] and the COREL collection. Compared with the strong baselines and the state-of-the-art methods, the experimental results show that our method improves the region labeling accuracy significantly. The rest of the paper is organized as follows: Section 2 reviews some related work. Section 3 presents our model design. Section 4 describes details of the parameter estimation and the inference procedure. Section 5 presents the experiments and our discussions, and in Section 6 we conclude the paper.

2 Related Work

In recent years, many research efforts have been devoted to the problem of image region labeling. Athanasiadis *et al.* [1] propose a method which involves region labeling and image segmentation simultaneously. Bi-Layer sparse coding is presented by Liu *et al.* [10], where they construct the semantic regions by selecting atomic pathes in each layer and propagating the labels from images to regions. Spatial contextual information ("label smoothing") has shown great usefulness in region labeling problem. Shotton *et al.* [14] utilize multi-local evidences to represent the region and also use a CRF to model the spatial correlation between adjacent pixels. Rubinstein *et al.* [13] propagate labels via dense image correspondences as well as the pixel correlations in an image. Corso *et al.* [2] propose a supervised approach for combining image segmentation and region labeling. In their model, pixels are constructed hierarchically and labels are shifted between neighboring sites on the same layer. Due to the difficulty of only exploiting the local visual observation for region labeling, Toyoda *et al.* [17] propose to employ the observation of multi-scales to reduce the inaccuracy produced by classifying the regions just based on the simple local evidence. Lin *et al.* [9] proposed a method based on CRFs to solve the region labeling problem by utilizing visual features of both the local region and the whole image while there is no semantic context in their model.

However, very few work of region labeling explores the high level semantic information. Han *et al.* [6] propose a sparse coding based method, in which they construct a graph of regions to introduce both the visual relationships and the semantic correlations into the framework. Zheng *et al.* [21] present a novel multi-layer hierarchical dictionary learning framework for region tagging when the tag taxonomy is known. Different from their work of exploiting tag taxonomy, we propose to integrate the semantic contextual information into image region labeling. We model the spatial context and the semantic context simultaneously by a hybrid graph of CRF model, and with the release of their reinforcement power, we achieve good labeling performance.

3 Conditional Random Fields with Region Graph and Concept Graph

In this section, we present our CRF model and the details of the site potential and edge potential of the CRF subsequently.

3.1   Conditional Random Fields

In Conditional Random Field (CRF) [8], discriminative models are employed for modeling the site potential and the interactions between sites. Let $G = (S, E)$ denote a graph with site set $S = \{1, 2, ..., m\}$ and edge set E. $\mathbf{y} = \{y_1, y_2, ..., y_m\}$ is a set of variables indexed by S, $X = \{x_1, x_2, ..., x_m\}$ is the observed data and x_i is the data from site i. Then $(\mathbf{y}, X)$ is said to be a conditional random field if, when conditioned on X, the random variable y_i obeys the Markov property with respect to the graph: $P(y_i|X, y_{s-\{i\}}) = P(y_i|X, y_{N_i})$ where $s - \{i\}$ is the set of all nodes in G except node i, N_i is the set of neighbors of node i in G, and $\mathbf{y}_{\Omega}$ represents the set of labels on nodes in the set Ω.

For modeling the visual context and semantic context simultaneously, we adopt a hybrid graph which consists of a region (grid) sub-graph and a concept sub-graph. In the region sub-graph, the sites represent region grids of an image. For the neighborhood of one site, only the adjacent 4 nearest grids are considered in our work. As the label of a grid could be one of the concepts in set $C = \{c_1, c_2, ..., c_K\}$, the labeling can be regarded as a K-class classification problem. We train K different sets of parameters, one for each concept. Therefore the conditional distribution over the labels of sites $\mathbf{y}^k$ given X is defined as follows:

$$P(\mathbf{y}^k|X)$$
$$= \frac{1}{Z} \exp\left(\sum_{i \in S} -A(y_i^k, X) + \sum_{i \in S}\left(\sum_{j \in N_i^R} -I_R(y_i^k, y_j^k, X) + \sum_{p \in N_k^C} -I_C(y_i^k, y_I^p, X)\right)\right),$$
$$(1)$$

where Z is a normalizing constant called the partition function, and A and I are the site potential and edge potential respectively. Notice that in this paper, we only consider cliques of size up to two. The superscript k on $\mathbf{y}$ indicates that the CRF is constructed for the concept c_k, and y_i^k denotes the presence or absence of the concept c_k in region i, and y_I^p denotes the presence or absence of the concept c_p in image I which contains region i. N_i^R denotes the neighboring regions of region i, and N_k^C denotes the neighboring concepts of the concept c_k.

3.2   Site Potential

In our CRF framework, $\mathbf{y}^k = \{y_i^k\}_{i \in S}$ represents the labels of semantic concept c_k for sites in S, and $X = \{x_i\}_{i \in S}$ represents the corresponding feature vectors. The site potential $A(y_i^k, X)$ is used to model the association between the site and the corresponding concept given the observation data. $y_i^k \in \{-1, +1\}$ represents the absence/presence of concept c_k for the region i. And $x_i \in X$ represents the feature vector of region i. The site potential is defined as follows:

$$A(y_i^k, X) = y_i^k(\lambda^k + \alpha^k P(y_i^k = 1|x_i)),$$
$$(2)$$

where λ^k, α^k are the parameters on the site potential, and $P(y_i^k = 1|x_i)$ is the probability that the concept presents in the site i. The probability $P(y_i^k = 1|x_i)$ is defined as:

$$P(y_i^k = 1|x_i) = \frac{1}{1 + \exp((\gamma_{k0} + \gamma_k^T x_i))}. \tag{3}$$

We employ the logistic regression to generate $P(y_i^k = 1|x_i)$, which is the conditional probability that the local region is labeled with the concept c_k.

3.3 Edge Potential with Region Graph and Concept Graph

To utilize spatial context and semantic context simultaneously in our CRF, we design two types of edge potentials. Both region edge and concept edge are modeled with linear discriminative models. The edge potential for spatial context modeling is defined as:

$$I_R(y_i^k, y_j^k, X) = \beta_{ij}^k y_i^k y_j^k P(y_j^k = 1|x_j) sim(x_i, x_j), \tag{4}$$

where

$$sim(x_i, x_j) = \exp(-dist(x_i, x_j)), \tag{5}$$

and β_{ij}^k is the parameter associated with edge(i, j) in the region graph. $P(y_j^k = 1|x_j)$ is the conditional probability of label y_i^k for the concept c_k given observation x_j. $sim(x_i, x_j)$ is used to measure the similarity of the neighboring regions given the corresponding observation visual features. When $\beta_{ij}^k < 0$, the edge potential of this type will favor the case that the neighboring regions have the same label which can raise the conditional probability in Eq.1 and penalize the case of different label values.

Similarly, the edge potential for semantic context modeling is defined as:

$$I_C(y_i^k, y_I^p, X) = \xi^{kp} y_i^k y_I^p P(y_I^p = 1|X) cor(k, p), \tag{6}$$

and

$$cor(k, p) = \mu \cdot occur(k, p), \tag{7}$$

where ξ^{kp} is the parameter associated with edge(k,p) in the concept graph. $P(y_I^p = 1|X)$ is the conditional probability of assigning the label y_I^p of the concept c_p to the image I which contains the region i given the observation X. $cor(k, p)$ is used to measure the correlation between the concept c_k and the concept c_p, $occur(k, p)$ is the co-occurrence frequency of the two concepts and μ is a smoothing parameter.

To measure the correlation between the concept c_p and the image I, we define the conditional probability of assigning the label y_I^p of the concept c_p to the image I given the observation X as:

$$P(y_I^p = 1|X) = \frac{1}{\sum_{j \in I} \sigma(y_j^p = 1)} \sum_{j \in I} \sigma(y_j^p = 1) P(y_j^p = 1|x_j), \tag{8}$$

where the indicator function $\sigma(\cdot)$ is 1 if the condition in the brackets is true, and otherwise zero.

Different from the edge potential in Eq.4, this edge potential is utilized to favor identical labels at a pair of neighboring concepts. When $\xi^{kp} < 0$, the neighboring concepts having the same label in one image will raise the conditional probability in Eq.1, while different values will cause the penalty.

3.4   Concept Graph

In our model, the construction of the concept graph is based on the co-occurrences of concepts in the training set $\mathcal{T} = \{X_I^n, \mathbf{y}_I^n\}_{n=1}^N$. X_I denotes the features of an image, and N denotes the number of training images. $\mathbf{y}_I = \{y_I^1, y_I^2, ..., y_I^K\}$ is the labels of image I, and K is the number of semantic concepts. We use the concept co-occurrences to decide wether there is an edge between two concepts:

$$edge(c_k, c_p) = \begin{cases} 1, & if \ \exists c_k, c_p \in C, \ |y_I^k \cap y_I^p| > \tau_k \ or \ |y_I^k \cap y_I^p| > \tau_p \\ 0, & otherwise, \end{cases} \tag{9}$$

where $|y_I^k \cap y_I^p|$ is the co-occurrence frequency of the concept c_k and the concept c_p in image I, and $\forall c_k \in C$,

$$\tau_k = \frac{1}{\sum_{c_q \wedge q \neq k} \sigma(|y_I^k \cap y_I^q| > T_0)} \sum_{c_q \wedge q \neq k} \sigma(|y_I^k \cap y_I^q| > T_0)|y_I^k \cap y_I^q| \tag{10}$$

Where T_0 is a predefined threshold constant. τ_k is the average co-occurrence frequency for the concept c_k considering its neighboring concepts. In our experiments, we set an average co-occurrence frequency for each concept because the specific co-occurrence frequency can present different importance for different pairs of concepts, and using different thresholds is helpful to extract more reasonable neighbors for different concepts.

4   Parameters Estimation and Model Inference

4.1   Parameters Estimation

For parameter estimation, we employ pseudo-likelihood [15] to make the joint label probabilities of the training objective decoupled to a serial of conditional probabilities over single variables and then the partition function can be computed efficiently. We rewrite the parameters to be estimated of the CRF for the concept c_k as $\theta_k = (\lambda^k, \alpha^k, \beta_{ij \forall j \in N_i^R}^k, \xi^{kp \forall p \in N_k^C})$, and the training data of the site i for the concept c_k as $v_i^k = (1, P(y_i^k = 1|x_i), y_j^k P(y_j^k = 1|x_j)sim(x_i, x_j)_{\forall j \in N_i^R}, y_I^p P(y_I^p = 1|X)cor(k, p)_{\forall p \in N_k^C})$. Then the negative log pseudo-likelihood is written as:

$$L_k(\theta_k) = \sum_{i=1}^{M_k} \{\log(1 + e^{2\theta_k^T v_i^k}) - (1 - y_i^k)\theta_k^T v_i^k\} + \frac{\| \theta_k \|^2}{2\sigma^2}, \tag{11}$$

We adopt a projected gradient algorithm to minimize Eq.11. The value of σ is chosen empirically and we set it to the same value for each CRF model. We denote θ_{kt} be the t_{th} element of θ_k, the the update rule of the parameters θ_k is

$$\theta_{kt\_new} = \theta_{kt\_old} - \eta_{old} \frac{\partial L_k(\theta_{k\_old})}{\partial \theta_{kt}}, \tag{12}$$

where η is the learning rate, which is normally designed to decrease as the iteration proceeds.

4.2   Inference Procedure

In this section, we introduce the inference procedure whose goal is to find the most probable label configuration of the sites. First, the probabilities $P(y_I^k|X)$ in Eq.8 and the corresponding labels y_I^k are updated. Second, for the CRF of the concept c_k we employ the iterated conditional mode (ICM) algorithm for inference in each CRF to find the optimal configuration which maximizes the local conditional probabilities sequentially. Third, after the inference procedure is finished for all CRFs, we refresh the label to the region by the following rule,

$$k \leftarrow \arg\max_k \{p(y_i^k|y_{N_i^R}^k, y_I^{N_k^C}, X)\}_{k=1}^K, \tag{13}$$

And the algorithm repeats the above steps until all the region labels **y** tend to be stable.

5   Experiments

5.1   Datasets

The MSRC-v2 [14] is a commonly used benchmark for region labeling. It contains 591 images and 23 semantic concepts. As usual, we select 21 categories for our experiments, since "horse" and "mountain" have very few positive samples and are finally filtered out from the dataset. We use 471 images with $471 \times 16 \times 12$ region grids for training and the remains for testing. **COREL-100** [10] is a subset of COREL collection. It contains 100 images and 7 concepts. We use 60 images with $60 \times 18 \times 12$ region grids for training and the remains for testing.

In our experiments, we extract three kinds of descriptors for all regions: HSV, SIFT [12] and Region Moments [3]. For both datasets, we set $T_0 = 0$ and $g_0 = 8$ and the prior probability in Eq.3 is computed by using logistic regression.

5.2   Experimental Results

For performance comparison, we implement two strong baselines: (1) Binary logistic regression (LR), which handles the K-class classification problem with k binary classification. (2) Conditional Random Fields (CRFs) with only spatial contextual information considered. Some state-of-the-art methods, for instance, Bi-Layer [10], G^2SRRT [6], and $MSDL$ [21] are also selected for comparison. Since different accuracy metrics are adopted in these previous work, we follow their rules to perform the comparison. That is, in comparison with Bi-Layer[10] we use the pixel level labeling accuracy and in comparison with G^2SRRT [6] and $MSDL$ [21], the region level labeling accuracy is chosen.

Since our method outputs grid labeling, we adopted a mapping approach to assign labels to image pixels after image grid labeling, we further segment the testing images by a graph based segmentation algorithm [4]. Then, for each segmented image region, the most frequent label of the included grids is assigned to the corresponding region. So all the pixels of one segmented region own the same label after label mapping.

We split our datasets into 5 parts and conduct the 5-fold cross validation. The final performance accuracy is the average over the five runs. Table 1 shows the labeling performance of the baseline algorithms (LR, CRF), Bi-Layer [10] and our proposed algorithm(CRFC) on MSRC-v2 and COREL-100 in pixel level accuracy. We can see that our method has the best result, which achieves accuracy of 0.653 and 0.638 on the two datasets respectively. It demonstrates that the integration of the contextual information of visual spatial grids and the semantic concepts brings significant improvement. Table 2 shows the labeling performance in the semantic region level accuracy on the two datasets. We can see that in MSRC-v2, our proposed method achieves the best result of 0.661.

Figure 3 shows the detailed accuracies of all concepts in each dataset. We notice that CRFC achieves higher accuracies on 17 categories on MSRC-V2 and 5 categories on COREL-100 compared with CRF. Our method can improve the accuracy on categories with large object size such as "water" and "airplane", and can also improve the accuracy on categories with small object size such as "bird", "cat", "dog", "boat" and so on.

Table 1. Pixel Level Accuracies on MSRC-v2 and COREL-100

Methods	MSRC-v2	COREL_100
LR	0.597	0.604
CRF	0.617	0.629
$Bi-Layer$ [10]	0.630	0.610
CRFC(ours)	**0.653**	**0.638**

Table 2. Semantic Region Level Accuracies on MSRC-v2

Methods	MSRC-v2
$G^2SRRT(kNN + Tag)$ [6]	0.533
LR	0.609
CRF	0.625
$MSDL$ [21]	0.634
CRFC(ours)	**0.661**

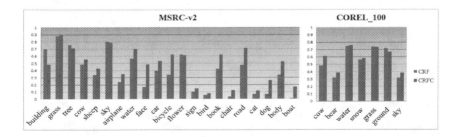

Fig. 3. The detail pixel level accuracies on MSRC-v2 and COREL-100

In Figure 4, we show some labeling results on MSRC-v2 and COREL-100. From the results of MSRC-v2, we notice that there is a large amount of grids misclassified by CRF. For instance, in the first row of the MSRC-v2 results, most grids of 'boat' are misclassified to 'car' by CRF, while in the corresponding CRFC result, almost all the wrong labels are corrected thanks to the semantic

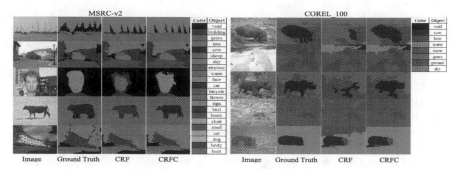

Fig. 4. Example results of CRF and CRFC on two datasets

contextual information, namely that the probability of assigning 'boat' to the grids should be decreased since its co-occurring semantic labels do not appear in the image. The same phenomenon can also be seen in the COREL-100 part. For instance, in the second row, the misclassified label 'bear' (by CRF) is finally corrected by our CRFC method.

6   Conclusion

We present a novel image region labeling method by exploring the high level semantic context modeling. Our method is based on the conditional random fields, in which the contextual information of the visual spatial and the high level semantics are integrated into a hybrid graph. We compared our method with some strong baselines and the state-of-the-art methods on two commonly used datasets. The experimental results demonstrates the significant improvement and effective performance of the proposed algorithm.

Acknowledgement. The authors thank Yu Xiang for his valuable help and discussions. And this work was partially supported by NSFC Grant No.61370157, No.61373106 and Shanghai Science and Technology Development Funds(13dz2260200, 13511504300).

References

1. Athanasiadis, T., Mylonas, P., Avrithis, Y.: A context-based region labeling approach for semantic image segmentation. In: Avrithis, Y., Kompatsiaris, Y., Staab, S., O'Connor, N.E. (eds.) SAMT 2006. LNCS, vol. 4306, pp. 212–225. Springer, Heidelberg (2006)
2. Corso, J.J., Yuille, A., Tu, Z.: Natural image labeling by dynamic hierarchical computing. In: CVPR (2008)
3. Doretto, G., Yao, Y.: Region moments: Fast invariant descriptors for detecting samll image strctures. In: CVPR (2010)
4. Felzenszwalb, P.F., Huttenlocher, D.P.: Efficient graph-based image segmentation. IJCV 59(2), 167–181 (2004)

5. Feng, S., Manmatha, R., Lavrenko, V.: Multiple bernoulli relevance models for image and vedio annotation. In: CVPR (2004)
6. Han, Y., Wu, F., Shao, J., Tian, Q., Zhuang, Y.: Graph-guided sparse reconstruction for region tagging. In: CVPR (2012)
7. Ji, C., Zhou, X., Lin, L., Yang, W.: Labeling images by integrating sparse multiple distance learning and semantic context modeling. In: Fitzgibbon, A., Lazebnik, S., Perona, P., Sato, Y., Schmid, C. (eds.) ECCV 2012, Part IV. LNCS, vol. 7575, pp. 688–701. Springer, Heidelberg (2012)
8. Lafferty, J.D., McCallum, A., Pereira, F.C.N.: Conditional random fields: Probabilistic models for segmenting and labeling sequence data. In: ICML (2001)
9. Lin, Z., Chan, W., He, K., Zhou, X., Wang, M.: Conditional random fileds for image region labeling with global observation. In: Huet, B., Ngo, C.-W., Tang, J., Zhou, Z.-H., Hauptmann, A.G., Yan, S. (eds.) PCM 2013. LNCS, vol. 8294, pp. 586–597. Springer, Heidelberg (2013)
10. Liu, X., Cheng, B., Yan, S.: Label to region by bi-layer sparsity priors. In: MM (2009)
11. Llorente, A., Motta, E., Rüger, S.: Exploring the semantics behind a collection to improve automated image annotation. In: Proceedings of the 10th Workshop of the Cross Language Evaluation Forum (2010)
12. Lowe, D.: Distinctive image features from scale-invariant keypoints. IJCV 60(2), 91–110 (2004)
13. Rubinstein, M., Liu, C., Freeman, W.T.: Annnotaion propagation in large image databases via dense image correspondence. In: Fitzgibbon, A., Lazebnik, S., Perona, P., Sato, Y., Schmid, C. (eds.) ECCV 2012, Part III. LNCS, vol. 7574, pp. 85–99. Springer, Heidelberg (2012)
14. Shotton, J., Winn, J.M., Rother, C., Criminisi, A.: Textonboost: Joint appearance, shape and context modeling for multi-class object recognition and segmentation. In: Leonardis, A., Bischof, H., Pinz, A. (eds.) ECCV 2006, Part I. LNCS, vol. 3951, pp. 1–15. Springer, Heidelberg (2006)
15. Sutton, C., McCallum, A.: An introduction to conditional random fields. Foundations and Trends in Machine Learning 4, 267–373 (2011)
16. Tang, J., Hong, R., Yan, S., Chua, T., Qi, G., Jain, R.: Image annotation by knn-sparse graph-based label propagation over noisily-tagged web images. ACM Transactions on Intelligent Systems and Techology 2 (2011)
17. Toyota, T., Hasegawa, O.: Random field model for integration of local information and global information. IEEE Tranaction on PAMI 30(8), 1483–1489 (2008)
18. Wang, H., Huang, H., Ding, C.: Image annotation using bi-relational graph of images and semantic labels. In: CVPR (2011)
19. Xiang, Y., Zhou, X., Chua, T., Ngo, C.: A revisit of generative model for automatic image annotation using markov random fields. In: CVPR (2009)
20. Xiang, Y., Zhou, X., Chua, T., Ngo, C.: Semantic context modeling with maximal margin conditional random fields for automatic image annotation. In: CVPR (2010)
21. Zheng, J., Jiang, Z.: Tag taxonomy aware dictionary learning for region tagging. In: CVPR (2013)

Manifold Regularized Multi-view Feature Selection for Web Image Annotation

Yangxi Li[1], Xin Shi[2], Lingling Tong[1,*], Yong Luo[2], Jinhui Tu[3],
and Xiaobo Zhu[1]

[1] National Computer Network Emergency Response Technical Team/Coordination
Center of China (CNCERT/CC)
[2] Key Laboratory of Machine Perception, Peking University, China
[3] China Construction Bank, Guangzhou, China
{liyangxi,shixinnk,yluo180}@gmail.com, tongling300@sina.com

Abstract. The features used in many multimedia analysis-based applications are frequently of very high dimension. Feature selection offers several advantages in highly dimensional cases. Recently, multi-task feature selection has attracted much attention, and has been shown to often outperform the traditional single-task feature selection. Current multi-task feature selection methods are either supervised or unsupervised. In this paper, we address the semi-supervised multi-task feature selection problem. We first introduce manifold regularization in multi-task feature selection to utilize the limited number of labeled samples and the relatively large amount of unlabeled samples. However, the graph constructed in manifold regularization from a single feature representation (view) may be unreliable. We thus propose to construct the graph using the heterogeneous feature representations from multiple views. The proposed method is called manifold regularized multi-view feature selection (MRMVFS), which can exploit the label information, label relationship, data distribution, as well as correlation among different kinds of features simultaneously to boost the feature selection performance. All these information are integrated into a unified learning framework to estimate feature selection matrix, as well as the adaptive view weights. Experimental results on a real-world web image dataset demonstrate the effectiveness and superiority of the proposed MRMVFS over other state-of-the-art feature selection methods.

Keywords: Feature selection, multi-task, multi-view, manifold regularization.

1 Introduction

Automatic image annotation lies at the heart of many multimedia analysis-based applications, such as web image search, online picture-sharing, etc [6]. In these applications, where high dimensional features are usually utilized, a compact features selected from the original features is helpful to reduce the computational

---

* Corresponding author.

W.T. Ooi et al. (Eds.): PCM 2014, LNCS 8879, pp. 103–112, 2014.
© Springer International Publishing Switzerland 2014

cost, save the storage space, and reduce the chance of over-fitting. Although traditional feature selection methods usually select features from a single task [9,5], more recently there has been a focus on joint feature selection across multiple related tasks [11,10,14,7,12]. This is because joint feature selection can exploit task relationship in order to establish the importance of features, and this approach has been empirically demonstrated to be superior to feature selection on each task separately [11].

Current multi-task feature selection methods are conducted either in a supervised [11,10] or unsupervised [14,7,12] manner, in terms of whether the label information is utilized to guide the selection of useful features. Supervised feature selection methods always require a large amount of labeled training data, and it may fail to identify the relevant features that are discriminative when the number of labeled samples is small. On the other hand, unsupervised feature selection methods are often unable to identify the discriminative features since the label information is ignored [13]. Therefore, given the high labeling cost and the large amount of unlabeled data are usually available, we focus on the semi-supervised multi-task feature selection in this paper.

To make use of both the limited labeled data and the abundant unlabeled data, we employ the idea of manifold regularization [1] into multi-task feature selection. The performance of manifold regularization relies much on the constructed graph Laplacian. However, the graph constructed using the features from a single view may be unreliable. For example, the different labels cannot be properly characterized by a single feature representation in image classification [8]. We thus further extract different kinds of features from the image, and weightedly combine the graphs constructed using the different feature representations. The combined graph is able to approximate the underlying data distribution more accurately then the single graph.

The proposed method is called manifold regularized multi-view feature selection (MRMVFS). To the best of our knowledge, little progress has been made in semi-supervised multi-view feature selection. There are some recent works on unsupervised multi-view feature selection [4,12], and we mainly differ from them in that the label information is incorporated. The proposed MRMVFS integrates four kinds of information, i.e., label information, label relationship, data distribution, as well as correlation among different views, to select the most representative feature components from the orginal multi-view features. In particular, a regression model is adopted to exploit the label information contained in the labeled samples. The $l_{2,1}$-norm penalty on the feature selection matrix is enforced to joint feature selection across multiple labels, and thus explore the label relationship. Meanwhile, the visual similarity graphs of different views are constructed and combined to model the geometric structure of the underlying data distribution. Besides, a set of non-negative view weights is learned to leverage the correlation among different views, and establish a reliable regularization term along the data manifold to smooth the prediction function. Finally, we integrate all these information into a unified learning framework. Based on this framework, we can simultaneously estimate feature selection matrix, as well as the adaptive view weights. In the experiments, we apply MRMVFS to automatic image annotation on a challenge web image dataset, NUS-WIDE-OBJECT [3], and compare it with several state-of-the-art feature selection methods. Experimental results demonstrate the effectiveness and superiority of the proposed MRMVFS.

2   Manifold Regularized Multi-view Feature Selection

In this section, we present the proposed manifold regularized multi-view feature selection (MRMVFS) method in detail.

2.1   Notations

We firstly introduce some important notations used in the rest of this paper. A matrix is represented by a capital letter, e.g., X. X_{ij} is the (i, j)th element of X, and $X_{i:}$ indicates the elements in the ith row of X. The bold lower case letter $\mathbf{x}$ indicates a vector and x indicates a scalar. Superscript indicates the view of data, e.g., $X^{(v)}$ is the vth view of data X. Subscript is used to denote if the data is labeled, for example, X_L is the labeled data, whereas X_U is the unlabeled data. $\|X\|_F$ denotes the matrix X's Frobenius norm. Specifically, for a matrix $X \in \mathbb{R}^{p \times q}$, its $l_{2,1}$-norm is defined as:

$$\|X\|_{2,1} = \sum_{i=1}^{p} \sqrt{\sum_{j=1}^{q} X_{ij}^2} \tag{1}$$

2.2   Problem Formulation

Given a set of l labeled samples $\mathcal{D}_L = \{(\mathbf{x}_i, y_i)_{i=1}^{l}\}$ and a relatively large set of u unlabeled samples $\mathcal{D}_U = \{(\mathbf{x}_i)_{i=l+1}^{n=l+u}\}$, we suppose each sample is represented by m different views, i.e., $\mathbf{x}_i = [\mathbf{x}_i^{(1)}; \mathbf{x}_i^{(2)}; \ldots; \mathbf{x}_i^{(m)}]$. The view we refer to here is a certain kind of feature or modality. Then the feature matrix of the vth view can be represented as $X^{(v)} = [\mathbf{x}_1^{(v)}, \mathbf{x}_2^{(v)}, \ldots, \mathbf{x}_n^{(v)}] \in \mathbb{R}^{d_v \times n}$, and feature matrix of all the views is $X = [X^{(1)}; X^{(2)}; \ldots; X^{(m)}] \in \mathbb{R}^{d \times n}$, where $d = \sum_{v=1}^{m} d_v$ and d_v is the feature dimension of the vth view.

To select the compact and representative feature components from raw features, we propose to integrate four kinds of information, i.e., the label information contained in labeled data, label relationship, data distribution, as well as correlation among different views of both labeled and unlabeled data, into the learning framework.

We firstly introduce how to integrate label information of labeled data into MRMVFS. Given the labeled feature matrix $X_L = [\mathbf{x}_1, \mathbf{x}_2, \ldots, \mathbf{x}_l] \in \mathbb{R}^{d \times l}$ and the corresponding label matrix $Y_L = [\mathbf{y}_1, \mathbf{y}_2, \ldots, \mathbf{y}_l]^T \in \mathbb{R}^{l \times c}$ with each

$\mathbf{y}_i = [y_i^1, y_i^2, \ldots, y_i^c]^T$, we can learn the prediction functions $f^p(\mathbf{x}), p = 1, \ldots, c$ by minimizing the prediction error over the labeled data in training set:

$$\min_{\{f^p\}} \sum_{p=1}^{c} \sum_{i=1}^{l} \mathcal{L}(f^p(\mathbf{x}_i), y_i^p), \tag{2}$$

where $\mathcal{L}$ is some pre-defined convex loss, $y_i^p = 1$ if the pth label is manually assigned to the ith sample, and -1 otherwise. We assume each $f^p(\mathbf{x})$ is a linear transformation with $f^p(\mathbf{x}_i) = (\mathbf{w}^p)^T \mathbf{x}_i$. Let $\mathbf{f}(\mathbf{x}) = [f^1(\mathbf{x}), f^2(\mathbf{x}), \ldots, f^c(\mathbf{x})]^T$ we have $\mathbf{f}(\mathbf{x}) = W^T \mathbf{x}$, where $W = [\mathbf{w}^1, \mathbf{w}^2, \ldots, \mathbf{w}^c] \in \mathbb{R}^{d \times c}$ is the transformation matrix. To make it suitable for feature selection, a $l_{2,1}$-norm regularization of W is added to (2) to ensure that W is sparse in rows. Thus the optimization problem becomes

$$\min_{W} \sum_{i=1}^{l} \mathcal{L}(\mathbf{f}(\mathbf{x}_i), \mathbf{y}_i) + \alpha \|W\|_{2,1}. \tag{3}$$

This is a general formulation of multi-task feature selection [11]. In this formulation, the importance of an individual feature is evaluated by simultaneously considering multiple tasks. In this way, different tasks help each other to select features assumed to be shared across tasks.

In many practical applications, the number of labeled samples l is quite small, and thus the learned W is often unreliable. Considering the data samples may lie on a low-dimensional manifold embedding in a high dimensional space, we propose to utilize the large amount of unlabeled samples to help learning W under the theme of manifold regularization (MR) [1]. MR has been widely used for capturing the local geometry and conducting low-dimensional embedding. In MR, the data manifold is characterized by a adjacency graph, which explores the geometric structure of the compact support of the marginal distribution. The geometry is then incorporated as an additional regularizer to ensure that the solution is smooth with respect to the data distribution. In our method, the regularization term is given by

$$\frac{1}{2} \sum_{p=1}^{c} \sum_{i,j=1}^{n} (f_i^p - f_j^p)^2 A_{ij} = \frac{1}{2} \sum_{i,j=1}^{n} A_{ij}(\mathbf{f}_i^T \mathbf{f}_i + \mathbf{f}_j^T \mathbf{f}_j - 2\mathbf{f}_i^T \mathbf{f}_j)$$
$$= \text{tr}(F^T(D - A)F) \tag{4}$$
$$= \text{tr}(F^T L F),$$

where $F = [\mathbf{f}(\mathbf{x}_1), \mathbf{f}(\mathbf{x}_2), \ldots, \mathbf{f}(\mathbf{x}_n)]^T \in \mathbb{R}^{n \times c}$ is the predictions over all the data (labeled and unlabeled). Here, A is the adjacency graph constructed using all the data, and each element A_{ij} indicates the similarity between sample $\mathbf{x}_i$ and $\mathbf{x}_j$; D is a diagonal matrix with $D_{ii} = \sum_{j=1}^{n} A_{ij}$ and $L = D - A$ is the graph Laplacian matrix. The regularization term guarantees that if two samples are similar in the feature space, then their predictions will be close. In this way, the data geometric structure existed in high dimensional space is preserved and the data distribution information is well explored.

In general, there are two ways to construct the graph: k-nearest neighbor method and ϵ-ball method. The former one is adopted in this paper. In semi-supervised problem, label information of labeled data can be utilized in graph construction. For those labeled data, feature similarity is measured as:

$$A_{ij} = \begin{cases} 1 & \text{if } \mathbf{x}_i \in \mathcal{N}_k(\mathbf{x}_j) \text{ or } \mathbf{x}_j \in \mathcal{N}_k(\mathbf{x}_i), \text{ and } y_i = y_j \\ 0 & \text{otherwise} \end{cases} \tag{5}$$

where $\mathbf{x}_i \in \mathcal{N}_k(\mathbf{x}_j)$ indicates $\mathbf{x}_i$ is $\mathbf{x}_j$'s k nearest neighbor, and y_i is $\mathbf{x}_i$'s label. For those unlabeled data, the feature similarity is measure as:

$$A_{ij} = \begin{cases} \exp(-\frac{\|\mathbf{x}_i - \mathbf{x}_j\|^2}{2\sigma^2}) & \text{if } \mathbf{x}_i \in \mathcal{N}_k(\mathbf{x}_j) \text{ or } \mathbf{x}_j \in \mathcal{N}_k(\mathbf{x}_i) \\ 0 & \text{otherwise} \end{cases} \tag{6}$$

The adjacency graph matrix A is usually symmetrized by setting $A \leftarrow (A + A^T)/2$.

By introducing the regularization term in (3), we obtain the following optimization problem:

$$\min_W \sum_{i=1}^l \mathcal{L}(\mathbf{f}(\mathbf{x}_i), \mathbf{y}_i) + \alpha\|W\|_{2,1} + \beta\mathrm{tr}(F^T L F). \tag{7}$$

In this formulation, the features from multiple views are concatenated to calculate the graph Laplacian. However, such a strategy ignores the complementary property of the different views. We thus further propose to weightedly combine the graph Laplacians calculated from the different views and learn the combination coefficients. Thus the optimization problem becomes

$$\min_{W,\lambda} \sum_{i=1}^l \mathcal{L}(\mathbf{f}(\mathbf{x}_i), \mathbf{y}_i) + \alpha\|W\|_{2,1} + \beta\sum_{v=1}^m \mathrm{tr}(F^T \lambda_v L^{(v)} F) + \gamma\|\lambda\|_2^2,$$
$$\text{s.t. } \sum_{v=1}^m \lambda_v = 1, \lambda_v \geq 0, \tag{8}$$

where the regularization term $\|\lambda\|_2^2$ is to avoid the variable λ overfitting to a trivial solution. By choosing the least square loss for $\mathcal{L}$ and considering that $\mathbf{f}(\mathbf{x}) = W^T \mathbf{x}$, we have the following formulation for the proposed manifold regularized multi-view feature selection (MRMVFS) framework,

$$\min_{W,\lambda} \|X_L^T W - Y_L\|_F^2 + \alpha\|W\|_{2,1} + \beta\mathrm{tr}(W^T X_{LU} \sum_{v=1}^m \lambda_v L^{(v)} X_{LU}^T W) + \gamma\|\lambda\|_2^2,$$
$$\text{s.t. } \sum_{v=1}^m \lambda_v = 1, \lambda_v \geq 0,$$
$$\tag{9}$$

where $X_{LU} \in \mathbb{R}^{d \times n}$ is the feature matrix of both the labeled and unlabeled data. According to the definition of $l_{2,1}$-norm, many rows of W will shrink to (or close to) zeros. Therefore, $\widetilde{\mathbf{x}} = W^T \mathbf{x}$ can be seen as the selected features consist of most representative dimensions of raw data $\mathbf{x}$. That is to say, we can rank raw features according to $\|W_{i:}\|$ in descending order and only preserve those top ranked feature components.

2.3   Optimization Algorithm

Problem (9) is a nonlinearly constrained nonconvex optimization problem. We adopt the alternating optimization to solve this problem. The variable λ is initialized as $\lambda_v = \frac{1}{m}$, and W is initialized as a random matrix. Then we fix one variable and update the other variable, alternatively and iteratively.

Fix λ, Optimize W. With a fixed λ, the problem (9) becomes:

$$\min_W \mathcal{L}(W) = g(W) + \alpha \|W\|_{2,1}, \tag{10}$$

where

$$g(W) = \|X_L^T W - Y_L\|_F^2 + \beta \mathrm{tr}(W^T X_{LU} \sum_{v=1}^{m} \lambda_v L^{(v)} X_{LU}^T W).$$

By setting $\frac{\partial \mathcal{L}(W)}{\partial W} = 0$, we have:

$$W = (P + \alpha Q)^{-1}(X_L Y_L), \tag{11}$$

where $P = X_L X_L^T + \beta X_{LU} \sum_{v=1}^{m} \lambda_v L^{(v)} X_{LU}^T$, and Q is a diagonal matrix with $Q_{ii} = \frac{1}{2\|W_{i:}\|_2}$. This means that Q is dependent on W. Fortunately, the solution can be obtained efficiently by repeating the following two steps until convergence,

(a) $W^{\tau+1} = (P + \alpha Q^\tau)^{-1}(X_L Y_L)$;
(b) update the diagonal matrix Q using $W^{\tau+1}$.

In practical, $\|W_{i:}\|_2$ may close to zero. Therefore, we can regularize it with $Q_{ii} = \frac{1}{2\sqrt{W_{i:}^T W_{i:} + \epsilon}}$, where ϵ is a tiny constant. Following [10], we can prove the convergence of this iteration algorithm.

Fix W, Optimize λ. With a fixed W, the problem (9) degenerates to:

$$\min_\lambda h^T \lambda + \gamma \|\lambda\|_2^2, \quad \text{s.t.} \sum_{v=1}^{m} \lambda_v = 1, \ \lambda_v \geq 0, \tag{12}$$

where $h = [h_1, \ldots, h_m]^T$ with $h_v = \beta \mathrm{tr}(W^T X_{LU} L^{(v)} X_{LU}^T)$. To solve this problem, we use a coordinate descent-based algorithm. The Lagrange of problem (12) is

$$\mathcal{L}(\lambda, \xi) = h^T \lambda + \gamma \|\lambda\|_2^2 - \xi(\sum_v \lambda_v - 1). \tag{13}$$

Algorithm 1. The optimization procedure of the proposed MRMVFS algorithm

---

Input: Labeled training data $D_L^{(v)} = \{(\mathbf{x}_i^{(v)}, y_i)_{i=1}^l\}$ and unlabeled training data $D_U^{(v)} = \{(\mathbf{x}_i^{(v)})_{i=l+1}^{l+u}\}$ form different views and $v = 1, \ldots, m$ is the view index.
Algorithm parameters: α, β, γ, and k
Output: The feature selection matrices W and the view combination coefficients $\{\lambda_v\}$.
Initialization: W is a random matrix, and $\lambda_v = \frac{1}{V}, v = 1, \ldots, V$. Set $t = 0$.
  1: **Iterate**
  2:   Update W by iterating the two steps (a) and (b) presented in the optimization
       section until convergence;
  3:   Update$\{\theta_v\}$ via (17);
  4:   $t = t + 1$.
  5: **Until convergence**

Randomly select two variables to update, i.e.,

$$\frac{\partial \mathcal{L}(\lambda, \xi)}{\partial \lambda_i^*} = 0, \Rightarrow h_i + 2\gamma \lambda_i^* - \xi = 0, \tag{14}$$

$$\frac{\partial \mathcal{L}(\lambda, \xi)}{\partial \lambda_j^*} = 0, \Rightarrow h_j + 2\gamma \lambda_j^* - \xi = 0, \tag{15}$$

We can eliminate the Lagrange multiplier λ by subtracting (15) with (14), which leads to

$$2\gamma(\lambda_i^* - \lambda_j^*) + h_i - h_j = 0. \tag{16}$$

By further considering that $\lambda_i^* + \lambda_j^* = \lambda_i + \lambda_j$, we have the following solution for updating λ_i and λ_j,

$$\lambda_i^* = \frac{2\gamma(\lambda_i + \lambda_j) + h_j - h_i}{4\gamma}, \ \lambda_j^* = \lambda_i + \lambda_j - \lambda_i^*, \tag{17}$$

To satisfy the constraint $\lambda_v \geq 0$, we set

$$\begin{cases} \lambda_i^* = 0, \lambda_j^* = \lambda_i + \lambda_j, \text{ if } 2\gamma(\lambda_i + \lambda_j) + h_j - h_i \leq 0, \\ \lambda_j^* = 0, \lambda_i^* = \lambda_i + \lambda_j, \text{ if } 2\gamma(\lambda_i + \lambda_j) + h_i - h_j \leq 0. \end{cases}$$

The optimization procedure is summarized in Algorithm 1. When the local optimum solution of W is obtained, we sort all the feature components according to the value of $\|W_{i:}\|$, the top ranked components are selected.

3 Experiments

In this section, we evaluate the effectiveness of proposed MRMVFS by applying it to automatic image annotation on a challenge web image dataset, NUS-WIDE-OBJECT [3].

3.1   Dataset and Settings

NUS-WIDE-OBJECT is a multi-view image dataset consisting of 31 categories and 30,000 images in total (17,927 for training and 12,703 for test). Six different kinds of features, namely 500-D bag of visual words, 64-D color histogram, 144-D color auto-correlogram, 73-D edge direction histogram, 128-D wavelet texture, and 225-D block-wise color moments are provided in [3] to represent each image.

In our experiments, we randomly select $s = \{5, 10, 20\}$ labeled samples for each category from the training set to construct the labeled set; 5,000 images are used as the unlabeled set. In the test set, 2,000 images are used for validation. The parameters performing the best on the validation set are used for test.

In particular, the following methods are compared:

- **AllFea**: the baseline in which all features are normalized and concatenated.
- **MTFS** [11]: a popular multi-task feature selection algorithm. The trade-off parameter γ is set in the range $\{10^{-3}, 10^{-2}, \ldots, 10^3\}$.
- **RFS** [10]: an efficient and robust multi-task feature selection algorithm that utilizes the $l_{2,1}$-norm for both the least squares loss and the regularization term. The trade-off parameter γ is chosen from the set $\{10^{-3}, 10^{-2}, \ldots, 10^3\}$.
- **MRMVFS**: the proposed feature extraction method. The parameters α, β and γ are tuned on the set $\{10^{-3}, 10^{-2}, \ldots, 10^3\}$. The nearest neighbor parameter k is set as 5 empirically.

All the 1,134 features (of different views) are sorted in descending order according to the value $\|W_{i:}\|_2, i = 1, \ldots, d$ and then the r top-ranked features are selected as the input of an SVM classifier. We apply both linear SVM (linSVM) and nonlinear SVM (kerSVM) to the selected features. For SVM training, we use the libSVM [2] toolbox[1] and tune the penalty factor C on the set $\{10^{-3}, 10^{-3}, \ldots, 10^3\}$ in linear SVM. We use the RBF kernels with the bandwidth parameter σ optimized over the set $\{2^{-8}, 2^{-7}, \ldots, 2^3, 2^4\}$ for nonlinear SVM.

3.2   Experimental Result and Analysis

The experimental results using linear and nonlinear SVM are shown in Figure 1 and Figure 2 respectively. Annotation accuracies at different selected feature dimensionality $r = \{20, 50, 100, 200, 500, 800, 1000\}$ are reported. It can be seen from the results that: 1) the accuracies of all the compared methods improve with an increase of labeled samples; 2) annotation using the feature subset selected by the feature selection algorithms can outperform the baseline that using all the features at some r. The reason is that feature selection methods discard redundant and noisy feature components; 3) the proposed MRMVFS is superior to the other two feature selection algorithms significantly in most cases, especially when the number of labeled samples is small. This indicates that the data distribution information is well explored by utilizing the large amount of

---

[1] http://www.csie.ntu.edu.tw/~cjlin/libsvm/

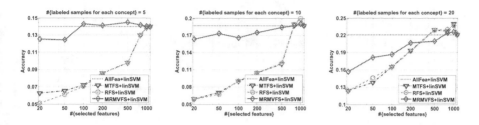

Fig. 1. Annotation performance using linear SVM vs. the number of selected features

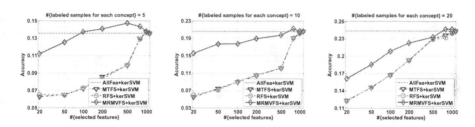

Fig. 2. Annotation performance using nonlinear SVM vs. the number of selected features

unlabeled data in the proposed method; 4) When the number of labeled samples increases, the improvements of MRMVFS compared to the other methods (including the baseline) becomes small. This is because that when more labeled samples are used, the underlying data distribution is better explored, and thus the significance of the using the unlabeled data decreases; 5) The improvements also becomes small with an increasing r. The reason is that the overlap of features selected by different algorithms becomes large when r is large. In particular, only 100 features selected by MRMVFS is needed to achieve the performance of using all the features when the number of labeled samples is 5 for each category. This demonstrates the advantage of the proposed method in handling the small-labeled sample size problem.

4 Conclusion

In this paper, we propose a manifold regularized multi-view feature selection (MRMVFS) for semi-supervised problem. In our method, four kinds of vital information from data, i.e., label information contained in labeled samples, label relationship, data distribution, and data correlation among different views of both labeled and unlabeled samples, are integrated into an unified learning framework. Experiment on a challenge web image annotation task demonstrates the superiority of the proposed method.

References

1. Belkin, M., Niyogi, P., Sindhwani, V.: Manifold regularization: A geometric framework for learning from labeled and unlabeled examples. Journal of Machine Learning Research 7, 2399–2434 (2006)
2. Chang, C.C., Lin, C.J.: Libsvm: a library for support vector machines. ACM Transactions on Intelligent Systems and Technology 2(3), 1–27 (2011)
3. Chua, T.S., Tang, J., Hong, R., Li, H., Luo, Z., Zheng, Y.: NUS-WIDE: a real-world web image database from national university of singapore. In: ACM International Conference on Image and Video Retrieval (2009)
4. Feng, Y., Xiao, J., Zhuang, Y., Liu, X.: Adaptive unsupervised multi-view feature selection for visual concept recognition. In: Lee, K.M., Matsushita, Y., Rehg, J.M., Hu, Z. (eds.) ACCV 2012, Part I. LNCS, vol. 7724, pp. 343–357. Springer, Heidelberg (2013)
5. He, X., Cai, D., Niyogi, P.: Laplacian score for feature selection. In: Advances in Neural Information Processing Systems, pp. 507–514 (2005)
6. Li, J., Wang, J.Z.: Real-time computerized annotation of pictures. In: ACM Multimedia, pp. 911–920 (2006)
7. Li, Z., Yang, Y., Liu, J., Zhou, X., Lu, H.: Unsupervised feature selection using nonnegative spectral analysis. In: AAAI Conference on Artificial Intelligence (2012)
8. Luo, Y., Tao, D., Xu, C., Xu, C., Liu, H., Wen, Y.: Multiview vector-valued manifold regularization for multilabel image classification. IEEE Transactions on Neural Networks and Learning Systems 24(5), 709–722 (2013)
9. Molina, L.C., Belanche, L., Nebot, À.: Feature selection algorithms: A survey and experimental evaluation. In: International Conference on Data Mining, pp. 306–313 (2002)
10. Nie, F., Huang, H., Cai, X., Ding, C.: Efficient and robust feature selection via joint l2,1-norms minimization. In: Advances in Neural Information Processing Systems 23, pp. 1813–1821 (2010)
11. Obozinski, G., Taskar, B., Jordan, M.: Multi-task feature selection. In: ICML Workshop on Structural Knowledge Transfer for Machine Learning (2006)
12. Tang, J., Hu, X., Gao, H., Liu, H.: Unsupervised feature selection for multi-view data in social media. In: SIAM International Conference on Data Mining, pp. 270–278 (2013)
13. Xu, Z., King, I., Lyu, M.T., Jin, R.: Discriminative semi-supervised feature selection via manifold regularization. IEEE Transactions on Neural Networks 21(7), 1033–1047 (2010)
14. Yang, Y., Shen, H.T., Ma, Z., Huang, Z., Zhou, X.: l 2, 1-norm regularized discriminative feature selection for unsupervised learning. In: International Joint Conference on Artificial Intelligence, pp. 1589–1594 (2011)

Semantic Concept Annotation of Consumer Videos at Frame-Level Using Audio

Junwei Liang, Qin Jin*, Xixi He, Gang Yang, Jieping Xu, and Xirong Li

Multimedia Computing Lab, School of Information, Renmin University of China
{leongchunwai,qjin,xxlanmi,yanggang,xjieping,xirong}@ruc.edu.cn

Abstract. With the increasing use of audio sensors in user generated content (UGC) collection, semantic concept annotation using audio streams has become an important research problem. Huawei initiates a grand challenge in the International Conference on Multimedia & Expo (ICME) 2014: Huawei Accurate and Fast Mobile Video Annotation Challenge. In this paper, we present our semantic concept annotation system using audio stream only for the Huawei challenge. The system extracts audio stream from the video data and low-level acoustic features from the audio stream. Bag-of-feature representation is generated based on the low-level features and is used as input feature to train the support vector machine (SVM) concept classifier. The experimental results show that our audio-only concept annotation system can detect semantic concepts significantly better than random guess. It can also provide important complementary information to the visual-based concept annotation system for performance boost.

Keywords: Semantic Concept Annotation, Video Content Analysis, Soundtrack Analysis.

1   Introduction

With the explosion of user generated content (UGC) on current social network sites, there has attracted tremendous research interest in developing automatic technologies for organizing and indexing multimedia content [1]. Semantic concept analysis, which consists of annotating and searching a multimedia collection for user-defined concepts, is one of the fundamental analysis tasks in multimedia content understanding. The outcomes of such analysis processes are "high-level" concepts for describing, indexing and searching consumer media [2, 3].

Traditional semantic concept analysis techniques have focused largely on the visual domain. Owing to the fact that vision is the highest bandwidth sensor for humans, it makes sense that machines would also be able to extract significant semantic information from image and video data. However, audio also conveys significant information that can semantically interpret the video content. Audio is extremely useful in certain situations when other sensors such as visual sensor fail to provide reliable

---

* Corresponding author.

W.T. Ooi et al. (Eds.): PCM 2014, LNCS 8879, pp. 113–122, 2014.

information. For example, when the object is occluded or is in bad illumination, the audio sensors are the key sensors in detecting the presence of objects assuming the objects make sound. Therefore, in the context of multimedia semantic concept analysis applications, audio stream (the soundtrack of a video) can provide important complementary information to visual stream [5, 6]. In this paper, we focus on video concept annotation based only on audio stream.

Most of the state-of-the-art semantic concept frameworks were conducted toward the videos with loose structures such as sports videos [7], surveillance videos [6], or medical videos etc. [8]. In recent years, the consumer generated videos are getting more research attention such as in TRECVID evaluations [3]. Consumer generated videos are unstructured compared to professional contents like films. It brings a lot of technical challenges to analyze them. Huawei organized a grand challenge in the International Conference on Multimedia & Expo (ICME) 2014: Huawei Accurate and Fast Mobile Video Annotation Challenge [9]. The goal of this task is to analyze UGC videos and annotate their contents automatically. The labels to be annotated are 10 concept classes, covering objects (e.g. "car", "dog", "flower", "food" and "kids"), scenes (e.g. "beach", "city view" and "Chinese antique building") and events ("football game" and "party"). The semantic concept annotation within the Huawei challenge is required to be at the frame-level. That means for each frame, we need to make a binary decision about the presence of a specific concept in the frame. Comparing to the semantic concept annotation task at the video level or supra segmental level in previous research, this task requires annotation with finer resolution and is a more challenging task.

The remainder of this paper is organized as follows. Section 2 presents the related work. Section 3 introduces the Huawei grand challenge dataset. Section 4 describes audio semantic concept annotation system. Experimental results are presented in Section 5. Some discussions follow in Section 6 and conclusions are made in Section 7.

2   Related Work

We summarize the prior work in soundtrack analysis from three different dimensions based on the type of data that researchers have focused on:

- First dimension: the quality of the audio data. Early work on audio event classification was largely done on sound databases [10] and clean broadcast or television program audio data [11]. Typical high quality database or broadcast data can be extremely clean, and "foreground" sounds are generally easy to distinguish from "background" sounds due to studio recording conditions. The growing popularity of video sharing services such as YouTube, Dailymotion, Youku in China and so on enables the vast increasing of user generated videos. Analyzing such consumer videos is more challenging.

- Second dimension: number of sound classes. Much early works focused on detecting or distinguishing between a small number of sound classes such as speech, music, silence, noise, or applause. This was solved using various traditional machine learning and signal processing approaches [11-13].

- Third dimension: the granularity of the audio processing. We can roughly categorize the soundtrack analysis work into two categories: sub-soundtrack classification or entire soundtrack classification. Distinguishing between a small numbers of sound classes can be considered as a sub-soundtrack classification problem. It produces annotations of input data according to a fixed number of classes for which one has trained models. Such sound classes can be aforementioned speech/music etc. There also have been efforts to classify short audio clips with respect to the environment in which they were recorded [14]. The multimedia event detection (MED) using soundtrack is the entire soundtrack classification problem [4]. Modeling the event based on sub-soundtrack classification results has been one type of approaches in such tasks [15, 16]. Though the semantic indexing (SIN) task in TRECVID [4] has a subtask of localizing concepts on frame-level since 2013, few work have used auditory method to help achieve the goal. Similar to the SIN subtask, the Huawei grand challenge can be categorized as a sub-soundtrack classification problem and sound classes are aforementioned 10 concept classes.

3   Data Description

Our experiments are conducted on the development data of the Huawei Accurate and Fast Mobile Video Annotation Challenge (MoVAC 2014). All the videos are collected from the Internet (e.g. Youku and Youtube) and converted to mpeg4 format. The selected 10 semantic concepts cover objects (e.g. "car", "dog", "flower", "food" and "kids"), scenes (e.g. "beach", "city view" and "Chinese antique building") and events ("football game" and "party"). There are also some extra videos as background videos which contain none of the predefined 10 concepts.

Table 1. Number of positive and negative frames for each concept in the dataset

Concept	#pos	#neg	%pos
beach	664793	8391067	6.6%
car	1161116	8843824	11.6%
chinese_antique_building	772805	772805	7.7%
city_view	801583	9203357	8.1%
dog	524560	9480380	5.2%
flower	1082986	8921954	10.8%
food	378046	9626894	3.7%
football_game	161873	8391067	16.1%
kids	1525771	8479169	15.2%
party	780240	9224700	8.0%
MEAN	**1030577**	**8974362**	**10.3%**

The development dataset contains 2,666 videos. The video resolution ranges between 640x480 and 1280x720. The videos are normally taken by mobile devices. The recording frame per second (fps) varies among all the videos. Some videos have been post-edited, such as a video on flowers with only music audio. Some videos (16 out of 2666) have no soundtrack. We divide the dataset into a training set (which contains 1764 videos) and a test set (which contains 886 videos). The ground truth label files with manual annotations are in the format of three columns: <concept>\t<start frame index>\t<end frame index>, such as in the following example:

```
Car   1      568
Car   93     1165
Car   1386   1423
Kids  1      1423
```

The detailed information about the amount of positive and negative examples in each concept class is listed in Table 1. As we can see from the table, the amount of negative examples is overwhelmingly larger than the amount of positive examples.

4   System Description

Our semantic concept annotation system using audio only contains the following key components (Figure 1): audio data pre-processing, audio features extraction, concept annotation models and post-processing.

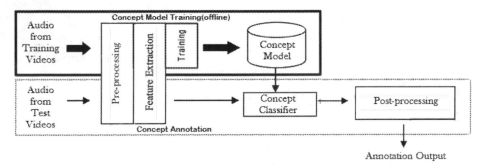

Fig. 1. Audio-only Concept Annotation System Components

Pre-processing. In order to detect concept on frame-level, we chunk the audio stream into small segments with overlap, extract audio features and apply concept detection on those segments. The length of the segment/chunk is a tunable parameter.

Audio Features. The codebook model is a common technique used in the document classification (bag-of-words) [18] and image classification (bag-of-visual words) [19]. The similar bag-of-audio-words model has also been applied in the sound track analysis work [5, 17]. In our system, we use bag-of-audio-words model to represent each audio segment. It is represented by assigning low level audio features to a discrete set of codewords in the vocabulary (codebook) thus providing a histogram of codeword's counts. These codewords are learnt via unsupervised clustering. The discriminative power of such a codebook is governed by the size of the codebook and by the assignment of features to codeword's [14]. In this paper we apply this model to the low level MFCC features. The MFCC features are computed every 25ms with 10ms shift and are in 39 dimensions (13 MFCC + 13 delta + 13 ddelta). The vocabulary is learnt by applying kmeans clustering algorithm with K=4096 on the whole training dataset. Each audio segment is then represented as a distribution over these 4096 codewords' by using soft-assignment of MFCC features to these codewords' (as shown in Figure 2).

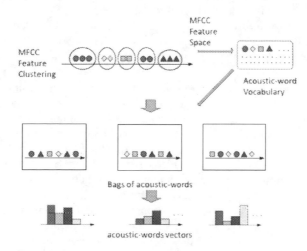

Fig. 2. Data-driven MFCC based Bag-of-Audio-Words model

Concept Annotation Models. After we calculate all the bag-of-audio-words features, we train two-class SVM classifiers for each of the 10 concepts. As shown in table 1 that the training data is overwhelmed by negative examples, we train classifiers by the Negative Bootstrap algorithm [20].   The algorithm takes a fixed number (N) of positive examples and iteratively selects negative examples which are most misclassified by current classifiers. The algorithm randomly samples $10xN$ number of negative examples from the remaining negative examples as candidates at each iteration. An ensemble of classifiers trained in the previous iterations is used to classify each negative candidate examples. The top N most misclassified candidates are selected and used together with the N positive examples to train a new classifier. In order to improve the efficiency of the training process, we use Fast intersection kernel SVMs (FikSVM) as reported in [21].

Post-processing. Intuitively, if a concept occurs within a video, it is usually not an instantaneous appearance. It normally lasts for certain duration. Therefore, we conduct boundary padding and cross-segment smoothing over the raw annotation results. We expand the beginning and ending of the detected segments. We also merge two detected segments if they belong to the same concept and the gap between them is below a certain threshold.

5   Experimental Results

We use the average precision to evaluate the concept annotation performance for each concept class:

$$AP = \frac{1}{R}\sum_{j=1}^{n} I_j \times \frac{R_j}{j} \qquad (1)$$

where R is total number of relevant segments of that concept, n as the total amount of segments, $I_j=1$ when the jth segment are relevant otherwise $I_j=0$ and R_j is the number of relevant segments in the first j segments.

We experiment with different length of chucking in the pre-processing step: 3sec-Chunk+1sec-Shift; 5sec-Chunk+2sec-Shift; 7sec-Chunk+3sec-Shift. The results are very close. Longer chunk only improves less than 1% on mean AP. Longer chunk takes more time to process, we therefore choose the 3sec-Chunk+1sec-Shift setup as our best setup. We choose $N=3000$ in the Negative Bootstrap Training. The second column in Table 2 shows the annotation performance of each concept class using audio only with the best setup. We achieve a mean average precision of 30% on all 10 concepts. Some concept classes, which are acoustically easy to distinguish such as "football game", "dog", "kids", clearly achieve much better performance than others, with AP of 72.8%, 47.9%, and 40.5% respectively. From the results, we can see that the concept annotation based on audio stream only achieves significantly better performance than random guess. Since significant semantic information is conveyed in the visual stream, we also develop the concept annotation system using visual stream. Intuitively, the audio and visual streams contain complementary information for interpreting a semantic concept. We then explore to combine these two systems. The linear fusion weights of the two systems are tuned on a held out dataset. As shown in Table 2, although the visual concept annotation system achieves much better performance than the audio system, combining them achieves certain improvement over all of the 10 concept classes.

As we inspect the data closely, we find out that since the manual annotation of videos is produced mainly according to visual evidence, there are a certain amount of audios that are unrelated to the concepts they are labeled. Therefore, we try to clean these acoustically false "positive" examples by listening to them and deciding whether we human can identify the concept(s) based on audio only. We only focus on three concepts, "kids", "football game", and "dog", which intuitively can be easily distinguished using acoustic cues. We conduct the following procedure to clean the data. The videos satisfy either one of the following criteria will be excluded: 1) the videos with music-only audio that may has nothing to do with the visual content and the music only relates to the editor's taste which is random; 2) the videos with loud background noise that we cannot hear any of the labeled concept(s), such as a dog swimming in a coming wave with strong wind blowing at the beach; 3) the videos with no target concept's sound at all, for examples, a sleeping dog or a kid quietly sitting on the chair. In the end, we exclude 10-30% amount of annotations for these three concepts, both in the training set and the test set. We then conduct the annotation experiment on this cleaned data set with the same setup. Table 3 compares the annotation performance of our audio-only system on the original data and on the cleaned one. We can see from the table that, there is obvious improvement for each concept class.

This suggests that in the future work we can look for better ways to address the video annotation problem from the audio sensor point of view, both manually and automatically. Some work in [22] can be related to this task.

Table 2. Perofrmance of audio-only, visual-only and fusion systems

Concept	Audio Only	Visual Only	Fusion	Audio Weight
beach	12.8%	69.7%	70.0%	0.14
car	24.5%	77.7%	77.8%	0.15
chn_anti_bldg	19.9%	75.2%	76.2%	0.21
city_view	22.1%	73.3%	74.9%	0.27
dog	47.9%	60.6%	68.2%	0.39
flower	26.0%	80.8%	81.4%	0.21
food	7.0%	59.7%	59.8%	0.06
football_game	72.8%	98.2%	98.5%	0.22
kids	40.5%	53.9%	60.7%	0.50
party	25.8%	85.0%	86.1%	0.22
MEAN	**29.9%**	**73.4%**	**75.3%**	-

Table 3. Performance comparison (original data vs cleaned data)

Concept	Original	Cleaned
dog	47.9%	55.3%
football_game	72.8%	75.6%
kids	40.5%	43.1%
MEAN	**53.7%**	**58.0%**

The evaluation criteria in the Huawei grand challenge is defined as follows (we call it Huawei accuracy):

$$\begin{cases} accuracy = sign(score \geq th) \\ sign(x) = \begin{cases} 1 & if\ x\ is\ true \\ 0 & else \end{cases} \\ score = \frac{\sum_{i=1}^{10} Interval_{ip} \cap Interval_{ig}}{\sum_{i=1}^{10} Interval_{ip} \cup Interval_{ig}} \end{cases} \quad (2)$$

$Interval_{ig}$ means the ground truth annotation interval(s) for concept i, and $Interval_{ip}$ means the prediction interval(s) for the same concept. An extra threshold will be used to get a discrete score of 1 or 0 for each prediction (the threshold is set 0.5 in the evaluation). Finally, all the scores will be summed up to get the Huawei accuracy. We achieve 19.2% Huawei accuracy (on original data) with our audio-only annotation system (with 3sec padding to all segments in the post-processing). Meanwhile, visual-only annotation system achieves 63.1% Huawei accuracy, and the fusion of both systems improves the Huawei accuracy to 65.6%.

6   Discussions

From our system annotation results, we discover the salient co-occurrence of "car" and "city view", "football game" and "car". We then check the ground truth (the manual annotation data) and notice that the "kids" concept do often co-occur with other concept such as "football game" and "beach", the "car" concept often co-occurs with "city view" and "football game" （Figure 3）. In Figure 3 the number (in thousands) in the table cell indicates the number of co-occurrence frames between the concept on X axis and the one on Y axis. Such co-occurrence information can potentially help with automatic annotation. In future work, we will consider using the co-occurrence information (such as co-training) to improve the system.

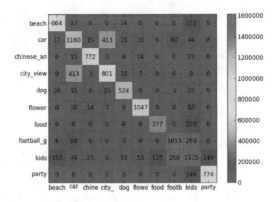

Fig. 3. Concepts' frame-level co-occurrence in the manual annotation

As shown in table 2, we can see that the performance of audio-only system varies among different concepts. This is not surprising, because although some concepts are visually consistent, they hardly have acoustic consistency. For example, the audios from "food" videos vary from incredibly noisy to completely silent. Some concepts cannot be an audio-concept at all, such as "flower". Most of the audios from flower videos are post-edited music. Even if we get a reasonable result on this dataset, we are not modeling the flower concept, rather the editor's music taste for flower. If the music taste changes, the model will fail. Therefore, more attention should be paid to the consistency of the audio data.

We can also see from results in table 2 that combining audio-only and visual-only systems benefits in general. It is even more beneficial on certain concepts, such as "kids" and "dog". In extreme cases when visual evidence is not available, such as the object is hidden but its sound is heard, audio-only system is the only solution for concept annotation in this case. For example, for a snap shot in the video tr0213.mp4 with kids talking behind the camera at a circus performance (Figure 4 (a)), audio-only system successfully detects the "kids" concept while visually is impossible. For another snap shot in the video tr0209.mp4 with kids talking behind the camera about

(a) tr0213.mp4   (b) tr0209.mp4

Fig. 4. Snap shot with kids' voice in the background

the chicken (Figure 4 (b)), audio-only system also successfully detects "kids", while visual-only system may only be able to detect "chicken" or "animal". These examples indicate that for acoustically salient concepts when visual system fails to detect, audio annotation systems will be the best solution.

7   Conclusions

This paper presents our semantic concept annotation system using audio stream only for the Huawei grand challenge. The system uses bag-of-audio-words representation based on the low-level features and negative bootstrap SVM concept classifier. The experimental results on the challenging Huawei UGC video data show that our audio only concept annotation system can detect semantic concepts significantly better than random guess. When combining with visual-only concept annotation system, it helps in general and more significantly on certain concepts. In the future work, we will explore different low-level features to build the acoustic vocabulary, since MFCC may not be the best feature to distinguish non-vocal sound. We will study the impact of audio data consistency on annotation performance and explore the potential of utilizing the concept co-occurrence property. We will also study building audio annotation systems on more acoustically salient concepts that visual method might fail.

Acknowledgements. This work is supported by the Fundamental Research Funds for the Central Universities and the Research Funds of Renmin University of China (No. 14XNLQ01), the Beijing Natural Science Foundation (No. 4142029), NSFC (No. 61303184), and SRFDP (No. 20130004120006).

References

1. Snoek, C., Worring, M.: Concept-based Video Retrieval. Foundations and Trends in Information Retrieval (2009)
2. Chang, S.F., Ellis, D., Jiang, W., Lee, K., Yanagawa, A., Loui, A.C., Luo, J.: Large-Scale Multimodal Semantic Concept Detection for Consumer Video. In: International Workshop on Multimedia Information Retrieval (MIR) (2007)

3. Naphade, M.R., Smith, J.R., Tesic, J., Chang, S.F., Hsu, W., Kennedy, L., Hauptmann, A., Curtis, J.: Large-Scale Concept Ontology for Multimedia. IEEE Journal MultiMedia 13(3) (2006)
4. Over, P., Awad, G., Michel, M., Fiscus, J., Sanders, G., Kraaij, W., Smeaton, A.F., Quéenot, G.: TRECVID 2013 – An Overview of the Goals, Tasks, Data, Evaluation Mechanisms and Metrics. In: Proceedings of TRECVID. NIST, USA (2013), http://www-nlpir.nist.gov/projects/tvpubs/tv13.papers/tv13overview.pdf
5. Lee, K., Ellis, D.P.W.: Audio-Based Semantic Concept Classificationfor Consumer Video. IEEE Transactions on Audio, Speech, and Language Processing 18(6) (2010)
6. Atrey, P.K., Kankanhalli, M.S., Jain, R.: Information Assimilation Framework for Event Detection in Multimedia Surveillance Systems. In: Multimedia Systems, pp. 239–253 (2006)
7. Kolekar, M.H., Sengupta, S.: Semantic concept extraction from sports video for highlight generation. In: International Conference on Mobile Multimedia Communications (MobiMedia) (2006)
8. Luo, H., Fan, J.: Building Concept Ontology for Medical Video Annotation. In: ACM Multimedia (2006)
9. ICEM 2014 Huawei Accurate and Fast Mobile Video Annotation Challenge, http://www.icme2014.org/huawei-accurate-and-fast-mobile-video-annotation-challenge
10. Wold, E., Blum, T., Keislar, D., Wheaten, J.: Content-based Classification, Search, and Retrieval of Audio. IEEE Multimedia 3(3) (1996)
11. Saunders, J.: Real-time Discrimination of Broadcast Speech/Music. In: ICASSP (1996)
12. Scheirer, E., Slaney, M.: Construction and Evaluation of a Robust Multifeature Speech/Music Discriminator. In: ICASSP (1997)
13. Williams, G., Ellis, D.P.W.: Speech/Music Discrimination Based on Posterior Probability Features. In: Eurospeech (1999)
14. Ma, L., Milner, B., Smith, D.: Acoustic Environment Classification. ACM Transactions on Speech and Language Processing 3(2) (2006)
15. Eronen, A., Peltonen, V., Tuomi, J., Klapuri, A., Fagerlund, S., Sorsa, T., Lorho, G., Huopaniemi, J.: Audio-based Context Recognition. IEEE Trans. on Audio, Speech, and Language Processing 14(1) (2006)
16. Brown, L., et al.: IBM Research and Columbia University TRECVID-2013 Multimedia Event Detection (MED), Multimedia Event Recounting (MER), Surveillance Event Detection (SED), and Semantic Indexing (SIN) Systems. In: TRECVID Workshop (2013)
17. Jin, Q., Schulam, F., Rawat, S., Burger, S., Ding, D., Metze, F.: Categorizing Consumer Videos Using Audio. In: Interspeech (2012)
18. Xue, X.B., Zhou, Z.H.: Distributional Features for Text Categorization. IEEE Transactions on Knowledge and Data Engineering 21(3) (2008)
19. Philbin, J., Chum, O., Isard, M., Sivic, J., Zisserman, A.: Object retrieval with large vocabularies and fast spatial matching. In: CVPR 2007 (2007)
20. Li, X., Snoek, C., Worring, M., Koelma, D., Smeulders, A.: Bootstrapping Visual Categorization With Relevant Negatives. IEEE Transactions on Multimedia 15(4) (2013)
21. Maji, S., Berg, A., Malik, J.: Classification using international kernel support vector machines is efficient. In: CVPR 2008 (2008)
22. Zha, Z.-J., Wang, M., Zheng, Y.-T., Yang, Y., Hong, R., Chua, T.-S.: Interactive Video Indexing with Statistical Active Learning. IEEE Transactions on Multimedia 14(1), 17–27 (2012)

Text Detection in Natural Scene Images with Stroke Width Clustering and Superpixel

Shuang Liu, Yu Zhou, Yongzheng Zhang, Yipeng Wang, and Weiyao Lin

Institute of Information Engineering, Chinese Academy of Sciences,
Beijing 100093, China
zhangyongzheng@iie.ac.cn

Abstract. Text information in natural scene images is important for various kinds of applications. In this paper a novel method based on stroke width to detect text in unconstrained natural scene images is proposed. Firstly, we use the stroke width transform to generate a rough estimation of stroke width map, then use K-Means clustering and the elbow method to find some specific stroke width values that are both dominant and consistent. Secondly, in order to generate better edge detection and gradient direction results we use these specific stroke width values as the size parameters in the superpixel algorithm to generate smooth and uniform region boundaries. Finally, we try to refine the stroke width map and recover valid edge pixels by applying stroke width regularized constraints on the improved edge detection and gradient direction results computed from these region boundaries. Our method was evaluated on three benchmark datasets: ICDAR 2005, 2011 and 2013, and the experimental results show that it achieves state-of-the-art performance.

Keywords: Text localization, superpixel, stroke width clustering.

1 Introduction

The widespread availability of intelligent mobile devices equipped with cameras leads to the explosive growth of natural scene images. Detecting text information from these images plays an important role in a variety of applications: image retrieval with text information, intelligent driving assistance, reading aid for visually impaired people and scene understanding, etc. Consequently, the research of text detection in scene images has received much attention from the computer vision community. Although numerous algorithms have been proposed, scene text detection is still an open problem. The difficulties are mostly caused by cluttered background, occlusion, blur and variations of font type, size, color, etc. As a specific kind of object detection, it also suffers from the common challenges such as lighting conditions, perspective distortions, etc.

Existing methods for text detection in scene images can be classified into two categories: the sliding window based methods and the connected component based methods. Sliding window based methods exploit the textural property of texts by scanning the image with sliding windows at different scales. However, to

W.T. Ooi et al. (Eds.): PCM 2014, LNCS 8879, pp. 123–132, 2014.

cope with the large variation in appearance of the text in natural scene images, window descriptors have to be extracted and classified at numerous locations and scales. The high computational complexity of these methods is a major disadvantage compared with connected component base methods. Connected component based methods, on the other hand, detect text by analyzing the shapes of the edges detected or components extracted from the images. The computational complexity of connected component based methods are only subjected to the number of edge pixels detected, which makes it much faster than sliding window based methods. However, these methods are very dependent on the quality of edge detection or component extraction.

The stroke width based methods are a type of connected component based methods. These methods are mainly based on the assumption that in a single text group the stroke width values of the characters are always nearly constant, which is independent of the language type and computationally efficient. The stroke width transform (SWT) was originally introduced by Epshtein et al[3], in which the stroke width value is computed by measuring the distance between the edge pixel pairs. Text is detected by grouping and filtering components extracted from the stroke width map, which is generated by detecting edges in images and shooting pixel rays from edge pixels following the gradient directions of the edge pixels. Therefore, the qualities of the edge detection and the gradient direction computation have critical influences on the performance. Since then, there are several major notable attempts to improve the SWT algorithm. Huang et al[5] proposed an algorithm called stroke feature transform (SFT) to include the stroke color continuity constraint during the pixel pair searching so that possible valid edges missed by the canny edge detector can be recovered. In the stroke color constraint the threshold decreases linearly with respect to the number of the pixels in the current ray. It will not generalize well for text of different sizes and images of different resolutions. Chen et al [2] first use an edge enhanced maximally stable extremal region (MSER) for component detection, then compute stroke width on the distance transform map by applying a flood-fill-like algorithm to propagate the largest distance values to their neighbors. It has been shown their method is more accurate than conventional SWT. However, because the distance transform is very sensitive to noise, their method suffers more severely from the distortion caused by noise than conventional SWT.

To remedy these weaknesses, in this paper we propose to first use the SWT algorithm to generate stroke width maps, then find consistent and dominant stroke width values by using K-Means clustering and the elbow method. We observe that the shape of the text can be precisely preserved by the boundaries of superpixels whose sizes are close to the stroke width values of the text. In order to detect edge pixels and compute gradient directions with higher accuracy, we represent the original images with superpixels to ignore the noise within the text regions. Among various superpixel algorithms, the Simple Linear Iterative Clustering (SLIC) algorithm [1] has the highest speed and lowest under-segmentation error. It also allows us to configure the size of the superpixels by choosing a step size. Thus, we use each of these stroke width values to define the size of the superpixels to generate a specific superpixel image. Then, we use stroke width constraints regularized by these specific values to recover valid edge pixels missed by the edge detector and refine the stroke width map. Finally, geometrical constraints are enforced to filter out invalid components and aggregate valid components into text lines.

The remainder of this paper is organized as follows. In Section 2, we describe the individual steps of our text detection algorithm. Section 3 demonstrates the robust performance of the proposed method and Section 4 concludes the paper.

2   The Proposed Method

The flow diagram of our system is shown in Fig. 1. First, we use the SWT algorithm to generate the stroke width map. In the stroke width map, we use the K-Means algorithm and the elbow method to find several consistent and dominant stroke values, which serve as clues for the following superpixel algorithm to select its parameter value. Next we use the superpixel algorithm to generate several superpixel images with smooth and uniform region boundaries. A stroke width map is generated again for each superpixel image with SWT, and connected components are extracted with some stroke width regularized constraints. Finally, we filter the extracted components with stroke width, color and geometrical constraints, and the text lines are constructed and filtered. In the following subsections, we will describe each step in detail.

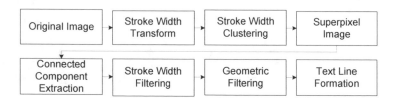

Fig. 1. System flow diagram

2.1   Stroke Width Clustering and Superpixel Generating

We first use conventional SWT to generate a map of stroke width. In the SWT algorithm, rays are emitted from edge pixels along the direction of their gradients computed from the grayscale input image. The reliability of the SWT algorithm is compromised by the poor quality of the edge detection and the gradient direction. Nevertheless, it does generate an estimation of the stroke width distribution, and according to this distribution the dominant stroke width values can be obtained. Thus we use K-Means clustering to discover these parts by finding consistent and dominant stroke value clusters. The number of the clusters is determined automatically using the elbow method. This procedure is outlined in Algorithm 1. When considering the validity of the clusters, we consider both the occurrences and the deviation of the stroke width values. The validity of a cluster is measured by the ratio of its stroke width values to the standard deviation of these values. This is because with small text regions the numbers of occurrences of their stroke width values are generally more consistent. How to effectively remove outliers is a very difficult problem. However, with this simple heuristic rule we are able to preserve most small text regions in experiments.

Algorithm 1. Algorithm for finding valid stroke width values

---

vc = validity threshold
ec = elbow threshold
m = maximal number of cluster
$v = 0$ ▷ sum of cluster variances
s = stroke width map
for $n = 1 \rightarrow m$ **do**
    $c = kmeans(s, n)$ ▷ cluster s into n cluster
    $cv = 0$ ▷ current sum of cluster vairances
    for $i = 1 \rightarrow n$ **do**
        $cv = cv + variance(c_n)$
    end for
    if $\|v - cv\| > ec$ **then**
        $v = cv$
    else
        break
    end if
end for
$sv = [\,]$ ▷ valid clusters
for $i = 1 \rightarrow n$ **do**
    if $\frac{count(c_n)}{variance(c_n)} > vc$ **then**
        $add(sv, c_n)$ ▷ add to the valid clusters
    end if
end for

After these stroke width value clusters have been found, we use the centroids of these clusters as step sizes to generate superpixels. Since all the pixels within a superpixel shares the same label, they are assigned with the same color, which makes the edge detection and gradient direction computation suffer less from noise.

SLIC is a classical superpixel algorithm, and we use SLIC to generate the superpixel image. SLIC algorithm is based on K-Means clustering, it uses geometrical constraints to reduce computational complexity. The nature of the SLIC algorithm is well suited for parallel implementation. We initialize and update the list centers that should be searched for every pixels on CPU, then use the list on GPUs to perform dense parallel computation. Only the iteration part is done on GPUs because CPU is better at random memory access, which is needed for maintaining the search list and enforcing connectivity. Our parallel implementation of the SLIC algorithm on GPUs allows us to achieve around 8x-16x speed up from the sequential implementation depending on the sizes of the images. This makes the SLIC algorithm a computationally trivial part of our method. Experimental results show that finding valid stroke clusters and generating superpixels takes no more than 400ms on average.

2.2 Stroke Width Constraints

In SWT, the stroke width is defined by the distance between edge pixel pairs. To find these pixel pairs pixel rays are shot along the gradient directions of the edge pixels. An edge pixel pair will be formed once the pixel ray encounters another edge pixel whose gradient direction is roughly opposite to the original edge pixel. It is apparent that the accuracy of SWT is highly dependent on the qualities of the edge detection and gradient direction computation. SFT extends SWT with color continuity constraints to recover valid edge pixels missed by the canny edge detector. A pixel will be considered as an edge pixel if the discrepancy between its color and the median color of the pixel ray is larger than a threshold. SFT does not address the problem of inaccurate gradient direction, and in the color continuity constraints the threshold only decreases linearly with respect to the number of pixels in the ray regardless of the resolution of the image and the font size of the text.

We first use canny edge detector to detect edge pixels on superpixcls obtained from the previous step. Then we search for pixel $\mathbf{q}$ following the gradient direction $\mathbf{d}_p$ of edge pixel $\mathbf{p}$ that satisfy either of these constraints:

(1) Firstly, $\mathbf{q}$ is an edge pixel. Secondly, $\mathbf{d}_q$ is roughly opposite to $\mathbf{d}_p$ ($\mathbf{d}_p = -\mathbf{d}_q \pm \pi/3$) and the distance between $\mathbf{q}$ and $\mathbf{p}$ is roughly the same as the specific dominant stroke width value.

(2) Firstly, the color discrepancy between the color $\mathbf{C}_q$ of $\mathbf{q}$ and the median ray color $\mathbf{C}_r$ satisfies $\|\mathbf{C}_q - \mathbf{C}_r\| > \mathbf{C} + T_c \times \|\mathbf{S} - \mathbf{D}\|$, in which $\mathbf{D}$ is the distance between $\mathbf{p}$ and $\mathbf{q}$, $\mathbf{S}$ is the stroke width, $\mathbf{C}$ and $\mathbf{T}_c$ are thresholds determined at the training stage. Secondly, $\mathbf{d}_p = -\mathbf{d}_q \pm \pi/6$. This constraint is adjusted according to the absolute difference between $\mathbf{D}$ and $\mathbf{S}$, which is different from the color continuity constraint in SFT. The strictness of this constraint is adjusted according to the absolute difference between $\mathbf{D}$ and $\mathbf{S}$, which increases the probability of recovering valid edge pixel pairs whose distances are close to the specific dominant stroke width value and decrease the interferences from noisy pixels with strong color discontinuities.

(3) Firstly, the gradient strength $\mathbf{G}_q$ satisfies $\mathbf{G}_q > \mathbf{G} + \mathbf{T}_g \times \|\mathbf{S} - \mathbf{D}\|$, in which $\mathbf{G}$ and $\mathbf{T}_g$ are thresholds determined at training stage. Secondly, $\mathbf{d}_p = -\mathbf{d}_q \pm \pi/6$. This constraint checks for pixels with strong gradient, because gradient strength around the boundaries of the text regions are usually strong. Similar to (2), it is regularized by the specific dominant stroke value.

We avoid searching pixels whose distance from the edge pixels are too large or too small than the expected stroke width, and search for valid pixel $\mathbf{q}$ by shoot pixel rays within the range of gradient direction $\mathbf{d}_s = \mathbf{d}_q \pm \pi/6$.

2.3   Component Filtering and Text Line Aggregation

Connected component analysis are performed on the stroke width map obtained from previous step to extract individual components. Following most state-of-the-art methods, we perform geometrical checks on these components to filter out non-text components. These checks include stroke width consistencies, aspect ratios, number of intersections with other components.

As text is usually in straight lines, we further group the remaining components by their height, aspect ratio, stroke width, color and distance between one another. Improbable text lines with less than two valid individual components or unlikely elongation ratio are filtered out. Also, since in natural scene images bars and windows cause a lot of false positives because they both have consistent width and similar color, we perform and operation on the binary masks of each components within text lines to reject invalid text lines with highly repetitive components.

3   Experimental Results

The proposed method was trained and tested on the publicly available ICDAR 2005, 2011 and 2013 datasets [7,10,6]. There are 509, 484 and 462 color images in each dataset respectively. We measure the *recall (R)*, *precision (P)* and *F-measure (F)* to evaluate the performance of each method. The results are shown in Table 1, 2 and 3. Compared with stroke width based methods such as SWT and SFT, better results in F-measure and recall is achieved. Our system also outperformed some other methods [13,8,4,9,14] by a wide margin. On these three datasets the average processing time of our implementation was 2.12 sec., which is only about 1 second slower than the original SWT implementation (0.94 sec.). These three datasets contain images with very different backgrounds, which is the main cause of the performance discrepancies. Nevertheless, our method showed consistent improvements.

Careful examination of the experimental results reveals several major improvements: (1) From Fig. 2 and Fig. 4 we can observe that the quality of the edge detection result on the superpixels is significantly improved when the stroke width is close to the step size. (2) The shape of the characters is well persevered on the superpixels, while the noise that interferes with the gradient direction computation is eliminated. On the superpixel images, the pixels within a superpixel are assigned with the same color, which makes the adjustment in color between region boundaries smooth and uniform. Since the shape of the text remains unchanged and the noise is averaged out, convolving gradient computation kernel on the superpixel images will generate more accurate results. (3) The proposed stroke width regularized constraints generalize well to images of different resolutions and text of different sizes. Compared with the color continuity constraints of SFT, in which the threshold only decrease from a fixed number with respect to the number of pixels in the pixel ray, our constraints are adjusted according to the stroke width values of the text that are expected to be found.

It can be observed in Fig. 3 that as long as the initial stroke width estimation generated by SWT is reasonably accurate, our method is likely to outperform SWT and SFT. (4) Knowing the stroke width values of the text that are expected to be found, it is easy to avoid unnecessary searching on pixels whose distances are not close to the expected stroke width. It reduces the computational complexity and allows us to search pixels within a wider range of the gradient direction, which significantly increase the chance of recovering valid edge pixels.

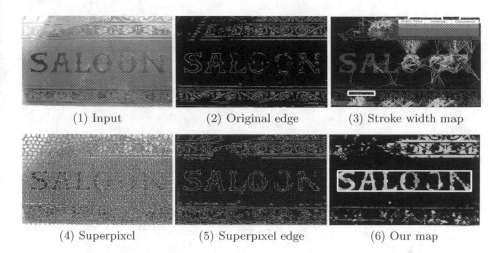

    (1) Input   (2) Original edge   (3) Stroke width map

    (4) Superpixel   (5) Superpixel edge   (6) Our map

Fig. 2. Comparing our method and the SWT algorithm where the gradient direction is inaccurate. The white bounding boxes are the detection results on the stroke width maps. Stroke value clusters in the charts of (3) are marked with green and red, which indicates their validities. It is clear that the result is greatly improved on superpixels that are generated with step size close to the stroke width.

Table 1. Evaluation on ICDAR 2005

Algorithm	Precision	Recall	f
Our system	**0.81**	0.67	**0.73**
Our system[§]	**0.85**	**0.77**	**0.77**
Huang[§][5]	**0.81**	0.74	0.72
Epshtein[3]	0.73	0.60	0.66
Fabrizio[4]	0.46	0.39	0.43
Yao[13]	0.69	0.66	0.67

[§] In [5], the f-measure is reported as the average of the f-measures of all the images and the precision and recall is measured on pixel level instead of bounding box level, which leads to results different from the standard ICDAR method. For a fair and comprehensive comparison, we use both two of measurement to evaluate our method and mark results with the measurement different from the standard ICDAR method with [§].

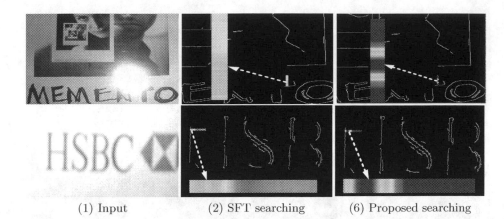

(1) Input   (2) SFT searching   (6) Proposed searching

Fig. 3. Comparing our stroke width regularized constraints with SFT. The likelihood of a pixel in the pixel ray being considered as valid is presented in jet colormap (unnormalized). The white arrow is pointing at the correct location of valid edge of the magnified searching pixel ray. Without stroke width information the heuristic constraint in SFT does not generalize well to text of different sizes.

Table 2. Evaluation on ICDAR 2011

Algorithm	Precision	Recall	f
Our system	0.80	0.69	**0.74**
Our system[§]	0.79	**0.77**	**0.74**
Huang[§][5]	**0.82**	0.75	0.73
Neumann[9]	0.73	0.65	0.69
Yi[14]	0.76	0.68	0.67
Neumann [8]	0.67	0.58	0.62

Table 3. Evaluation on ICDAR 2013

Algorithm	Precision	Recall	f
Our system	0.84	**0.75**	**0.79**
Epshtein[3]	0.81	0.73	0.77
Yin[15]	**0.88**	0.66	0.76
Neumann[9]	**0.88**	0.65	0.74
Shi[12,11]	0.85	0.63	0.72

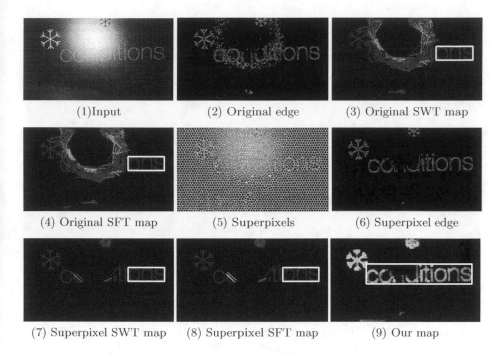

(1)Input　　　　　　　(2) Original edge　　　　(3) Original SWT map

(4) Original SFT map　　　(5) Superpixels　　　　(6) Superpixel edge

(7) Superpixel SWT map　　(8) Superpixel SFT map　　　(9) Our map

Fig. 4. Comparing our method to the SWT and the SFT algorithm. The white bounding boxes are the detection results on the stroke width maps. Evidently, the SWT and the SFT are rather inaccurate on the original image. But still, the SWT and the SFT algorithm fail on the superpixels that are generated with the step size close to the stroke width. This is because the SWT algorithm cannot recover missing edge pixels and the stroke width is too small to trigger the color constraint of the SFT algorithm.

4　Conclusion

We proposed a novel way of using stroke width information and superpixels to localize text lines in challenging natural scene images. Consistent and dominant stroke values on stroke width map are found and used as instructions to choose appropriate step sizes to generate superpixels. We observed that when the stroke width of the text expected to be found is close to the step size, it improves the qualities of edge detection and gradient direction computation. Furthermore, we proposed to use stroke width regularized constrains to recover valid edge pixels missed by the edge detector. Experimental results demonstrated that it achieved state-of-the-art performance on ICDAR 2005, 2011 and 2013 datasets.

Acknowledgment. This paper is partially supported by National Natural Science Foundation of China under Contract nos. 61303170, 61402472 and 61471235, and also supported by the National High Technology Research and Development Program of China (863 programs)under Contract nos. 2013AA014703 and 2012AA012803.

References

1. Achanta, R., Shaji, A., Smith, K., Lucchi, A., Fua, P., Süsstrunk, S.: SLIC superpixels. École Polytechnique Fédéral de Lausssanne (EPFL), Tech. Rep. 149300 (2010)
2. Chen, H., Tsai, S.S., Schroth, G., Chen, D.M., Grzeszczuk, R., Girod, B.: Robust text detection in natural images with edge-enhanced maximally stable extremal regions. In: 2011 18th IEEE International Conference on Image Processing (ICIP), pp. 2609–2612. IEEE (2011)
3. Epshtein, B., Ofek, E., Wexler, Y.: Detecting text in natural scenes with stroke width transform. In: 2010 IEEE Conference on Computer Vision and Pattern Recognition (CVPR), pp. 2963–2970. IEEE (2010)
4. Fabrizio, J., Cord, M., Marcotegui, B.: Text extraction from street level images. City Models, Roads and Traffic (CMRT) 3 (2009)
5. Huang, W., Lin, Z., Yang, J., Wang, J.: Text localization in natural images using stroke feature transform and text covariance descriptors. In: Proc. IEEE Int. Conf. Comp. Vis., pp. 1241–1248 (2013)
6. Karatzas, D., Shafait, F., Uchida, S., Iwamura, M., Mestre, S.R., Mas, J., Mota, D.F., Almazan, J.A., de las Heras, L.P., et al.: ICDAR 2013 robust reading competition. In: 2013 12th International Conference on Document Analysis and Recognition (ICDAR), pp. 1484–1493. IEEE (2013)
7. Lucas, S.M.: ICDAR 2005 text locating competition results. In: Proceedings of the Eighth International Conference on Document Analysis and Recognition 2005, pp. 80–84. IEEE (2005)
8. Neumann, L., Matas, J.: Text localization in real-world images using efficiently pruned exhaustive search. In: 2011 International Conference on Document Analysis and Recognition (ICDAR), pp. 687–691. IEEE (2011)
9. Neumann, L., Matas, J.: Real-time scene text localization and recognition. In: 2012 IEEE Conference on Computer Vision and Pattern Recognition (CVPR), pp. 3538–3545. IEEE (2012)
10. Shahab, A., Shafait, F., Dengel, A.: ICDAR 2011 robust reading competition challenge 2: Reading text in scene images. In: 2011 International Conference on Document Analysis and Recognition (ICDAR), pp. 1491–1496. IEEE (2011)
11. Shi, C., Wang, C., Xiao, B., Zhang, Y., Gao, S.: Scene text detection using graph model built upon maximally stable extremal regions. Pattern Recognition Letters 34(2), 107–116 (2013)
12. Shi, C., Wang, C., Xiao, B., Zhang, Y., Gao, S., Zhang, Z.: Scene text recognition using part-based tree-structured character detection. In: 2013 IEEE Conference on Computer Vision and Pattern Recognition (CVPR), pp. 2961–2968. IEEE (2013)
13. Yao, C., Bai, X., Liu, W., Ma, Y., Tu, Z.: Detecting texts of arbitrary orientations in natural images. In: 2012 IEEE Conference on Computer Vision and Pattern Recognition (CVPR), pp. 1083–1090. IEEE (2012)
14. Yi, C., Tian, Y., et al.: Text extraction from scene images by character appearance and structure modeling. Computer Vision and Image Understanding 117(2), 182–194 (2013)
15. Yin, X., Huang, K., Hao, H.: Robust text detection in natural scene images. IEEE Transactions on Pattern Analysis and Machine Intelligence 36(5), 970–983 (2013)

Sketch-Based Retrieval
Using Content-Aware Hashing

Shuang Liang[1,*], Long Zhao[1], Yichen Wei[2], and Jinyuan Jia[1]

[1] Tongji University, Shanghai, China
{shuangliang,9012garyzhao,jyjia}@tongji.edu.cn
[2] Microsoft Research Asia, Beijing, China
yichenw@microsoft.com

Abstract. In this paper, we introduce a generic hashing-based approach. It aims to facilitate sketch-based retrieval on large datasets of visual shapes. Unlike previous methods where visual descriptors are extracted from overlapping grids, a content-aware selection scheme is proposed to generate candidate patches instead. Meanwhile, the saliency of each patch is efficiently estimated. Locality-sensitive hashing (LSH) is employed to integrate and capture both the content and saliency of patches, as well as the spatial information of visual shapes. Furthermore, hash codes are indexed so that a query can be processed in sub-linear time. Experiments on three standard datasets in terms of hand drawn shapes, images and 3D models demonstrate the superiority of our approach.

Keywords: sketch-based retrieval, LSH, content-aware windows.

1 Introduction

Using sketch as input to retrieve visual information, such as hand drawn shapes, images and 3D models, has attracted a lot of research interests in recent years. A sketch is regarded as a collection of hand drawn strokes representing the contour or skeleton of an object, while detailed appearance information are lost. As a result, sketch-based retrieval algorithms are usually quite different from traditional image retrieval systems. One research direction for sketch-based retrieval aims to produce sketch-like edge maps from images or 3D models. Such works focus on removing edge noises such as background clutters in an image [10,20,21] or selecting suitable viewpoints for 3D models [8]. In this paper, we try to answer another question: *How to extract and compare features carried by a sketch (or a sketch-like edge map) in a proper way?* Therefore, in the following we briefly discuss previous works from the two viewpoints of interest in this paper: the ways to extract sketch features and methods to measure feature similarities.

Recently, segmentation-based methods [9,11,12] are proved effective for sketch retrieval and recognition systems with line drawings. Stroke segmentation is usually involved to lower the computational complexity. Topology/geometry attributes are then calculated from the extracted segments. However, an accurate

---

*** Corresponding author.

W.T. Ooi et al. (Eds.): PCM 2014, LNCS 8879, pp. 133–142, 2014.

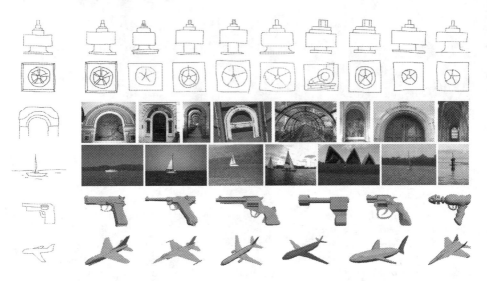

Fig. 1. Example results of our approach on the Magic Sketch Database [11] (top), TU Berlin Benchmark [7] (middle) and PSB Dataset [8] (bottom)

stroke segmentation is very hard to achieve. Especially, for natural images and 3D models where usually a lot of noises exist in their edge maps, there is almost no perfect segmentation methods. This limits segmentation-based methods unsuitable for generic sketch-based retrieval tasks.

Other works [2,7,8,14,21] originates from image retrieval. A sketch is divided into patches and visual descriptors are then extracted. Overlapping grids [2] or dense windows [14] are usually applied to describe the distribution of features over the whole shape. This is suitable for an image filled with details as colors and textures, but inappropriate for sparse edge maps, e.g., a sketch with a few strokes. In such cases, most of the content of a sketch is gathered into several salient patches while other patches are left almost empty. Due to this imbalance, it is really ineffective and even harmful when these empty patches are used for similarity computation, especially, when they are binarized before hashing.

Another issue in patch-based methods is that the similarity measurement for sketches is weak. As sketches are hand drawn artifacts with various line styles to represent objects other than colors and textures, they are different from images in two points: large intra-class differences (because of painters' subjective understanding) and small inter-class differences (due to the loss of visual information like colors and textures). Similar sketches may still have a lot of different patches. Therefore, the common concept (*images are similar when most patches are similar*) used in image retrieval and adopted by current patch-based methods is too strict and thus ineffective.

In this paper, we present a generic, content-aware sketch-based retrieval framework that overcomes the above issues in current patch-based methods by taking the sketch content distribution into consideration. Fig.1 shows some results of

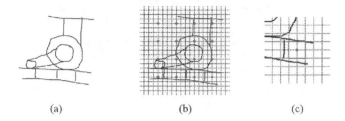

(a)   (b)   (c)

Fig. 2. Example of window selection. (a) input sketch; (b) an initial $n{\times}n$ grid and $m{\times}m$ uniformly sampled seeds; (c) the first $\delta w(x, y, 1)$ (magenta) and second $\delta w(x, y, 2)$ (green) rims of windows for seed (x, y). The ith $\delta w(x, y, i)$ is added to $w(x, y)$ iteratively.

our sketch-based retrieval on hand drawn shapes, images and 3D models. First, a window[1] selection scheme with regard to sketch content distribution is proposed to generate candidate patches. It ensures that features carried by a sketch are uniformly distributed in all patches. Second, we refine the common image retrieval concept in a more reasonable way: *sketches are similar when their most salient patches are similar*. Thus we introduce a salient window detection algorithm, which helps to compute shape similarity. We then follow hash-based retrieval approaches [2,16,23] and employ a locality-sensitive hashing (LSH) method, which combines the above cues together with structural information. To be applicable to large datasets, indexing is performed as well to enable sub-linear runtime performance. The rest of our paper is organized as follows. Sec. 2 presents the detailed algorithm, and our whole retrieval framework is introduced in Sec. 3. Finally, all experimental results are shown in Sec. 4.

2   Proposed Algorithm

Our proposed approach consists of three components: selecting candidate windows with regard to sketch content, detecting salient windows adaptively using key points, and combining these two features together with spatial cues into hash codes. Detailed algorithms of each step are described as follows.

2.1   Content-Aware Window Selection

Given a sketch, we begin by dividing it into a $n \times n$ grid. Then $m \times m$ seeds are uniformly sampled at the crossing points of the patches, as shown in Fig. 2(b). We define a set of windows $\delta w(x, y, i)$ as the ith-rim windows around the seed (x, y), as shown in Fig. 2(c). In order to generate a proper window, $\delta w(x, y, i)$ should be added to $w(x, y)$ iteratively until certain constraints are satisfied or $w(x, y)$ becomes invalid, e.g., it flows over the whole sketch or it is larger than a quarter of the sketch. We present two effective constraints as follows.

---

[1] We use "patches" and "windows" interchangeably in this paper.

Algorithm 1. Selecting candidate windows for the input sketch.

---
initialize $n \times n$ spatial grid for the input sketch
initialize $m \times m$ seeds uniformly sampled from the grid
initialize histogram $\boldsymbol{h}_i$ for each window w_i
initialize global histogram $\boldsymbol{\mathcal{H}} = \sum_{i=1}^{n^2} \boldsymbol{h}_i$
$W \leftarrow \{\}, H \leftarrow \{\}$
for $y = 1$ to m **do**
    for $x = 1$ to m **do**
        $w \leftarrow \{\}, \boldsymbol{h} \leftarrow \boldsymbol{0}$
        for $i = 1$ to $n/4$ **do**
            $w \leftarrow w + \delta w(x, y, i), \boldsymbol{h} \leftarrow \boldsymbol{h} + \boldsymbol{\delta h}(x, y, i)$
            if $\mathcal{F}_{app}(\boldsymbol{h}) \geq k_{app} \times \mathcal{F}_{app}(\boldsymbol{\mathcal{H}})$ **then**
                if $\mathcal{F}_{var}(\boldsymbol{h}) \leq k_{var} \times \mathcal{F}_{var}(\boldsymbol{\mathcal{H}})$ **then**
                    break
                end if
            end if
        end for
        $W \leftarrow W + w, H \leftarrow H + \boldsymbol{h}$
    end for
end for
return W and H

Since the histogram of oriented gradients (HoG) [5] is known to perform well in object detection and image retrieval problems, we use its unnormalized version to describe the feature inside windows for fast computation. Let $\boldsymbol{h} = \{b_1, \ldots, b_n\}$ denote the feature histogram of window w, $\boldsymbol{\delta h}(x, y, i)$ denote corresponding histograms of $\delta w(x, y, i)$ and $\boldsymbol{\mathcal{H}}$ denote the global histogram of the whole sketch. We define the *appearance constraint* C_{app} as

$$\mathcal{F}_{app}(\boldsymbol{h}) \geq k_{app} \times \mathcal{F}_{app}(\boldsymbol{\mathcal{H}}), \quad where \quad \mathcal{F}_{app}(\boldsymbol{h}) = \frac{1}{n} \sum_{i=1}^{n} b_i \qquad (1)$$

$\mathcal{F}_{app}$ is the appearance objective function, which essentially computes the mean value of $\boldsymbol{h}$. When $\mathcal{F}_{app}$ is high, it shows information carried by the window is relatively large, while low $\mathcal{F}_{app}$ indicates that the window is almost empty. Apparently, C_{app} constrains each window to contain enough information as expcted. Our second constraint is called *variety constraint* C_{var}, which is formulated as

$$\mathcal{F}_{var}(\boldsymbol{h}) \leq k_{var} \times \mathcal{F}_{var}(\boldsymbol{\mathcal{H}}), \quad where \quad \mathcal{F}_{var}(\boldsymbol{h}) = \frac{1}{n} \sum_{i=1}^{n} (b_i - \mathcal{F}_{app}(\boldsymbol{h}))^2 \qquad (2)$$

$\mathcal{F}_{var}$ is the variety objective function, which is the variance of $\boldsymbol{h}$ indeed. If $\mathcal{F}_{var}$ is low and $\mathcal{F}_{app}$ is high, it means that all bins in $\boldsymbol{h}$ have comparatively high values. As a result, windows satisfying C_{var} should have more diverse information, which is proved to be useful in sketch retrieval [14]. Note that parameters k_{app} and k_{var} control the global effects of $\boldsymbol{\mathcal{H}}$, and they are set to 0.8 and 1 experimentally. The whole algorithm is summarized in Alg. 1.

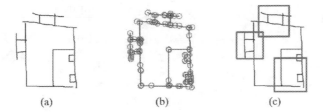

(a)   (b)   (c)

Fig. 3. Salient windows of a sketch. (a) input sketch; (b) salient points detected by Harris corner; (c) top 3 salient windows computed by Eq. 3 with overlapping area less than 20%.

We note that the objective function $\mathcal{F}_{app}$ of C_{app} is *incrementally computable*, for it satisfies the equation: $\mathcal{F}_{app}(\boldsymbol{h} + \boldsymbol{\delta h}) = \mathcal{F}_{app}(\boldsymbol{h}) + \mathcal{F}_{app}(\boldsymbol{\delta h})$. As the window grows, it is unnecessary to compute $\mathcal{F}_{app}$ across the entire histogram but sufficient to only update the affected part. Therefore, to evaluate C_{app} in each iteration is considerably fast. Though F_{var} is not incrementally computable, it does not need to be evaluated until C_{app} is satisfied. So our algorithm still has competitive computational speed. Experiments show that using candidate windows generated by our algorithm can improve retrieval results achieved by many traditional visual descriptors such as SIFT descriptor [13], HoG [5] and GALIF [8]. Detailed results are shown in Sec. 4.1.

2.2   Salient Window Detection

For local feature detection, keypoint-based detectors, such as difference of Gaussian (DoG) [13], Hessian operator [1] and Harris-Laplace detector [15], are more suitable for sketch feature extraction than region-based methods [17,24]. This is because a sketch usually includes separate lines and points other than continuous areas. These keypoint-based methods are mainly designed for finding salient points, which correspond to the corners and end points in sketches. Harris corner detector is employed in our method, for it achieves better performance than other keypoint-based methods as shown in [14].

For each window $w_i \in W$, where W is a set of candidate windows output by Alg. 1, we compute k_i to measure the saliency of w_i in a sketch as

$$k_i = 1 + \frac{Number(S_i)}{\sqrt{Area(w_i)}} \tag{3}$$

where $Number(S_i)$ denotes the number of salient points in w_i detected by Harris corner detector and $Area(w_i)$ is the pixel-based area of w_i. It has the intuitive interpretation that *a window is salient when it is small while contains many salient points*. Note that we use the square root of the area to make it less sensitive to scales: the measure remains stable across different window resolutions. See Fig. 3 for visual examples of salient windows detected in a sketch.

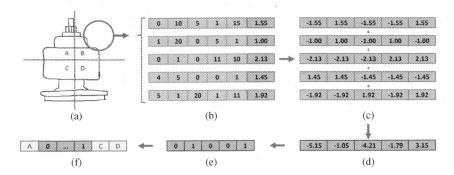

Fig. 4. Example of hashing. (a) input sketch partitioned into 4 parts spatially; (b) feature vectors f_i (blue) and k_i (orange) of all windows in part B; (c) weighted binary vectors by $\bar{f}_i \times k_i$; (d)-(f) vectors are summed, binarized and appended into the final hash code.

2.3   Hashing from Window Content and Saliency

Let f_i denote the feature vector extracted from w_i using certain visual descriptor (HoG in this paper). f_i is then binarized into $\bar{f}_i$ by setting the highest 40% values to 1 and the rest to -1. Note that thanks to the two constraints proposed in Sec. 2.1, features are evenly distributed and message loss can be reduced in this step. Then we follow the manner of Sim-hash [4] to produce locality-sensitive hashing for each sketch via $\bar{f}_i$ and k_i, where window saliency k_i is used as a weighting term for $\bar{f}_i$. Spatial information is shown to be helpful to sketch retrieval systems [6]. In order to capture spatial information of local features, we partition each sketch image into four parts. Other than hashing patch features in the entire sketch space, we do it in each part and finally append them together (a window is considered to belong to a certain part when its center is inside that part). Detailed steps are explained in Fig. 4.

3   Retrieval Framework

Before feature extraction and hashing, images and 3D models are converted into edge maps. Given an image, we generate a coarse edge map E_c by Canny algorithm [3], which yet contains many erroneous detected edges from background clutter. In order to find the most salient area in the image, we use [24] to get its saliency map S. We also apply maximum filter $\mathcal{MF}$ on S to enlarge salient regions to avoid the degenerate case when object contours are missing due to segmentation errors. Then the final salient edge map E is computed as $E = E_c \times \mathcal{MF}(S)$, where E is binarized by threshold 0.5 empirically. Fig. 5 presents our final edge maps. Given a 3D model, we follow the pipeline in [8] to get its projection edge maps.

Both sketches and edge maps are cut out of their minimum square bounding boxes and resized into 160×160 resolution to be translation and scale invariant.

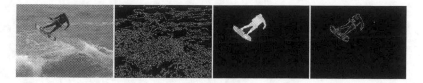

Fig. 5. Example of an image edge map. From left to right: input image, Canny lines [3], saliency map detected by [24] and our final edge map.

Algorithm 2. Pipeline of the retrieval framework.

Index:
1. produce edge maps E for all images/3D models in datasets
2. generate candidate windows w_i for E according to Alg. 1
3. compute HoG feature f_i and saliency k_i for each w_i according to Sec. 2.2
4. hash f_i and k_i into h according to Sec. 2.3
5. follow [16] to generate index I for all h

Query:
1. follow the step 2-4 of **Index** to generate h_s for input sketch S
2. return similarity ranking R by h_s from I as query results

---

Following Alg. 1, a sketch is divided into a 80×80 grid, and 15×15 seeds are uniformly sampled. In each patch, we compute the unnormalized HoG with 8 orientations and produce a 8-bin histogram. After all candidate patches are generated, they are resized to 16×16. The HoG descriptor is applied to extract features in each patch, for its unnormalized version has already been calculated.

Given a query, we find the most similar edge maps via the hash code proposed in Sec. 2.3 based on their *Hamming distance*. For all hash codes are binary, their Hamming distances can be calculated by several bit *xor* and *shift* operations, which is competitively fast even without indexing. In order to gain better runtime performance, we further follow the indexing strategy proposed by [16], which can be combined with our approach directly. It enables our retrieval process to be performed in sub-linear time. The pipeline of our retrieval framework is summarized in Alg. 2.

4   Experimental Results

In this section, extensive experiments are conducted to compare the proposed algorithm with other state-of-the-arts for sketch-based retrieval tasks. All experiments are done on the following three standard datasets. 1) *Magic Sketch Database*: Liang et al. [11] establish this database with a total of 1100 sketches from 55 classes drawn by 10 people. We use it to evaluate variants of our approach. 2) *TU Berlin Benchmark*: This is a sketch-based image retrieval benchmark introduced by Eitz et al. [7] which consists of 31 subjects, each one including 1 sketch and 40 corresponding test images. We use this benchmark to evaluate our approach for sketch-based image retrieval tasks. 3) *PSB Dataset*:

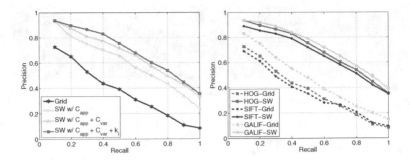

Fig. 6. Left: Evaluation of each component in our method. Right: Improvement of different visual descriptors caused by our method.

Eitz et al. [8] perform a large-scale experiment to collect 1,914 sketches for all 3D model categories in the Princeton Shape Benchmark (PSB) [19]. Our approach is evaluated on it to show the performance in sketch-based 3D model retrieval tasks.

4.1 Evaluation of Our Approach

Our method spends about 1.87 seconds to retrieve a sketch on the Magic Sketch Database, tested on a desktop computer with Intel 3.39GHz Quad-core CPU and 16GB memories and implemented by MATLAB without parallelized.

To evaluate the effectiveness of selecting windows with constraints C_{app} (Eq. 1) and C_{var} (Eq. 2), we implement a baseline using overlap grid (Grid) according to [2]. Then results via selected windows (SW) with C_{app} only and both of them are compared. To evaluate the performance of window saliency k_i (Eq. 3), we equally set window saliency by 1 in previous results for comparison. Results in Fig. 6(left) shows that our two constraints are complementary, and removing any of these three components would cause performance drop.

To further understand our method, we evaluate results using traditional visual descriptors such as SIFT [13] and GALIF [8] in addition to HoG [5]. Fig. 6(right) shows the performance of their grid-based version and the other one with our constrainted selected windows (SW). All of them are improved by our approach. We note that although GALIF achieves the best performance due to its multi-scale sampling strategy other than histogram-based method, it requires more time to be calculated. Therefore, HoG is employed in our approach.

4.2 Comparison with State-of-the-Art

On the TU Berlin Benchmark, we compare our algorithm with other methods using Bag-of-Words (BW) [7], key shapes (KS) [18] and Min-hash (MH) [2]. We use the benchmark scores introduced by [7] to evaluate the performance. On the Magic Sketch Database, methods via biased SVM (BSVM) [11], spatial relations

Table 1. Results on the TU Berlin Benchmark [7]

Method	BW[7]	KS[18]	MH[2]	Ours
Score	0.277	0.289	0.336	**0.352**

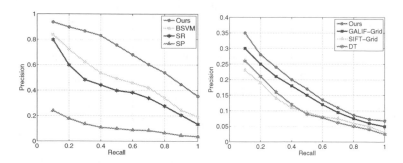

Fig. 7. PR curves of different algorithms on the Magic Sketch Database [11] (left) and PSB Dataset [8] (right)

(SR) [12] and spatial proximity (SP) [9] are compared with ours. On the PSB Dataset, we compare ours with methods via diffusion tensor (DT) [22], grid-based SIFT (SIFT-Grid) and GALIF (GALIF-Grid) [8]. All results on these two datasets are evaluated by standard Precision-Recall (PR) Curves. Both results in Table 1 and Fig. 7 show that our approach outperforms other state-of-the-arts. Some visual retrieval results of our approach are shown in Fig. 1.

5   Conclusion

In this paper, we present a novel sketch-based retrieval framework using content aware hashing. Feature window patches are selected with appearance and variety constraints to account for content distribution. A saliency value of each window is then calculated. For feature extraction, hashing is performed to combine content, saliency and spatial information of visual features and enables efficient retrieval in sub-linear time. Experimental results on sketch, image, and 3D model datasets demonstrate our proposed algorithm performs favorably against other alternatives for sketch-based retrieval systems. Future work includes exploring more effective visual descriptors for sketches.

Acknowledgement. This research work was supported by the National Science Foundation of China (No.61272276, No. 61305091), the National Twelfth Five-Year Plan Major Science and Technology Project of China (No.2012BAC11B01-04-03), Special Research Fund of Higher Colleges Doctorate (No. 20130072110035), the Fundamental Research Funds for the Central Universities (No.2100219038), and Shanghai Pujiang Program (No.13PJ1408200).

References

1. Bay, H., Tuytelaars, T., Van Gool, L.: Surf: Speeded up robust features. In: Leonardis, A., Bischof, H., Pinz, A. (eds.) ECCV 2006, Part I. LNCS, vol. 3951, pp. 404–417. Springer, Heidelberg (2006)
2. Bozas, K., Izquierdo, E.: Large Scale Sketch Based Image Retrieval Using Patch Hashing. In: Bebis, G., et al. (eds.) ISVC 2012, Part I. LNCS, vol. 7431, pp. 210–219. Springer, Heidelberg (2012)
3. Canny, J.: A Computational Approach to Edge Detection. IEEE Transactions on PAMI Pattern Analysis and Machine Intelligence 8(6), 679–698 (1986)
4. Charikar, M.S.: Similarity estimation techniques from rounding algorithms. In: STOC (2002)
5. Dalal, N., Triggs, B.: Histograms of oriented gradients for human detection. In: CVPR (2005)
6. Eitz, M., Hays, J., Alexa, M.: How Do Humans Sketch Objects? In: SIGGRAPH (2012)
7. Eitz, M., Hildebrand, K., Boubekeur, T., Alexa, M.: Sketch-Based Image Retrieval: Benchmark and Bag-of-Features Descriptors. IEEE Transactions on Visualization and Computer Graphics 17(11), 1624–1636 (2011)
8. Eitz, M., Richter, R., Boubekeur, T., Hildebrand, K., Alexa, M.: Sketch-Based Shape Retrieval. In: SIGGRAPH (2012)
9. Fonseca, M., Ferreira, A., Jorge, J.: Sketch-based Retrieval of Vector Drawings. In: Sketch-based Interfaces and Modeling, pp. 181–201. Springer, London (2011)
10. Furuya, T., Ohbuchi, R.: Visual saliency weighting and cross-domain manifold ranking for sketch-based image retrieval. In: Gurrin, C., Hopfgartner, F., Hurst, W., Johansen, H., Lee, H., O'Connor, N. (eds.) MMM 2014, Part I. LNCS, vol. 8325, pp. 37–49. Springer, Heidelberg (2014)
11. Liang, S., Sun, Z.: Sketch retrieval and relevance feedback with biased SVM classification. Pattern Recognition Letters 29(12), 1733–1741 (2008)
12. Liang, S., Sun, Z., Li, B.: Sketch Retrieval Based on Spatial Relations. In: CGIV (2005)
13. Lowe, D.: Distinctive image features from scale-invariant keypoints. IJCV 42(3), 145–175 (2001)
14. Ma, C., Yang, X., Zhang, C., Ruan, X., Yang, M.H.: Sketch Retrieval via Dense Stroke Features. In: BMVC (2013)
15. Mikolajczyk, K., Schmid, C.: Scale and Affine Invariant Interest Point Detectors. IJCV 60(1), 63–86 (2004)
16. Norouzi, M., Punjani, A., Fleet, D.J.: Fast Search in Hamming Space with Multi-Index Hashing. In: CVPR (2012)
17. Perazzi, F., Kr, P., Krahenbuhl, P., Pritch, Y., Hornung, A.: Saliency Filters: Contrast Based Filtering for Salient Region Detection. In: CVPR (2012)
18. Saavedra, J.M., Bustos, B.: Sketch-based image retrieval using keyshapes. In: Multimedia Tools and Applications, pp. 1–30. Springer, US (2013)
19. Shilane, P., Min, P., Kazhdan, M., Funkhouser, T.: The Princeton Shape Benchmark. In: Shape Modeling International (2004)
20. Sun, X., Wang, C., Xu, C., Zhang, L.: Indexing Billions of Images for Sketch-based Retrieval. In: ACM Multimedia (2013)
21. Wang, C., Cao, Y., Zhang, L.: MindFinder: A Sketch-based Image Search Engine based on Edgel Index. In: CVPR (2011)
22. Yoon, S.M., Scherer, M., Schreck, T., Kuijper, A.: Sketch-based 3D model retrieval using diffusion tensor fields of suggestive contours. In: ACM Multimedia (2010)
23. Zhang, L., Zhang, Y., Tang, J., Lu, K., Tian, Q.: Binary Code Ranking with Weighted Hamming Distance. In: CVPR (2013)
24. Zhu, W., Liang, S., Wei, Y., Sun, J.: Saliency Optimization from Robust Background Detection. In: CVPR (2014)

Pixel Granularity Template Matching Method for Screen Content Lossless Intra Picture

Lixin Feng, Pin Tao, Daxin Guang, Jiangtao Wen, and Shiqiang Yang

Computer Science and Technology Depa., TNList,
Tsinghua University, Beijing, China

Abstract. The-state-of-art High Efficiency Video Coding(HEVC) is designed towards the natural picture. Screen Content picture contains many similarities in one picture which can be used to improve the intra picture compression ratio. We propose sample rearrangement and template matching method by exploiting the similarities in the screen content picture. The 21 pixels template and the high efficiency multiple hash table are designed. Experiment results show that our proposal method can improve the lossless compression ratio by up to 4.23 times than HEVC Range Extensions.

Keywords: Screen Content Coding, High Efficiency Video Coding (HEVC), template matching, lossless coding.

1 Introduction

With popularity of smart devices and networks, video content is increasingly ubiquitous categorized as screen content video and camera-captured video recently. The generation method differs between screen content video and camera captured video. Screen Content(SC) commonly refers as the computer display screen such as text, slide, web pages, graphics, video game, and camera-captured video content usually is produced by video captured devices. Driven by the application in many scenarios such as remote desktop sharing, wireless display, game broadcasting, digital operation room, screen content coding attracts a great deal of attention in recent years. Unlike the camera-captured video coding which aims at the most maximum compression while achieving an acceptable distortion, screen content video coding needs high fidelity because human visual system cares more contrast. Therefore, the coding efficiency of screen content needs to be higher than camera-captured video.

The start of art video coding standard, High Efficiency Video Coding(HEVC) mainly focuses on the camera-captured video [1]. In response to these challenges, HEVC has a specifically range extension. For increasing the fidelity of screen content, many methods has been proposed categorized as two approaches according to the modified extent of the HEVC standard. The first kind of methods adopted in current HEVC standard for SC are the modification of existing HEVC coding framework, such as the coefficient level updates [2], transform skipping modifications [3], non-interpolation intra-picture prediction [4] and extended cross

W.T. Ooi et al. (Eds.): PCM 2014, LNCS 8879, pp. 143–152, 2014.

component deceleration [5]. The second kind of methods employed new ways to take advantage of the SC video, for example, exploiting spatial correlation such as clustering the color and getting the index map [6], catching the action of selecting background of text [7], following image histogram concept and prediction basing on color cluster [8] and reducing the pattern repetitive such as intra motion compensation [9], combining current HEVC coder with 1-D dictionary coding method [10] or 2-D dictionary coding method [11].

Intra-picture prediction plays important role in the intra-picture video coding. HEVC uses closed up intra prediction like H.264 by according to the above or left coding units(CU) edge pixel values [1]. Template matching applied in video coding can improve the coding efficiency[12–15] even been treated as a special case of the sparse prediction [16, 14]. All above processing unit is the block or sub block. Pixel level of intra prediction such as sample-based angular intra prediction [17] and pixel based template matching [18] can enhance the coding efficiency too. But template prediction method also has some disadvantage when searching out all available position. When processing unit is based on every pixel, it also leads to high accurately prediction picture than other prediction method. First, it needs more computation resource and extra space with respect to the high-dimensional template space which cases [12, 13, 18] only used three neighboring pixel and limited range. Second, the prediction error of the neighbor pixel diffuses very quickly, it will make the prediction error of following pixels becoming worse and worse.

This paper employs a fast template prediction method for screen content within rate-distortion optimization basing on current HEVC range extension(HM 12.0-RExt4.1). The rearrangement is used for exploiting the original picture template spatial correlation and avoiding pixel prediction error spreading. For the purpose of reducing the storage space and searching time, a multiple level hash function was adopted. 70%~80% of the residual pixel is zero after the predictive procedure. Unfortunately, the standard HEVC entropy coding cannot compress the residual picture efficiency. With a novel minimization path coding and rate-distortion optimization selection, our proposal method could further improve the coding performance on screen content video.

The rest of sections are organized as follows. Section 2 is the detail about our fast template prediction method. Section 3 gives the experimental results. Section 4 is the conclusion of this paper.

2   Proposed Template Matching Method

Firstly, this section gives the coding architecture of our proposal method, then describes the details of pixel granularity template matching method following by rearrangement, template matching, multiple hash table and new entropy coding method. The coding method processes on the rearrangement picture, but the template matching is based on the original picture.

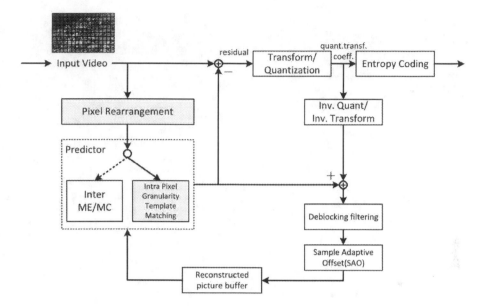

Fig. 1. The coding architecture. (the gray blocks stand for the new method)

2.1   The Algorithm Architecture

Like the formal video coding standard, HEVC is also a hybrid coding to achieve multiple goals. Coding tree unit(CTU) similar to macro-block in H.264 supports quadtree-like partition to gain high coding efficiency. The CTU size can be 64×64, 32× 32 or 16 × 16. To adapt to the HEVC intra-picture coding, the proposal method are employed as a new pixel level intra picture coding method as shown in Figure 1.

Our method doesn't use the standard intra prediction scheme in order to take full advantages of similarity in the SC picture. The standard intra prediction uses the neighbor pixels to predict the whole block. Even though the block is as small as 4×4 size, the farthest pixel in the block is 4 pixels away from the prediction pixels. It is hard for the prediction algorithm to achieve the accuracy prediction. Therefore, we propose to use the pixel based prediction to predict the pixel one by one. In order to overcome the prediction error spreading, the rearrangement operation is used. Then the prediction operation has four options: template matching, copy from above CTU, copy from left CTU and sample-based weighted prediction [15]. Finally, The Entropy coding has two options: content adaptive binary arithmetic coding(CABAC) and the new entropy coding method which will be described in following section. By rate-distortion optimization, the current CTU chooses the best option.

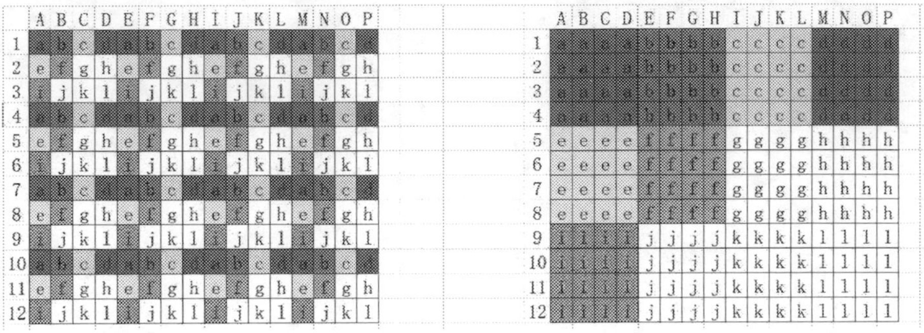

Fig. 2. The rearrangement example. (left:original right:rearrangement)

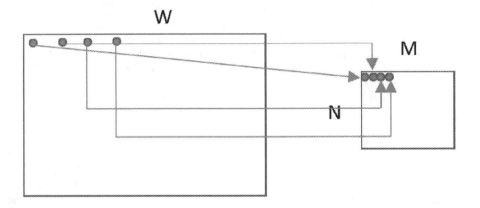

Fig. 3. Rearrangement (left:original picture, right:first rearrangement picture CTU)

2.2 Rearrangement

Direct using template matching method in pixel level causes the spreading of prediction error in every block. And that predict pixel far from the boundary pixel usually is hard to predict. Using sub-block such as 4×4 level decreases the error spreading problem [12, 13] but can't avoid the problem. Rearrangement the original picture solves the problem as shown in Figure 2. It is a special example to show the idea with coding unit of 4×4. The following part describes the rearrangement idea.

As shown in Figure 3, the rearrangement procedure is applied the full picture with $W \times H$ which W and H stands for the width and height of the picture. The rearrangement picture has the same width and height with different mapping order. The rearrangement consists of many blocks from left to right and top to bottom which size is $M \times N$ (generally take M = N). The corresponded position relation is indicated by

$$R_x = O_x \div StepR_x + O_x \bmod StepR_x * StepO_x \qquad (1)$$

$$R_y = O_y \div StepR_y + O_y \bmod StepR_y * StepO_y \qquad (2)$$

Which R means rearrangement position coordinate and O means original position coordinate. The StepR means the distance between original picture order and StepO usually refers as the CTU size in the equation formula 1-2. Based on template matching prediction for $M \times N$ pixel blocks, beginning from the top left corner pixel, the search order is from left to right, and top to bottom. The rearrangement picture that is coded by our method guarantees that every pixel in original picture is coded after above and left pixel is reconstructed, so the template matching method could avoid the spreading of error and maximize total method efficiency.

2.3 Template Matching

K(in this paper is 21) pixels around each pixel represent the template, the current pixel has the closest similarity by searching the most similar template location in the match index table. The template selecting is one of the key factors for pixel level template matching. Within perfect case, every pixel has an optimal template but the cost is very high for compression efficiency. Currently, we use the template showed in table 1 as a typical configuration. We only consider the nearby 21 pixels of pixel "×" as shown in table 1, the total 21 points as 8 groups as follows from G_1 to G_8.

- G_1:{ 1 }
- G_2:{ 2 }
- G_3:{ 3 }
- G_4:{ 4, 5 }
- G_5:{ 6, 7, 13, 14 }
- G_6:{ 8, 15, 16, 17 }
- G_7:{ 9, 10, 18, 19 }
- G_8:{ 11, 12, 20, 21 }

Obviously the pixel near current pixel "×" is more important than other far pixels. Therefore, pixel group G_1,G_2,G_3 only contains one pixel, and G_5 to G_8 contains 4 pixels as those pixels have minor significance. This template is empirical selected from the above and left image pixels according to current pixel for screen content picture. But straight searching and updating operation in the available space require many time and storage space. Hence, the multiple hash table is used which will be introduced in the next section.

Table 1. The 21-pixels Template

16	17	18	19	20	21
15	8	9	10	11	12
14	7	2	3	4	5
13	6	1	×		

2.4   Multiple Hash Table

The average value of each pixel group $G_i(i=1...8)$ are used as the feature value of the template and the template is represented by K bits(K means bit depth per pixel, usually is 8) value between 0 to 2^K-1. The average operation in the pixel group can suppress the impact of noise issued by pixel far from current pixel "×" in the template matching procedure. Every bit in the same position of G_i constitutes a hash key. The multiple hash table is organized by multiple level hash key value from left to right which indicted the address storing template match pixel position.

To each pixel in the picture, we should find the best match template with above multiple hash table. Then the pixel of best match template is used as the prediction of current pixel. This is the prediction procedure. It is almost same to the hash table construction procedure but has a slight difference.

The template and the hash key calculation are same as the hash table construction for the current pixel. With the hash key, our method finds the corresponding template list in the hash table. By comparing the current template and those ones in the list, it calculates the similarity of them which represents by how many pixels of total 21 in the template are same. The prediction procedure will not stop until all nodes in the list are tried and then the best one which has the most same pixels is selected. This is different from hash table construction. In hash table construction procedure, if one similar template is found, e.g. 19 pixels are same for all 21 pixels, the procedure is stopped to accelerate the procedure and keep the list not being too long.

For screen content picture, about 80% or more pixels will get the exactly same prediction value as the original pixel. That means the prediction residual of more than 80% pixels are zero. It is the good hints for following entropy compression.

2.5   New Entropy Method

The position of non-zero residual picture coding can use many kinds of entropy coding method. Currently, this paper use Golomb code of zero to encode the position of non-zero residual picture and residual value. The position of non-zero residual is organized by the changed distance of width and height.

The M×N residual block mostly has almost 90% zero so the nonzero value is small. But the nonzero absolute value is generally higher than the standard HEVC, so the traditional transform/quantization method will gain bad compression efficiency. Recording position using the first non-zero residuals, then the residual data is encoded non-zero, better compression efficiency can be achieved for sparse residual data blocks. The observation about the nonzero residual value show that it is not evenly distributed between -255 and 255 and is not a Gaussian distribution centered at zero or other distributions only related different characteristics according to the image non-linear distribution. Therefore, we use adaptive entropy coding algorithm through the establishment of a length of 512 residuals probability table recording different residuals prior probability M×N blocks. Before the new M×N blocks zero residuals coding, sorting the residual

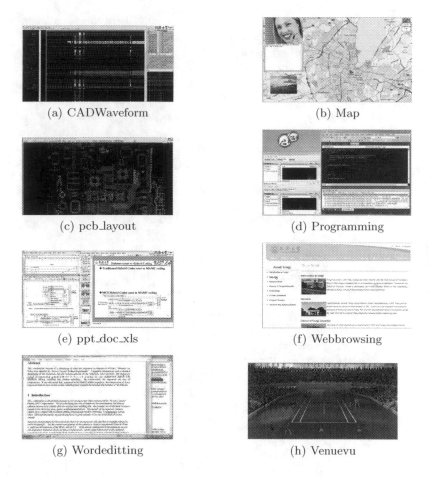

(a) CADWaveform

(b) Map

(c) pcb_layout

(d) Programming

(e) ppt_doc_xls

(f) Webbrowsing

(g) Wordeditting

(h) Venuevu

Fig. 4. Screen Content testing video sequences

probability table to make sure of the highest probability for the shortest code length encoding. The higher the probability of occurrence of the code length of the residuals is smaller cause in higher entropy coding compression ratio.

3   Experimental Result

Eight SC video sequences are used, shown in Figure 4. The format of these sequences is RGB 4:4:4 which has the components order from Green, Blue and Red. The anchor encoder is HEVC Test Model HM12.0-RExt4.1 which is the special range extension for screen content coding. The new method have been implemented into the anchor encoder. For lossless coding, the *TransquantBypassEnableFlag* and *CUTransquantBypassFlagForce* have both set to 1 .

The first picture of these sequences is used to do the lossless intra coding. The lossless compression ratio of the anchor and our proposal method are shown in Table 2. Experiment results show that our proposal method can achieve a better compression ratio than the anchor encoder. The best one is *pcblayout* sequence, our proposal method achieves 4.23 times than the anchor method on compression ratio. The worst case is *Venuevu* sequence, our method still has 1.9 times better than the anchor. The average compression ratios improvement is 2.86 times than current HEVC HM12RExt4.1. The results show our proposal method can do better on the lossless intra coding of SC sequences. It can also be seen that sequences within the simple texture such as *Wordeditting* could be compressed better than sequences within the complex texture such as *Venuevu*. The result also indicates proposal method could handle the simple texture better than HEVC.

Table 2. Experimental result

Sequence	Methods	Bits	CR	CR Ratio
cadwaveform	HM12_R4.1	3558528	13.99	2.87
	Proposal Method	1239550	40.15	
Map	HM12_R4.1	8338392	2.65	1.7
	Proposal Method	4918370	4.5	
pcblayout	HM12_R4.1	4138432	12.03	4.23
	Proposal Method	979078	50.83	
programming	HM12_R4.1	4385432	5.04	1.82
	Proposal Method	2408663	9.18	
ppt_doc_xls	HM12_R4.1	6556344	7.59	2.24
	Proposal Method	2923726	2.24	
Webbrowsing	HM12_R4.1	3131760	7.06	2.3
	Proposal Method	1359637	16.27	
WordEditing	HM12_R4.1	5528816	4.00	3.42
	Proposal Method	1616294	13.68	
Venuevu	HM12_R4.1	23641856	2.11	1.9
	Proposal Method	12416917	4.00	

4   Conclusions and Future Work

In this paper, we proposed a template matching method codec for screen content intra coding. By taking full advantages of the similarity in screen content picture, proposal method can achieve about 2.85 times improvement in lossless compression ratio than the-state-of-art HEVC range extension software. Furthermore, we believe the proposal method has future potential to get the better performance by improving the template design, hash table design and better entropy encoder.

Acknowledgment. This work was supported by the National Science Fund for Distinguished Young Scholars of China (Grant No. 61125102) and the National Significant Science and Technology Projects of China under Grant No. 2013zx01039-001-002-003.

References

1. Woo-Jin, H., Sullivan, G.J., Ohm, J., Wiegand, T.: Overview of the high efficiency video coding (hevc) standard. IEEE Transactions on Circuits and Systems for Video Technology 22, 1649–1668 (2012)
2. Budagavi, M.: Ahg8: Coefficient level criceparam updates for screen content coding. Document JCTVC-M0316, ITU-T/ISO/IEC Joint Collaborative Team on Video Coding (JCT-VC) (April 2013)
3. Gao, W., Cook, G., Yang, M., Song, J., Yu, H.: Near lossless coding for screen content. Document JCTVC-F564, ITU-T/ISO/IEC Joint Collaborative Team on Video Coding (JCT-VC) (July 2011)
4. Saxena, A., Chen, H., Felix, F.: Rce3: Nearest-neighbor intra prediction for screen content video coding. Document JCTVC-O0049, ITU-T/ISO/IEC Joint Collaborative Team on Video Coding (JCT-VC) (October 2013)
5. Nguyen, T., Khairat, A., Marpe, D.: Ahg5 and ahg8: extended cross-component decorrelation for animated screen content. Document JCTVC-P0097, ITU-T/ISO/IEC Joint Collaborative Team on Video Coding (JCT-VC) (January 2014)
6. Cuiling, L., Guangming, S., Feng, W.: Compress compound images in h.264/mpge-4 avc by exploiting spatial correlation. IEEE Transactions on Image Processing 19(4), 946–957 (2010)
7. Xiulian, P., Jizheng, X., Feng, W.: Exploiting inter-frame correlations in compound video coding. In: 2011 IEEE Visual Communications and Image Processing (VCIP), pp. 1–4 (2011)
8. Xu, J.-Z., Lu, Y., Li, S., Guo, X., Li, B., Wu, F.: Ahg8: Major-color-based screen content coding. Document JCTVC-O0182, ITU-T/ISO/IEC Joint Collaborative Team on Video Coding (JCT-VC) (November 2013)
9. Budagavi, M., Kwon, D.-K.: Ahg8: Video coding using intra motion compensation. Document JCTVC-M0350, ITU-T/ISO/IEC Joint Collaborative Team on Video Coding (JCT-VC) (April 2013)
10. Tao, L., Peijun, Z., Shuhui, W., Kailun, Z., Xianyi, C.: Mixed chroma sampling-rate high efficiency video coding for full-chroma screen content. IEEE Transactions on Circuits and Systems for Video Technology 23(1), 173–185 (2013)
11. Zhu, J.X.W., Ding, W.: Screen content coding using 2-d dictionary mode. Document JCTVC-O0357, ITU-T/ISO/IEC Joint Collaborative Team on Video Coding (JCT-VC) (November 2013)
12. Tan, T.K., Boon, C.S., Suzuki, Y.: Intra prediction by template matching. In: 2006 IEEE International Conference on Image Processing, pp. 1693–1696 (2006)
13. Cuiling, L., Jizheng, X., Feng, W., Guangming, S.: Intra frame coding with template matching prediction and adaptive transform, pp. 1221–1224
14. Tiirkan, M., Guillemot, C.: Image prediction based on neighbor-embedding methods. IEEE Transactions on Image Processing 21(4), 1885–1898 (2012)
15. Wige, E., Yammine, G., Amon, P., Hutter, A., Kaup, A.: Sample-based weighted prediction with directional template matching for hevc lossless coding. In: Picture Coding Symposium (PCS), pp. 305–308 (2013)

16. Martin, A., Fuchs, J.-J., Guillemot, C., Thoreau, D.: Sparse representation for image prediction. In: Proc. European Signal Processing Conference (EUSIPCO), Poznan, Poland (2007)
17. Minhua, Z., Wen, G., Minqiang, J., Haoping, Y.: Hevc lossless coding and improvements. IEEE Transactions on Circuits and Systems for Video Technology 22(12), 1839–1843 (2012)
18. McCann, K.: Samsungs response to the call for proposals on video compression technology. Document JCTVC-A124, ITU-T/ISO/IEC Joint Collaborative Team on Video Coding (JCT-VC) (April 2010)

Music Interaction on Mobile Phones

Wilber Chao, Kuan-Ting Chen, and Yi-Shin Chen*

Institute of Information Systems and Applications, Department of Computer Science,
National Tsing Hua University, Taiwan
{cecol3500123,sanpeter992000,yishin}@gmail.com

Abstract. This paper proposes an interactive musical application for smart phones to increase the possible exposure of music. Users can adjust the expressions of music using gestures on smart phones. Our experimental results demonstrate user satisfaction with this system. The user study also demonstrates that user preference of music pieces could be improved after playing with the system.

1 Introduction

Classical music is not as popular as it was before, as suggested in a study by Kolb [4], which mentioned that there is a decline in audience demand for American and European orchestral music and that its audience is aging. The same study also indicated that such a declining and aging audience could be eased by increasing the probability of young people's early exposure to the music.

To ease the declining and aging audience issue, the most effective way is to create an environment that audience can interact with music easily. In the study by Peery and Peery [10], interacting with music can increase the understanding and appreciation of music. In the past, researchers have proposed various ways to interact with music. For example, in the system proposed by Chew et al. [3], users can adjust the tempo of the music using a driving interface. Many proposed conducting systems [1,7,6,8,11] also allow users to adjust music tempo and volume using different types of devices.

Even though these systems can help users interact with music, such interactions require special devices and the systems are usually located at a fixed position. Furthermore, users must understand music very well in order to operate these conducting systems. These systems offer little help in increasing musical exposure. In order to attract most people, regardless of their abilities, the interaction approach should be the most familiar one possible and the device should be popular and nearly ubiquitous. Based on observations at many popular music concerts, we have discovered that the audience has a tendency to jump up and down or wave their hands to the musical beats. Several early experiments [5,9] have also pointed out that musical tempo can influence audience preference. This has inspired us to design an interactive interface to adjust music tempo using gestures with a smart phone.

---

* Corresponding author.

W.T. Ooi et al. (Eds.): PCM 2014, LNCS 8879, pp. 153–162, 2014.

There are numerous types and brands of smart phones, each using different devices and systems. After comparing most phone models, we determined that the acceleration sensor is essential and is embedded in almost every phone. Our proposed interface analyzes the acceleration signals to capture gestures and adjust music tempo based on the analyzed results. In order to have robust system performance, our gesture recognition method extracts the repeated intentions from the gesture signals and aligns the repeated intention with the music tempo. Our experimental results demonstrate that the proposed system can provide accurate, robust, and smooth music interaction. The user study also demonstrates that user preference to music can be improved after playing with the system. The proposed system might be helpful in promoting the popularity of classical music.

The rest of this paper is organized as follows. An overview of the system framework and detailed methods are given in Section Methodology. The empirical study is reported in Section Performance Evaluation. Section Conclusion concludes the paper and discusses future work.

2   Methodology

In the proposed system, the user can select a music piece to interact with as illustrated in Figure 1. If the interaction mode is selected, the user should give two preparation beats for the music piece. The music then starts on the next beat. The music tempo should follow the gesture, except in one particular scenario. That is if the user stops the beating gesture, the music tempo will stay the same until the end of the piece.

To achieve low user overhead, our system has two goals. First, that users can operate the system without any prior user knowledge. Second, that users can use their own beating patterns (including all professional conducting gestures and newly invented behaviors) to interact with music. In other words, no matter if

Fig. 1. Interact with music on a smart phone

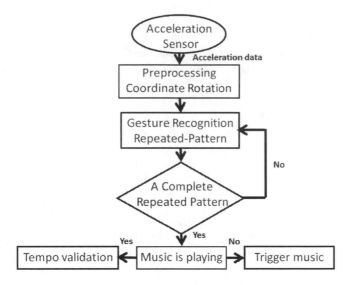

Fig. 2. System framework

users are using simple beating gestures or very complicated conducting gestures, they can smoothly adjust the tempo. The entire system framework is illustrated in Figure 2.

As illustrated in Figure 2, our system should recognize start gestures, that is, the gesture is detected by the acceleration sensor embedded in the smart phone. The gesture data contain the x-, y-, and z-axes of the Cartesian coordinates.

Initially, our system will load the input music and initialize a flag, called PlayStatus, which is set with a value equal to false if the input music has not yet been started. During the interaction mode, the system starts receiving the gesture data and waits for the start signal. Once the start gesture has been recognized, the system then sets the PlayStatus flag to be true and starts playing the input music. While the input music is playing, the gesture data will be analyzed until the end of music or until the system has been interrupted by user commands.

2.1 Data Preprocessing

Fortunately, smart phones not only contain an acceleration sensor, but they also contain a gravity and a geo-magnetic sensor, where the gravity sensor detects the direction of gravity and the geo-magnetic sensor can align the geomagnetic line. By utilizing the gravity matrix and geomagnetic sensor, the system then rotates the direction of the acceleration to fix it on the earth's coordinates.

2.2 Gesture Recognition

Although there are various types of beating gestures, most intended beating gestures have recurring patterns to indicate beat points. Beating gestures can be categorized into numerous beat patterns, in which each fragment represents a series of hand movements through a beat. Theoretically, the lapse of time for each beat should be nearly equivalent if the tempo remains constant. Each beat pattern contains several significant features, including a start point, a beat point, and an end point. The start point and end point usually go through the origin of beating gestures, where the acceleration values of these points are zero. The beat point, which is represented by a bouncing movement, is identified by the down turning point. In other words, the beat point is the lower stationary point (e.g., the circled points in Figure 3).

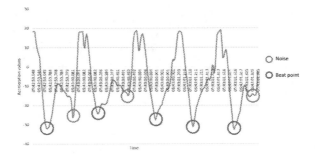

Fig. 3. Samples of acceleration data

If the lower stationary points are the only consideration, the inevitable hand trembles definitely affect the accuracy of beat detection. As illustrated in Figure 3, the circled points are the lower stationary points. However, the green circles denote falsely detected points, which are generated from unintentional hand trembles. To avoid falsely detecting beat points, one hypothesis is made based on the study by Cambouropoulos et al. [2]. Cambouropoulos et al. stated that people prefer music with a smooth tempo. Therefore, we can assume that the tempos of two consecutive beats are nearly the same. In other words, the time periods of two consecutive beats should be nearly equivalent. Furthermore, the consecutive beating gestures should have the repeated intention. Detecting repeated intentions should be able to avoid noise and maintain tempo smoothness.

The speed of each beat can be calculated as:

Definition 1. *Let* s, l, *and* e *denote a start point, a low stationary point, and an end point of a beat pattern, respectively.* $t_s < t_l < t_e$; *The speed* v *of this beat is*

$$v = \frac{t_e - t_s}{F} * 60$$

,where F *is the frequency of the sampled rate of the acceleration sensor. The speed* v *is in the format of bpm.*

After receiving the change of tempo command, the system then detects the existence of repeated intentions. If the newly issued tempo difference is smaller than a specified threshold, it is considered to be hand trembles and no action is taken. If the newly issued tempo difference is larger than a specified threshold, the system then checks the intention records. If there exists the same intention record (i.e., the user issued the same slow-down/speed-up command in the previous beat), the system will follow-up the intention. If no such intention exists (or even an opposite record exist), the system then clears the existing intention records and remains at the same speed. Algorithm 1 illustrates this in more detail.

Algorithm 1. CheckRepeatedIntention

Require: newTempo , currentTempo
  currentIntention = difference(currentTempo - newTempo)
  if $(absolute(currentIntention) \geq trembleThreshold)$ **then**
    if (intentionREC is empty) **then**
      Add (currentIntention) into intentionREC
    else
      preIntention = difference value in intentionREC
      if $(currentIntention * preIntention > 0)$ **then**
        ChangeTempo(newTempo)
        CLEAR(intentionREC)
      end if
    end if
  end if

The objective of Algorithm 1 can ensure that the tempo of the music is smooth. The cost of tempo smoothness is the latency. As is seen in Algorithm 1, the system waits for repeated intentions to be validated. Therefore, the latency of the system increases. However, such a compromise is essential.

3   Performance Evaluation

The objective of the proposed system is to attract people to listen to classical music and have more appreciation after the interaction. In this section, we first demonstrate that functions of our system can work accurately during interacting with music in Section System Performance. Then, the results of user study and the influence of the user preference are revealed in Section User Study.

3.1   System Performance

Experimental Setup. The experimental system was executed on a Galaxy Nexus containing a dual-core 1.2 GHz Cortex-A9 processor, Android 4.0, and 1 GB memory. The experiment involved 10 users of different music backgrounds.

The corresponding beating trajectories were recorded. Meanwhile, a think-aloud protocol used in the usability evaluation was adopted. In the think-aloud protocol, users were asked to tell the observer their intentions (e.g., speed up); this was synchronized with their beating movements during the experiments. The user's intention log was the standard used in validating the system output. After interacting with the system, a thorough interview was conducted to collect user opinions.

During the data collection procedures, the users were asked to perform a beating gesture according to the tempo generated by a metronome. The users first followed one fixed tempo, ranging from 60 *bpm* to 160 *bpm*. Afterward, the users were asked to interact with a song with a specified tempo for a while. After several measures, users were asked to speed-up/slow-down the tempo to a specified tempo. Finally, users were asked to perform gestures at a constant speed but intentionally perform errors, which would imitate trembles. Fifty errors in total were to be performed by each user.

Tempo Accuracy. Precisely interpreting the users' gestures with the music is an important index in expressive performance. Figure 4 illustrates the differences between the expected tempo and the tempo recognized by the system. It shows that the differences in all the speeds investigated are less than 1 bpm. The speed with the largest difference is 90 *bpm*.

Based on the interviews, many users mentioned that 90 *bpm* tempo is very difficult to follow since such a tempo is not natural for them. Conversely, the 60 *bpm* and 120 *bpm* tempos are easy to precisely follow. The possible reasons for this are that the time intervals between each beat are 1 s, half a second, and 0.6 s for 60 *bpm*, 120 *bpm*, and 90 *bpm*, respectively. The time interval 1 s is the most familiar to users since the tick-tock sound from clocks have the exactly same interval. Conversely, 0.6 s can be considered the most unfamiliar time interval. The tempo of 160 *bpm*, is very fast for subjects. They can only concentrate on tracking the speed of the metronome. Thus, the difference to expected tempo is smaller than the cases of 90 *bpm* tempo.

Time Latency. Table 1 shows the algorithm computing time of gesture recognition and tempo adjustment.

Owing to the requirement of checking repeated intentions, the system needs a latency of two beats to synchronize the music tempo to the gesture speed. The system can detect the tempo change within 0.5 s and follow up the music tempo

Table 1. Performance

	Gesture recognition	Tempo adjustment
Computing time(nanosecond)	105116	44080

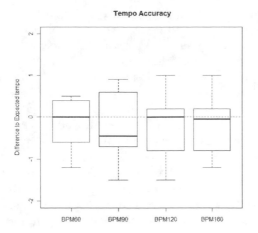

Fig. 4. Tempo accuracy

Table 2. Tempo modification latency

BPM(from-to)	120-150	120-180	120-90	120-60
System Latency(s)	0.4	0.3	0.5	0.5
Music Latency(s)	0.7	0.6	1	1.4

within 1.5 s, as shown in Table 2. Although we compromised between tempo smoothness and the speed of the music tempo change by a hand gesture, the system performs well regarding the latency of the system.

Tempo Smoothness. The performance of tempo smoothness is the most important factor in the user experience of music appreciation [2]. Therefore, we develop repeated intention detection to decrease the effect of hand trembling. Such a tempo validation technique can increase smooth expression. As shown in Table 3, 96 % of errors are filtered out by our method. This demonstrates that our system can provide good smoothness performance by eliminating the tremble effect and deviations in humans' motions.

Table 3. Tempo smooth result

Errors	Filtered out	Not filtered out
Count	48	2

3.2   User Study

Experimental Setup. There are two sets of experiments. In the first set of the experiments, data are collected from 15 users who are undergraduate or graduate students from the university. We designed a two-step test. In the pre-test, we asked the users to listen to certain classical music pieces and then to interact with any pieces they want to select. When they finished listening, we asked them about their preference regarding the music. After they interacted with music, we asked them about preference again. Half a month later, we began the post-test in which we asked the users to listen to two types of music, the same music as in the pre-test and, music with a different melody. Therefore, we can evaluate whether our music interaction system has an effect on all kinds of classical music. Finally, we asked them about their preferences regarding music pieces and the system using a 5-point scale, which indicated user preference level as very dissatisfied (1), dissatisfied, average, satisfied, and very satisfied (5). The entire experimental processes were recorded using a video camera. After interacting with the system, a thorough interview was conducted to collect user opinions.

In the first set of experiments, users select the music pieces based on their preferences. Such choices might mislead the conclusion; since users might select the songs they prefer and of course have higher preference values after interaction. To alleviate such concerns, the second set of experiments only allows users to interact with a selected music pieces. In the second set of the experiments, the data are collected from 25 users. Two-step tests are given. In the pre-test, users listen three classical music pieces and to interact with a selected one (i.e., Mozart K. 265/300e), Three days later, the post-test is given.

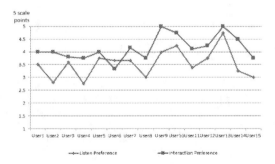

Fig. 5. Users' pre-test preferences

Preference Comparison Results. Among the 15 users that participated in this set of experiments, five can play musical instruments and seven are familiar with classical music. Figure 5 illustrates that after the interaction, all subjects' preference is significantly increased except for user 6, who wanted to listen to the music only rather than using the interaction system to enjoy music. We also utilized a paired t-test (which is a statistical hypothesis test) to see whether the improvement is significant after interaction. Our hypothesis for the t-test is that preference will be increased after users interact with the system. The hypothesis passed the t-test with a p-value of 3.542×10^{-05}, which indicates that the results are significant at the $p < 0.05$ level.

Even though the average preference after the post-test is higher than that of the pre-test, as illustrated in Table 4, the associated t-test shows no significant improvement after a half month. In the post-test, the p-values of the same music and the other music are 0.277 and 0.372, respectively. This may imply that period of the interaction is too short to have an impact on the users, and the interval between two tests is too long to be influenced. Therefore, the effect on preference is not statistically significant.

Table 4. Effect on preference

Pre-test preference	Post-test preference	
	Same music	Different music
3.541	3.674	3.655
	3.665	

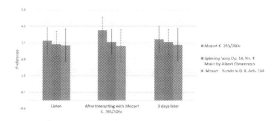

Fig. 6. Preference comparison for fixed selections

Figure 6 illustrates the preference comparison for the second set of experiments. As illustrated in Figure 6, the average preference after interaction is significantly increased. Our hypothesis for the t-test is the preference will be increase after users interact with the system. The hypothesis passed the t-test with the p-value 5.974×10^{-04}, which indicates that the results is significant at the $p < 0.05$ level. Even though the average preference of the post-test is more than that of the pre-test as illustrated in Figure 6, the associated t-test shows no significant improvement after 3 days. It may also imply that the period of the interaction is too short to impress the users and the interval between two tests is too long to remember the previous test. Therefore, the effect on the preference is not remarkable.

The improvement of preference in the post-test is not as significant as that of the pre-test, which might indicate that one can use our music interaction system more frequently to improve the long-term effect in the preference of music.

4   Conclusion

In this paper, we proposed an interaction between gestures and tempo to help users enjoy music on smart phones. Given the ubiquitous nature of smart phones, most people can have the opportunity to interact with music. Our experiment shows that our system can provide an accurate, robust, and smooth musical interaction experience. In the user study, we also demonstrate that preference can be improved and that people like classical music more after using the system.

References

1. Borchers, J., Lee, E., Samminger, W., Mühlhäuser, M.: Personal orchestra: a real-time audio/video system for interactive conducting. Multimedia Syst. 9(6), 594 (2004)
2. Cambouropoulos, E., Dixon, S., Goebl, W., Widmer, G.: Human preferences for tempo smoothness. In: Proceedings of the VII International Symposium on Systematic and Comparative Musicology, III International Conference on Cognitive Musicology, pp. 18–26. University of Jyväskylä, Jyväskylä (2001)
3. Chew, E., François, A., Liu, J., Yang, A.: Esp: a driving interface for expression synthesis. In: Proceedings of the 2005 Conference on New Interfaces for Musical Expression, pp. 224–227. National University of Singapore (2005)
4. Kolb, B.M.: The decline of the subscriber base: a study of the Philharmonia Orchestra audience. International Journal of Arts Management (2001)
5. LeBlanc, A., McCrary, J.: Effect of Tempo on Children's Music Preference. Journal of Research in Music Education 31(4), 283–294 (1983)
6. Lee, E., Grüll, I., Kiel, H., Borchers, J.: conga: a framework for adaptive conducting gesture analysis. In: Proceedings of the 2006 Conference on New Interfaces for Musical Expression, pp. 260–265. IRCAM &#8212, Centre Pompidou (2006)
7. Lee, E., Kiel, H., Dedenbach, S., Grüll, I., Karrer, T., Wolf, M., Borchers, J.: isymphony: an adaptive interactive orchestral conducting system for digital audio and video streams. In: CHI 2006 Extended Abstracts on Human Factors in Computing Systems, pp. 259–262. ACM (2006)
8. Liu, Y.-L., Chen, Y.-S.: iconduct: music control in the interactive conducting system. Technical report (2009)
9. Montgomery, A.P.: Effect of Tempo on Music Preferences of Children in Elementary and Middle School. Journal of Research in Music Education 44(2), 134–146 (1996)
10. Peery, J.C., Peery, I.W.: Effects of exposure to classical music on the musical preferences of preschool children. Journal of Research in Music Education 34(1), 24–33 (1986)
11. Toh, L.-W., Chen, Y.-S.: An interactive conducting system using kinect. In: IEEE International Conference on Multimedia and Expo, ICME 2005, 2013, pp. 1–6. IEEE (2012)

Semantically Enhancing Multimedia Lifelog Events

Peng Wang[1,*], Alan Smeaton[2], and Alessandra Mileo[3]

[1] Beijing Institute of System Engineering, Beijing, P.R. China, 100101
pengwangnudt@sina.com
[2] Insight Centre for Data Analytics
Dublin City University, Glasnevin, Dublin 9, Ireland
alan.smeaton@dcu.ie
[3] Insight Centre for Data Analytics,
National University of Ireland, Galway, Ireland
alessandra.mileo@deri.org

Abstract. Lifelogging is the digital recording of our everyday behaviour in order to identify human activities and build applications that support daily life. Lifelogs represent a unique form of personal multimedia content in that they are temporal, synchronised, multi-modal and composed of multiple media. Analysing lifelogs with a view to supporting content-based access, presents many challenges. These include the integration of heterogeneous input streams from different sensors, structuring a lifelog into events, representing events, and interpreting and understanding lifelogs. In this paper we demonstrate the potential of semantic web technologies for analysing lifelogs by automatically augmenting descriptions of lifelog events. We report on experiments and demonstrate how our results yield rich descriptions of multi-modal, multimedia lifelog content, opening up even greater possibilities for managing and using lifelogs.

Keywords: lifelogs, events, semantic web, semantic enhancement.

1 Introduction

Lifelogging is a user-controlled form of gathering personal multimedia information, though not necessarily for sharing with others [1]. Lifelogging is the process of sensing and digitally recording our everyday behaviour, capturing a person's experiences in the form of events, states and relationships. This rich pool of information is collected by wearable sensors by individuals to characterize their own contexts and activities and build applications that can enhance their quality of life.

An important premise for the use of lifelogs is the fact that *events*, discrete and often repetitive activities, are the atomic unit of interaction with lifelogs [2,3]. However, the identification of event boundaries alone is just a first step because we need to know what events actually are and how they relate to each

---

* Research funded by Science Foundation Ireland under grant SFI/12/RC/2289.

W.T. Ooi et al. (Eds.): PCM 2014, LNCS 8879, pp. 163–172, 2014.

other. In this work we discuss how semantic enrichment of lifelog events and the creation of semantic links between those descriptions and external knowledge can help bridge this data-to-knowledge gap and develop applications that can use lifelog data to determine user contexts and activities more effectively.

In Section 2 we discuss the shortcomings of current event enhancement approaches and the potential of using semantic web for modeling, linking and re-using knowledge about events are reported in Section 3. Experimental results are reported in Section 4, followed by conclusions and future work.

2   Background and Related Work

Lifelogging is concerned with digitally capturing media-rich representations of activities of everyday life and making use of them across a variety of use cases. Arbitrary sets of information including location or geo-tags, textual information, photos, audio and video clips from mobile sensors, etc. can characterize a particular scenario in everyday life.

Among all forms of information, visual information contains more semantics of events which can be used to infer other contexts like "Who", "What", "Where" and "When". Visual lifelogging is the term used to describe both image- and video-based lifelogging from wearables. The SenseCam, shown in Figure 1 and which we use in the work in this paper, is a sensor-augmented wearable camera designed for visual lifelogging by recording a series of images and a synchronised log of sensor data. It captures the view of the wearer from a fisheye lens and pictures are taken automatically at the rate of about one every 50 seconds. On-board sensors are also used to

Fig. 1. SenseCam worn by a user

trigger additional capture of pictures when sudden changes are detected in the environment of the wearer. SenseCam has been shown to be effective in supporting recall of memory from the past for individuals [5], as well as having applications in diet monitoring [6], activity detection [7], smoking cessation, sleep monitoring, etc. and these are representative of the kinds of applications to which multimedia lifelogs can be applied [1].

Once lifelogging hardware reached the stage of being useful for real world applications, the focus shifted towards mining deeper meanings from lifelogs i.e. eliciting their semantics. However, the state-of-the-art in enhancing lifelog events is almost all based on low-level feature matching. An everyday event enhancement technique with more comprehensive online resources is introduced in [8], in which SenseCam images are augmented with images from external publicly-available data sources such as Flickr, YouTube, etc. The shortcomings of this are obvious in that the task is performed based on image matching using the scalable colour MPEG-7 image descriptors and geotags. Though a number of approaches have been proposed for multimedia event detection, as shown in TRECVid [13] and MediaEval [14], and egocentric media summarization [15], etc., these approaches still lack capabilities in interpreting events from different contexts, which is more important for lifelogs obtained from sensing devices.

An example of work on enhancing automatically collected lifelogs is reported in [9] which describes the *DigMem* platform which tracks location and combines it with the wearer's heart rate and photos taken by the wearer from a tablet computer. This is turned into RDF and semantically enriched using Linked Data, in a similar way to what is reported later in this paper. What differentiates this work from ours is that our photo capture is true ambient life logging with no user involvement which makes the nature of the photos different with strong similarity among our photos when taken during the same event. Furthermore, our lifelog data is structured into spatio-temporal events which are enhanced semantically through links whereas in previous work there may be only one photo taken during an event thus enhancement is nearly impossible.

3   Semantically Enhanced Lifelogging

The goal of enhancing lifelog events is to inter-link knowledge from different contexts to enhance lifelog applications. We now describe the model and implementation of the EventCube system which does this.

3.1   Contexts and Events: Semantic Representation

Context information can be collected in lifelogging through the deployment of heterogeneous sensors. To model the semantics of events, contexts should be modeled in a way which allows each to be processed separately. That is because high uncertainty is often embedded in context processing due to data loss (e.g. GPS signal dropouts) or detection defects such as Bluetooth signal quality. The following concepts describing multimedia should ideally be included in lifelogs in which events are to be interpreted:

- **Event**: an occurrence as the intersection of time and space.
- **Location**: the geographical context of an event, as a recall cue.
- **Time**: the temporal context as a recall cue for an event.
- **Actor**: the human who carried out the event, normally the user.
- **Attendee**: the human/humans present and possibly involved in the event.
- **Image**: the class abstract for the image.
- **Annotation**: a class abstract for the textual description of events.

For consistency with the definition of an event as "a real-world occurrence at a specific place and time", we explicitly model the event class with spatial and temporal constraints in terms of OWL cardinality restrictions, as shown in Listing 1.1, formalized in the language Turtle [4]. The restriction `owl:cardinality` is used to enforce that one event has exactly one value for the properties of starting time and ending time. In Listing 1.1, the restriction `owl:minCardinality` is stated on the property `:hasLocation` with respect to event class, indicating that any event instance needs to be related to at least one GPS location. In lifelogging, there are cases when more than one GPS coordinate is needed to reflect the spatial characteristics of an event, such as "walking" or "driving".

Besides the raw-content of lifelog events captured directly by wearable sensors, there are three main external types of context, namely spatial, temporal and social for which there are already well-established ontologies to describe them. We investigated existing ontologies which may be reused and integrated into our context enhancing event ontology and chose the OWL-Time and GeoNames ontologies to model spatial and temporal contexts respectively. In our architecture, the people involved in an event including the actor and attendees are modeled by the FOAF (Friend Of A Friend) ontology which describes persons with their properties and relations. The visual information about events which answers the "What" question can be depicted by SenseCam images (addressed by the `FOAF:Image` class), or combined together with semantic annotations obtained from concept detectors.

```
:Event  rdf:type  owl:Class ;
        rdfs:subClassOf  time:TemporalEntity ,
[rdf:type owl:Restriction ;
 owl:onProperty :hasLocation ;
 owl:minCardinality "1"^^xsd:nonNegativeInteger],
[rdf:type owl:Restriction ;
 owl:onProperty :endAt ;
 owl:cardinality "1"^^xsd:nonNegativeInteger],
[rdf:type owl:Restriction ;
 owl:onProperty :beginAt ;
 owl:cardinality "1"^^xsd:nonNegativeInteger].
```

Listing 1.1. Definition of Ontological event classes

3.2   Semantics of Locations

Semantic enhancement of events in lifelogs needs an effective measure to locate the user because location can be a key facet in indexing multimedia lifelogs. Because GPS does not work inside buildings or where satellite signals are weak, we apply a location enhancement algorithm consisting of location clustering, reverse geocoding and an LOD (Linked Open Data) semantic query.

Location clustering is first explored on GPS coordinate records using k-means clustering [10] in which we use 100m as the clustering radius and a filtering time span of 10 minutes. Reverse geocoding is used to translate latitude/longitude pairs into semantically meaningful address names and returns the closest addressable location. Once the placename is obtained, we query all the semantics (properties and values) of the placename in DBpedia's RDF repository through a SPARQL query and provide links for the user to navigate the returned RDF graphs.

Current reverse geocoding web services label GPS coordinates with semantic tags by returning the nearest place names however due to GPS error margins and the sizes of what we define as places (city vs. streetname), the nearest place is not necessarily the correct answer for the target event. To address this, we provide nearby places as a ranked list for the user and we enhance selected places at the user's request. We rank the place list according to popularity, analyzed from Flickr social tags. The underlying assumption here is that the better-known places will be easier to recall when a user reminisces about the event and the most popular places are usually a benchmark of the region.

3.3   Semantics of Social Profiles

As social contexts, the actor and attendee concepts together reflect the agent aspect of a lifelog event. While these two contexts may answer the "Who" is carrying out the event and "Who" else is involved, enhancing social context aims to enrich the social profiles of these agents. In our implementation the FOAF profile and the lifelogger's personal information in Facebook are combined.

FOAF profiles are datasets in LOD and contain personal information modeled in RDF. While FOAF profiles contain information about millions of persons including persons relevant to the event, lifeloggers' social profiles like Facebook contain more semantics which have been customized and might have higher correlation with lifelog events. When a user re-experiences events, social information can improve the understanding of the "Who" aspect. The combination of FOAF profiles and Facebook involves transforming the XML feeds from Facebook to RDF, the FOAF profiles and Facebook are integrated in the same data model for which the same vocabularies like the FOAF ontology are needed for consistent semantic representation. Finally the RDF statements are populated in the event model thus enhancing the semantic description of the social context for the event.

3.4   EventCube: Conceptual Architecture

In designing an application for enhancing the representation of multimedia lifelog events, we mimic the traditional behavior of users in organizing personal digital photos. In [11], users state that the most important feature of photo organization is to automatically place photographs into albums and as shown in [12], albums are suggested to be more desirable for image organization and retrieval. There is also evidence from memory science that organising our lives into events assists with recall of the past, which is done through these events [2]. Motivated by this, we propose an event enhancement architecture – EventCube – to enhance the descriptions of events. The architecture of EventCube is shown in Figure 2.

For the application of this architecture to lifelogs, we employed SenseCam (shown in Figure 1) and Bluetooth-enabled handsets plus GPS trackers as context-sensing devices. The processing of raw lifelog sensor data into enhanced events is in three steps: first, the user uploads sensor readings to the database where SenseCam images and other sensor streams are segmented into events. A keyframe is automatically selected as a thumbnail for each event. Second, the

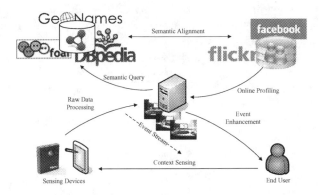

Fig. 2. Event enhancement architecture

recorded GPS coordinates are clustered (as described in Section 3.2) and stored together with Bluetooth proximity records. In this step, sensor readings are synchronized with segmented events. Finally, online knowledge bases and social profiles are accessed and combined to create the relevant semantics for current event contexts. These enhanced contents are provided as links for further use.

4   EventCube Experiments and Evaluation

4.1   Experiment Setup

SenseCam was used to collect image sequences while GPS recording and nearby Bluetooth device detection were implemented on a Nokia n810 internet tablet communicating with an external GPS module. GPS data was recorded every 10 seconds. The unique addresses and friendly device names of nearby Bluetooth devices were recorded every 20 seconds. All these sensor readings including Sense-Cam images were recorded with timestamps and synchronized. One subject wore the above recording devices for our initial event enhancement experiments. We processed logged data on a daily basis as the subject uploaded the SenseCam, GPS and Bluetooth readings after each day's continuous recording. For retrieval of physical information recorded by ambient sensing devices, we segmented each day's SenseCam data streams into individual events indexed by keyframes selected for their visual representation of events [3].

Besides local ambient information sources, the environment also includes the information space constructed by online semantic repositories and users' social profiles. Retrieval from online knowledge bases such as datasets in LOD enhances the "Who" and "Where" aspects of events. We used the web services listed in Table 1 to access the data sources.

4.2   Event-Centric Enhancement Application

We now present an event-centric enhancement application which includes event visualization and allows a user to browse through their own multimedia lifelog.

Table 1. Online data sources employed

Dataset	Web Service Endpoints	Event Aspects
DBpedia	http://dbpedia.org/sparql	Who, Where
GeoNames	http://www.geonames.org/export/	Where
Flickr	http://www.flickr.com/services/api/	Where
DBLP	http://dblp.l3s.de/d2r/	Who
Facebook	http://api.facebook.com/1.0/	Who

This is a browser-based application with a SenseCam event viewer, geospatial map and contextual enhancement embedded, shown in Figure 3.

The event viewer presents event keyframes allowing the user to view events on a day-by-day basis. The calendar on the left corner provides a navigation and a display option. After the user selects a target date, the event viewer will list all events for that day organized in chronological order. Figure 3 illustrates the temporal sequence when the user attended a presentation.

After the user picks an event s/he is interested in, the GPS location is queried and located on the map. To enhance location context, relevant place names are retrieved according to the event's GPS coordinates. The abstract information is shown in the browser as a brief description of the most relevant named place plus links from the web pages or RDF triple repositories. Social context is enhanced in the same manner with brief information and links to external semantics using DBLP and DBpedia. The temporal context is visualized with a timestamp indicating the start time, end time and duration of the selected event.

4.3 Assessing Context Enhancement

We use 25 consecutive days of lifelog data to evaluate our methodology for semantic enhancement of events. The final dataset includes 38,026 images, 327,244 GPS records and 45,898 Bluetooth detections involving 958 unique devices. We applied the location-clustering described in Section 3.2 to find significant places in the lifelog. Since user behavior is usually periodic over a relatively long period, images in the same cluster are more conceptually similar and can be well-represented using keyframes. GPS records were validated in order to filter invalid coordinates such as empty GPS records logged when satellite signals were not visible. Ultimately, 59,164 GPS coordinates were selected for location clustering and each day's locations were clustered with 2,400 coordinates on average.

For those detected significant places we enrich the location context by DBpedia and exemplar results are shown in Table 2, in which the abstracts (defined by predicate `dbpedia-owl:abstract`) and home pages (by `foaf:homepage`) are shown where available. After applying a SPARQL query, semantics about places can be retrieved from DBpedia. Besides abstracts and home pages, there may be dozens of other properties queried from DBpedia for location enhancement. Relevant properties about the target location could include the type of place,

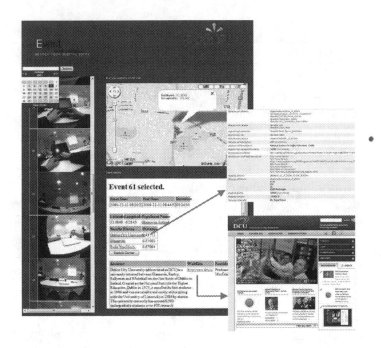

Fig. 3. Event enhancement interface (event viewer and map-enhancement browser)

the exact geospatial location information, affiliation, image, etc., which are all provided as links. As reflected in Table 2, we did not apply placename disambiguation before applying the enhancement. "The Spire"[1] is enhanced as a novel in Table 2, which is not the true interpretation of its meaning as a tourist attraction. However, the `dbpedia-owl:wikiPageDisambiguates` property allows users to navigate various options of resources with the same name as "The Spire" and to choose the correct one, which is described as "the Monument of Light ...on O'Connell Street in Dublin, Ireland".

For social context while most benchmark locations can be queried from DBpedia, not many persons involved in the event can be enhanced in this way so we enhance the social context by combining resources from DBpedia, DBLP and Facebook. Sophisticated approaches to identifying users are beyond the scope of this paper and in our application we allow the user to map their real friends' names to Bluetooth friendly names. The purpose of this is to focus on event enhancement issues rather than user identification. Social context is then enhanced by querying relevant information from the aforementioned data sources by interlinking the friends' names to those data sets.

Table 3 shows some samples of enhanced social descriptions for lifelog events. For simplicity, we only illustrate the person abstracts obtained from DBpedia. The column for DBLP shows the number of records in DBLP datasets reflected

---

[1] The Spire is a large, spire-shaped public monument in the centre of Dublin city.

Table 2. Enhanced samples for places

Place Name	Abstract	Home Page
Dublin City Univ.	a University situated between Glasnevin, Santry, Ballymun and ...	www.dcu.ie
Trinity College	formally known as the College of the Holy and Undivided Trinity of ...	www.tcd.ie
Glasnevin	a largely residential neighborhood of Dublin, Ireland ...	–
Baile Átha Cliath	capital and largest city of Ireland ...	www.dublincity.ie
Croke Park	the principal stadium and headquarters of the Gaelic Athletic Association ...	www.crokepark.ie
Merrion Square	a Georgian square on the southside of Dublin city centre ...	–
The Spire	a 1964 novel by ...	–

by the number of `dc:creator`/`foaf:maker` properties queried from DBLP. For the Facebook column we illustrate only the hometown defined in the aligned RDF models. Since these profiles are accessed from either the publicly-available LOD repository or the lifelogger's own Facebook account, there are no ethical issues when interlinking and using them in the enhancement application.

Table 3. Enhanced samples for social context

Bluetooth Name	DBpedia	DBLP	Facebook home town
NeilOHare-MacBook	–	13	Drogheda, Ireland
Alan Smeaton's MacBook Pro	Alan Smeaton is an author and academic at Dublin City University ...	227	Dublin, Ireland
cdvpmini-AlansOffice	Alan Smeaton is an author and academic at Dublin City University ...	227	Dublin, Ireland
cdvpminiColum	–	12	–
Pete	a British multimedia artist living in Newfoundland, Canada ...	23	–

A similar problem to that caused by a lack of name disambiguation is mis-enhancement for commonly-used names. For example, the recorded person 'Pete' (Peter as real name), who was a colleague of our subject, is incorrectly enhanced when querying DBpedia. The characteristics of Bluetooth also cause another artifact for social descriptor enhancement. Bluetooth has a range of about 10 meters and in some cases it can penetrate walls. In our enhancement experiment, we rank the Bluetooth records in terms of their frequency of occurrence during the time span of the selected event. In this way, accidentally logged device proximities can be ranked lower and have less chance to be enhanced.

Using Semantic Web techniques, our enhancement leverages information retrieved from public resources by applying SPARQL queries and aligning semantics to a standardized RDF model. The populated profiles provide a comprehensive tool to realize detailed aspects about lifelog events.

5   Conclusion and Future Work

This paper introduced a method for semantic enhancement of multimedia lifelog events to improve their interpretation by leveraging external knowledge. The approach we have taken has only recently become feasible because a critical mass of semantic descriptions of people, places, and activities has now been modeled and published as Linked Data. The effectiveness of detecting significant places and accessing billions of triples in the LOD knowledge bases for enhancing location and social contexts has been demonstrated. In future work we will explore how we can improve our understanding of lifelog events by integrating more semantic information from richer contextual views, and investigate how to use such richer semantic structure to enhance the characterization of events.

References

1. Gurrin, G., Smeaton, A.F., Doherty, A.R.: Lifelogging: Personal Big Data. Foundations and Trends in Information Retrieval 8(1), 1–125 (2014)
2. Doherty, A.R., Pauly-Takacs, K., Caprani, N., et al.: Experiences of Aiding Autobiographical Memory Using the SenseCam. HCI 27(1-2), 151–174 (2012)
3. Doherty, A.R., Smeaton, A.F.: Automatically segmenting lifelog data into events. In: WIAMIS 2008, pp. 20–23. IEEE Computer Society, Washington, DC (2008)
4. Beckett, D.: Turtle-Terse RDF triple language. W3C Technical Report (2007)
5. Silva, A.R., Pinho, S., Macedo, L.M., Moulin, C.J.: Benefits of SenseCam Review on Neuropsychological Test Performance. AJPM 44(3), 302–307 (2013)
6. O'Loughlin, G., Cullen, S.J., McGoldrick, A., et al.: Using a wearable camera to increase the accuracy of dietary analysis. AJPM 44(3), 297–301 (2013)
7. Wang, P., Smeaton, A.F.: Using visual lifelogs to automatically characterise everyday activities. Information Sciences 230, 147–161 (2013)
8. Doherty, A., Smeaton, A.F.: Automatically augmenting lifelog events using pervasively generated content from millions of people. Sensors 10(3), 1423–1446 (2010)
9. Dobbins, C., Merabti, M., Fergus, P., et al.: Exploiting linked data to create rich human digital memories. Computer Communications (2013)
10. Ashbrook, D., Starner, T.: Using GPS to learn significant locations and predict movement across multiple users. Personal UbiCom 7(5), 275–286 (2003)
11. Rodden, K.: How do people organise their photographs? In: Proceedings of the BCS IRSG Colloquium (1999)
12. Platt, J.C.: AutoAlbum: Clustering digital photographs using probabilistic model merging. In: CBAIVL 2000, pp. 96–100. IEEE Computer Society (2000)
13. Little, S., Jargalsaikhan, I., Clawson, K., et al.: Interactive Surveillance Event Detection at TRECVid2012. In: ICMR 2013, pp. 301–302 (2013)
14. Reuter, T., Papadopoulos, S., Petkos, G., et al.: Social Event Detection at MediaEval 2013: Challenges, Datasets, and Evaluation. In: MediaEval Workshop (2013)
15. Lu, Z., Grauman, K.: Story-Driven Summarization for Egocentric Video. In: CVPR, Portland, OR, USA, pp. 2714–2721 (2013)

Haar-Like and HOG Fusion Based Object Tracking

Chong Xia[1], Shui-Fa Sun[1,2,*], Peng Chen[1,2], Heng Luo[1], and Fang-Min Dong[1,2]

[1] Institute of Intelligent Vision and Image Information, China Three Gorges University,
HuBei, YiChang, 443002
[2] Hubei Key Laboratory of Intelligent Vision Based Monitoring for Hydroelectric Engineering,
China Three Gorges University, Yichang 443002, China
watersun1977@hotmail.com

Abstract. Only unitary feature for object is adopted in the conventional tracking system, making it difficult for robust tracking. Regarding the characteristic of both Haar-like and HOG features, a tracking algorithm fusing these two features is proposed: using the Haar-like features for the structure of the object and HOG features for the edge. A mixed feature pool is constructed with these two features. The On-line Boosting feature selection framework is adopted to select out the notable features, and update these features on line to realize the optimal selection. Four representative videos are used to test the performance of the proposed algorithm in the aspect of illumination change, tacking small targets, complex motion of the object, similar object interference during tracking and so on. Statistical analysis Results of the error show that the tracking system using the fused features outperforms the system using either of the two features.

Keywords: Object Tracking, On-line Boosting, Haar-like, HOG, fusion.

1 Introduction

The common method used in object tracking is to treat it as a binary classification problem. Effective tracking system should be able to distinguish target from complex background and have the ability to adjust to appearance change of target. In order to divide the target and background efficiently, researchers have put forward various features to denote the characteristic of target, such as Haar-like [1], histogram of oriented gradient (HOG) [4], EOH [7], SIFT [8]. Haar-like feature is introduced into tracking system in [2-3,5] and gets remarkable effect and runs in real-time. HOG feature, presenting local appearance and shape of target, has been used in tracking system such as [6, 9-10]. In [6], a tracking system based on On-line Boosting framework [2-3] is designed to select out outstanding HOG features, and realize stable tracking. To improve the performance of features, some researchers raised the idea of feature fusion [7,11-12], taking advantage of multiple features to represent the same target.

Haar-like feature can be calculated efficiently but is sensitive to illumination change. HOG feature is robust to illumination variation, but is not able to reflect the local texture information and computational expensive. This article uses Haar-like

---

* Corresponding author.

W.T. Ooi et al. (Eds.): PCM 2014, LNCS 8879, pp. 173–182, 2014.
© Springer International Publishing Switzerland 2014

features to depict the structure information of local area in the target and HOG features to account for the edge information. Based on the On-line Boosting feature selection framework, we merge these two kinds of features into our tracking system to enhance the performance. We have carried out some experiments, and the results show that when the Haar-like features perform poorly, by bringing the HOG features into the tracking system , the merged features can present target precisely, and vice versa. By fusing these two features, the reliability and stability of tracking system could be improved greatly.

The remainder of this paper is organized as follows: Haar-like and HOG features are analyzed briefly in Section 2; s the overall framework of our system and how to use the On-line Boosting framework to select out fused features are illustrated in Section 3; experiments including selection of effective features and contrastive analysis using different type of features are performed in section 4; finally this paper is summarized and some advices are put forward for further improvement in section 5.

2   Feature Analysis

2.1   Haar-Like Features

Haar-like feature uses simple luminance information or color information, and adopts Integral Image [1] to speed up the computation of feature values, making it a preferable selection for real-time tracking system. By generating sub-windows at random positions in the image and using random size, a huge number of Haar-like features is got. Through adopting Integral Image to scan the entire image, the sum of pixels in each region can be calculated conveniently, thus the features' value could be efficiently calculated.

2.2   HOG Features

Histogram of Oriented Gradient (HOG) feature [4] represents the intensity and orientation of the gradient of an image. The target's appearance and shape can be perfectly described by gradient or the oriented intensity of edge. Thus it is robust to illumination and geometrical change in local area.

In [6] the authors mentioned that using random size and width-height ratio to form a block. Using the randomized block and appropriate step to travel through the entire image, we can get plenty of HOG features.

The Integral Histogram Image [9] is brought in to speed up the calculation of HOG features. The procedure is somewhat like the method of Integral Image, and an Integral Histogram of Oriented Histogram (IHOG) image can be constructed.

Utilizing these feature templates based on the block to get the initial values of the features in the target as reference feature values. Then calculate the distance between feature values of the samples and the reference feature values. Euler distance is adopted to measure the value of the distance.

3 On-Line Boosting Tracking Algorithm Based on Feature Fusion

3.1 The Overall Framework of This Algorithm

For the purpose of enhancing the features' performance, this paper takes into account of the advantages of both Haar-like and HOG features. We build up a feature pool that contains both features, and use them in the weak classifiers. The On-line Boosting feature selection framework [3] is then adopted to select out some of the best weak classifiers to construct selectors and finally sums these weighted selectors to form a final strong classifier.

3.2 Design the Weak Classifier

Considering the computational complexity of Haar-like and HOG features differs from each other, when we build up the mixed feature pool, different weak classifiers need to be designed for these two kinds of features. Suppose the distribution of feature values of these samples yields Gaussian distribution. f(x) is the value of a Haar-like feature or the distance between HOG feature and the reference feature. The weak classifier of Haar-like or HOG feature is defined as:

$$h_i^{weak}(x) = p_j \cdot sign(f_j(x) - \theta_j) \qquad (1)$$

A weak classifier is in response with a Haar-like or HOG feature. θ_j is the threshold of a weak classifier and p_j the parity factor. For each weak classifier the threshold θ_j is set to $\theta_j = |\mu^+ + \mu^-|/2$, μ^+ represents the mean value of positive samples using the corresponding feature template. The parameter μ^- is acquired in the same manner.

3.3 Feature Selection

The number of feature is usually very large, and not all of them can perform well to distinguish the target from background. So, it is necessary to utilize a learning method to select out features that are capable to classify the unknown samples into positive or negative. In this paper, we use the On-line Boosting feature selection framework to select out prominent fused features.

The On-line Boosting algorithm framework (OAB) uses Discrete AdaBoost, and the output of weak classifier is a discrete value, namely {-1,+1}. The final strong classifier can be expressed as (2):

$$H(x) = \sum\nolimits_{n=1}^{N} \{-1, +1\} \cdot \alpha_n \qquad (2)$$

Algorithm 1. On-line Boosting for fused features selection
(The presentation below uses '*' to represent the data used by Haar-like or HOG)

- Given training example $< x, y >$, x belongs to the training samples, y is the label of the training sample, $y \in \{-1, +1\}$. Set the number of weak classifier to M, including M_1 Haar-like weak classifiers and M_2 HOG weak classifiers. Ser the number of selectors to N, which contains N_1 Haar-like selectors and N_2 HOG selectors.

- Initialize the importance weights of samples and the weights of weak classifiers $\lambda^* = 1$, $\lambda_{n,m}^{corr,*} = \lambda_{n,m}^{wrong,*} = 1$

- for   n=1,...,N
  - for   m =1,...,M

    for each training sample <x,y>, update the weak classifier $h_{n,m}^*$

    $\lambda_{n,m}^{corr,*}$, $\lambda_{n,m}^{wrong,*}$, as well as $e_{n,m}^*$

    end for

  - choose weak classifier with lowest error: $m_+^* = \arg\min_m (e_{n,m}^*)$

    $$e_n^* = e_{n,m^+}^* \qquad h_n^{sel \blacklozenge *} = h_{n,m^+}^*$$

  - if $e_n^* = 0$ or $e_n^* > 1/2$, then exit

  - calculate the weights   of selectors $\alpha_n^* = \frac{1}{2} \cdot \ln\left(\frac{1 - e_n^*}{e_n^*}\right)$

  - update importance weights   of samples λ^*
  - replace the worst weak classifier with a new one

  end for

- output of strong classifier:

$$H(x) = sign[\sum_{n=1}^{N_1} \alpha_n^{Haar} h_n^{sel,haar}(x) + \sum_{n=N_1+1}^{N_1+N_2} \alpha_n^{HOG} h_n^{sel,hog}(x)] \qquad (3)$$

3.4   Algorithm Analysis

Haar-like features can be calculated quickly but can only represent the rough information of image intensity. HOG features have advantage in target appearance and shape curving, but face the problem of computational complexity. By analyzing the On-line Boosting algorithm framework, we find out that the selector's error rate rises with the increasing number of selectors. To avoid this, we propose the idea of feature fusion on the selector level using Haar-like and HOG features' complementary nature. Then we can use the fused features to represent the same target. When one type of selectors degrades in performance, the other type of selectors would be able to represent the target accurately. So we can make use of this type of selectors to replace the first ones to avoid selecting redundant features. Thus keep both types of selectors a low error rate and achieve an overall performance increasing for the strong classifier.

4   Experiments and Analysis

4.1   Multiple Feature Selection

In order to verify the effectiveness of the fused features, we show the first six selectors of both Haar-like and HOG features. Fig 1(a ~ c),(d ~ f) shows the Haar-like and HOG features from frame 176~178 and 571~573 respectively. The yellow rectangles represent the selected HOG features while the rectangle with white and black stand for the selected Haar-like features.

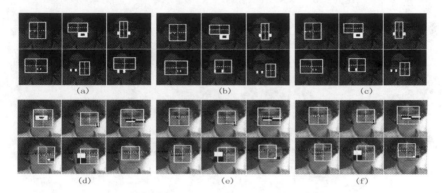

Fig. 1. Feature selection under different background

As shown in the Fig. 1, when the light is dim, Haar-like features can't represent the target effectively just as (a ~ c) shows. In (a ~ c), the first six features almost show the same region. But in this circumstance, HOG features can still perform well, and the selected features have diversity in type. This indicates that when the Haar-like features perform poorly, using HOG features can make up for the degradation. When the light increases, Haar-like and HOG features can all perform well on the target, such as (d ~ f), the selected features in each frame differs from each other in type and location.

4.2   Feature Fusion Experiments

To further evaluate the algorithm's performance, we use four representative test videos with Ground Truth (GT) to test the tracking system. The test video of David and Sylvester are from [5], EgTest video comes from the VIVID data set[1], Person video is from BoBoT data set[2]. The video of David is used for illumination change test and Sylvester for testing the algorithm when there exists rotation and size changes of

---

[1] http://vision.cse.psu.edu/data/vividEval/datasets/datasets.html.
[2] http://www.iai.uni-bonn.de/~kleind/tracking/index.htm.

target. EgTest targets at testing the algorithm's stability when tracking small object for a long time. Person used to measure the system's effectiveness when there is similar disturbing object. The following experiments compare our algorithm with the OAB using Haar-like[2] and HOG features[6] respectively and the other two state-of-the-art tracking algorithms MIL[5] and CT[13], which use unitary Haar-like feature to represent the target as well. The blue, yellow dashed and red rectangle show the result of On-line Boosting system using Haar-like, HOG and our fused features respectively. The white and magenta dashed rectangle represents the tracking results of MIL and CT, respectively.

(1) Experiment using Sylvester
The background of Sylvester is simple and the light changes little, all of the three features can represent the target appropriately. This experiment together with the system performance testing experiment shows the accuracy and stability of the proposed system.

Fig. 2. Tracking result of EgTest

From Fig 2 we can see that there exists little difference among the tracking results. But by comparing the distance between the center of tracking result and the GT center of the three algorithms, we can still find some fine shadow. As is shown in Fig 6(a) and table 1, we can obviously find out the accuracy and stability of these tracking systems.

(2) Experiment using EgTest
In this video the target is small, about 21*25 pixels, but the smallest HOG features used in this paper is 12*12 and the step size is 3*3. So we could not get plenty effective HOG features to represent the target. This experiment is used for verifying the system's stability using fused features when the HOG features perform poorly.

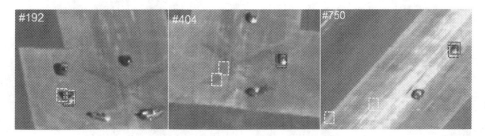

Fig. 3. Tracking result of EgTest

Since there is not plenty HOG features to represent the target, the tracker misses the target a few seconds later. The system using Haar-like features can tack the target for a long time, but it is vulnerable to the shadow of car, which makes the system suffers from accuracy degradation. Our system fused a small number of HOG features and a part of Haar-like features to avoid the influence of shadow and can track the target stably.

(3) Experiment using David
In this video there exist dramatic light changes, since Haar-like features is sensitive to illumination change, this experiment mainly test the system's stability using HOG features and fused features in the circumstance of dramatic illumination change.

Fig. 4. Tracking result of David

When the light is dim, the target is overwhelmed by the darkness. And Haar-like features can't represent the target accurately, causing the tracking system track target inaccurately. HOG features is robust to illumination change, the system using HOG features can track the target efficiently. The fused features can take advantage of HOG features to compensate for the drawback of Haar-like features in this situation, and thus achieve robust tracking in dim circumstance. When the light turns up, Haar-like features can effectively express the target and the system using fused features can get better tracking result than using HOG features alone.

(4) Experiment using Person
In this video there are illumination changes, target scale changes and similar object disturbing. This experiment aims at testing the three systems' performance on accuracy under complex environment.

Fig. 5. Tracking result of Person

Affected by the illumination change, when the position of the two people intersects, the system using Haar-like features tends to miss the target (shown in frame 341). And the system using HOG features also suffers from serious shift according to GT, this is caused by the fact that the tracking rectangle is fixed and the target changes in scale may bring in background information, which leads to an imprecise update of classifiers. By using the fused features, Haar-like features can be used to express structural information inside the target, while using HOG features to depict edge information of the target and overcome the influence of illumination. Thus getting a more accurate and stable result.

4.3 System Performance Test

In order to evaluate the three algorithms' performance more accurately, we calculated the Euler distance between tracking rectangle center and GT center, the average value and the standard deviation of the distance to estimate the tracking results. At Frame t, the Euler distance d_t between tracking rectangle center and GT center is defined as:

$$d_t = \sqrt{\left(x_o - x_g\right)^2 + \left(y_o - y_g\right)^2} \qquad (4)$$

(x_o,y_o)denotes the center of tracking rectangle, (x_g,y_g) denotes GT center.

The average value of the distance is defined as:

$$\mu = \frac{1}{N}\sum_{t=1}^{T} d_t \qquad (5)$$

N is the amount of frames. The standard deviation of the distance is as follows:

$$\sigma = \sqrt{\frac{1}{N}\sum_{t=1}^{T}(d_t - \mu)^2} \qquad (6)$$

From the definition above, we can see that the smaller d_t and μ , the more accurate the system is. And the system tends to be more stable if σ is smaller. In Fig 6(a ~ d) shows the trend of Euler distance between tracking result center and GT center of Sylvester, EgTest, David and Person respectively. The blue line, red and green dashed line represent the distance with the systems using fused, HOG and Haar-like features respectively, while the yellow and magenta dashed line represent the result of MIL and CT.

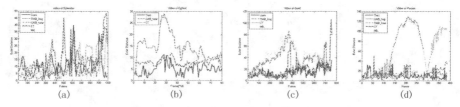

(a)   (b)   (c)   (d)

Fig. 6. Eula distance between tracking rectangle center and GT center

As is shown in Fig 2 that the tracking results are intuitional the same, Fig 6(a) shows the overall tracking result more accurately. And we can find out that the red line have many peak value, which means that in this video the system using HOG features is not as stable as the other two systems. The trend of green line is above the blue line on the whole, illustrating that the accuracy of the system using Haar-like features can't compare with the system using fused features. Similarly, we can see from Fig 6(b ~ d) that the system using fused features performs better in accuracy and stability than the system using either one of the features(In 6(b) using only HOG features fail to track the target, and 6(c) using only Haar-like features fail too, we do not show these two situation on the Fig).

Table 1. Comparison of accuracy and stability among three systems

Video sequence	Ours		OAB with Haar		OAB with Hog		CT		MIL	
	μ	σ	μ	σ	μ	σ	μ	σ	μ	σ
Sylvester	2.01	5.21	2.11	5.58	2.75	7.25	2.03	5.69	2.36	3.31
EgTest	6.30	2.01	8.96	3.72			14.50	5.54		
David	14.63	7.52			19.94	16.00	37.45	16.91	42.55	16.72
Person	9.54	5.93	27.79	33.90	13.68	8.96	48.86	45.79	45.46	45.58

The table above gives a data analysis of Fig 6, and we can get the overall tracking result from Fig6, table 1 lists the mean value and standard deviation of the Euler distance in each system using the four selected video sequence. A smaller mean value means a higher accuracy, and a smaller standard deviation means a more stable result.

5   Conclusion and Prospect

This paper uses multiple features for target representation and adopt On-line Boosting algorithm framework to select out outstanding features. Through randomly select Haar-like and HOG features' location and scale, we can generate plenty of features and utilize Integral Image and Integral Histogram to speed up the calculation of both features. By merging these two types of features, we can use the fused features to represent the target and overcome the drawback of Haar-like features that is sensitive to illumination change to some extent. Comparing with the system using only HOG features, our system could reduce the complexity of the algorithm. The comparison experiments show that our system performs well in the aspect of multiple feature selection, illumination change and complex circumstance. Considering the advantages of each feature, we could bring in the idea of Directional Texture Entropy [14] to speed up the selection of features. And make use of the idea of Cascade [15] to select out possible regions quickly so as to improve the accuracy and efficiency of the system.

Acknowledgments. This project is supported by National Natural Science Foundation of China (61102155, 61272237, and 61272236), the Young and Middle-aged Science Funding of Hubei Provincial Department of Education(Q20111205).

References

1. Viola, P., Jones, M.: Robust real-time object detection. International Journal of Computer Vision 57(2), 137–154 (2004)
2. Grabner, H., Bischof, H.: On-line boosting and vision. In: Proc. IEEE Computer Society Conference on Computer Vision and Pattern Recognition, pp. 260–267 (2006)
3. Grabner, H., Grabner, M., Bischof, H.: Real-time tracking via on-line boosting. In: British Machine Vision Conference, pp. 47–56 (2006)
4. Dalal, N., Triggs, B.: Histograms of oriented gradients for human detection. In: International Conference on Computer Vision and Pattern Recognition, pp. 886–893 (2005)
5. Babenko, B., Yang, M.-H., Belongie, S.: Robust Object Tracking with Online Multiple Instance Learning. IEEE Transactions on Pattern Analysis and Machine Intelligence 33(8), 1619–1632 (2011)
6. Sun, S., Guo, Q., et al.: On-line Boosting Based Real-time Tracking With Efficient HOG. In: International Conference on Acoustics, Speech and Signal Processing, pp. 2297–2301 (2013)
7. Ma, Y., Deng, L., et al.: Integrating Orientation Cue With EOH-OLBP-Based Multilevel Features for Human Detection. IEEE Transactions on circuits and systems for video technology 23(10), 1755–1766 (2013)
8. Zhou, H., Yuan, Y., et al.: Object Tracking using SIFT features and mean shift. Computer Vison and Image Understanding 113(3), 345–352 (2009)
9. Zhu, Q., Yeh, M.C., Cheng, K.T., Avidan, S.: Fast human detection using a cascade of histograms of oriented gradients. In: Proceedings of the IEEE Computer Society Conference on Computer Vision and Pattern Recognition, pp. 1491–1498 (2006)
10. Xu, F., Gao, M.: Human detection and tracking based on HOG and particle filter. In: International Congress on Image and Signal Processing, pp. 1503–1507 (2010)
11. Cai, Y.: Fusing multiple features to detect on-road vehicles. Computing Technology and Automation 32(1), 98–102 (2013)
12. Wang, H., Wang, J., et al.: A new robust object tracking algorithm by fusing multi-features. Journal of Image and Graphics 14(3), 489–498 (2009)
13. Zhang, K., Zhang, L., Yang, M.-H.: Real-time compressive tracking. In: Fitzgibbon, A., Lazebnik, S., Perona, P., Sato, Y., Schmid, C. (eds.) ECCV 2012, Part III. LNCS, vol. 7574, pp. 864–877. Springer, Heidelberg (2012)
14. Hu, S., Sun, S.-f., Lei, B.-j., Dan, Z.-p.: Haar-like feature based on-line boosting tracking algorithm with directional texture entropy. In: Huet, B., Ngo, C.-W., Tang, J., Zhou, Z.-H., Hauptmann, A.G., Yan, S. (eds.) PCM 2013. LNCS, vol. 8294, pp. 538–549. Springer, Heidelberg (2013)
15. Kalal, Z., Mikolajczyk, K., Matas, J.: Tracking-Learning-Detection. IEEE Transactions on Pattern Analysis and Machine Intelligence 34(7), 1409–1422 (2012)

Noise Face Image Hallucination via Data-Driven Local Eigentransformation

Xiaohui Dong[1], Ruimin Hu[1,2], Junjun Jiang[1], Zhen Han[1,2],
Liang Chen[1], and Ge Gao[1,2]

[1] National Engineering Research Center for Multimedia Software,
Computer School of Wuhan University, China
[2] Research Institute of Wuhan University in Shenzhen, China

Abstract. Face hallucination refers to inferring an High-Resolution (HR) face image from the input Low-Resolution (LR) one. It plays a vital role in LR face recognition by both manual and computer. The eigentransformation method based on Principal Component Analysis (PCA), which represents face image as a linear combination of the eigenfaces, has attracted considerable interests because of its simplicity and effectiveness. However, the face image observed is in a high-dimensional non-linear space, whose statistical properties cannot be captured by the PCA based linear modeling method. To this end, in this paper we advance a Data-driven Local Eigentransformation (DLE) method for face hallucination by exploiting the local geometry structure of data manifold and learning a specified eigentransformation model for each observed image. Experimental results show the effectiveness of the proposed approach for hallucinating face images especially with noise.

Keywords: Face hallucination, super-resolution, local eigentransformation, noise face image, video surveillance.

1 Introduction

With the deepening construction of Safe City, surveillance cameras are already prevalent in our daily life. They provide a powerful support for finding criminal suspects and tracing their routes. However, in many cases, the resolution of interested faces captured by the surveillance cameras is always low because of the long distance between camera and the object. The Low-Resolution (LR) of human face becomes a primary obstacle to face identification and recognition [1]. At the same time, the video captured usually contains lots of noise because of low-light, cheap imaging device or other factors, which make the image quality worse. In order to gain detailed facial features for recognition, it is necessary to infer High-Resolution (HR) face image from the LR one and this technique is called face hallucination [2].

1.1 Prior Work

Face hallucination is the technology which incorporates image super-resolution technique into facial image synthesis. It was firstly proposed by Baker *et al.* [2]

W.T. Ooi et al. (Eds.): PCM 2014, LNCS 8879, pp. 183–192, 2014.

in 2000. They learned the gradient distribution and then employed Bayesian theory to infer the HR face image. After that, a number of face hallucination techniques have been proposed in recent years [3].

For example, Wang *et al.* [4] developed a face hallucination algorithm through eigentransformation. Its core idea is to use a linear combination of the HR images in the training set to reconstruct the target HR image. They used Principal Component Analysis (PCA) [5] to fit the input face image as a linear combination of the LR training face images, and then the target HR image can be reconstructed by linear combination of those corresponding HR images in the HR training set. By maximizing the facial information from the LR face image, the reconstructed face image removes most of the noise distortion and retains most of the facial characteristics. But this method only focused on global structure information without paying attention to local details, so the results seem unclear and lacking detailed features, especially when the size of training set is small. Image patch based methods, which divide the global image to local image patches and thus achieve better representation ability, have attracted enormous amount of attention among scholars in field of image processing and analysis [6–12].

Chang *et al.* [7] assumed that the LR training patch space and HR training patch space share the same local geometry structure, and they proposed a Neighbor Embedding (NE) super-resolution method. In their method, the target HR patch can be obtained by K neighbor HR patches with the coefficients calculated in the LR space. By incorporating the position information, Ma *et al.* [8] proposed a position-patch based method. They obtained the optimal weights for each LR patch with the LR training image position-patches by Least Squares Representation (LSR). Yang *et al.* [9] employed the sparse coding [10, 11] to sparsely represent an LR input patch as a combination of the learned LR image patch dictionary, and the target HR patch is generated by the same combination coefficients and the corresponding HR image patch dictionary. Recently, Jiang *et al.* [12] proposed a Locality constrained Representation (LcR) method. By incorporating the locality term into the objective function, it can adaptively choose the neighbor patch samples to represent the input LR patch, thus leading to sparsity and locality of the patch representation and good face hallucination performances. These patch based methods are mostly designed for image without noise. In actual surveillance application, however, the degraded face images are almost inevitably affected by noise. Designing a face hallucination method robust to noise is the primary task of this paper.

1.2   Motivation

As discussed above, it may be the optimal choice for noise robust face hallucination to combine advantages of the noise robustness of PCA based methods and the good representation ability of local patch based methods. As a subspace based approach, however, PCA cannot capture the characteristics of high-dimensional non-linear space. It has been well highlighted that in the space of rectangular arrays of positive numbers, only a small subset has images of natural scenes, especially for those face image patches of the same position that share

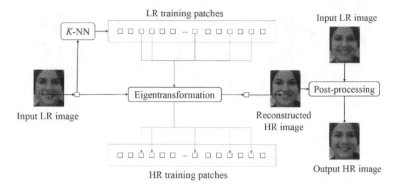

Fig. 1. Flowchart of the proposed method. Note that the red patches denote the K LR nearest neighbors of the input LR patch in the LR training set and K corresponding HR nearest neighbors in the HR training set, and we use the K LR and HR patch pairs to guide eigentransformation learning.

the similar structure [13]. The general solution is to identify this subset as a low-dimensional manifold and use its geometry to characterize images. Inspired by this, in this paper we propose to use the local geometry structure of the input LR patch to guide the eigentransformation learning, namely Data-driven Local Eigentransformation (DLE). As shown in Fig. 1, given an LR patch image, DLE firstly searches K nearest neighbors in the LR space, and secondly performs eigentransformation learning for the K nearest neighbors in the LR space and corresponding HR ones in the HR space.

2   Proposed Face Hallucination Method

2.1   Notations

The face hallucination problem for facial images can be formulated as the estimation of an HR image I_t^H from one input LR image I_t^L, given a training set of HR images and their corresponding LR versions, $I^H = \left\{I_i^H\right\}_{i=1}^m$ and $I^L = \left\{I_i^L\right\}_{i=1}^m$, where m is the number of training images. Human face is a kind of object in highly structured with allied five sense organs and outline, which makes the whole very similar with each other, too. Under these circumstances, the position prior of face plays an important role in the reconstruction of face image. For the patches at a certain position, they have the same position-patch structures and linear mapping. Divide each of the training images and the input LR image into small patches according to the location of face, $\{I_i^H(p,q)|1 \leq p \leq U, 1 \leq q \leq V\}_{i=1}^m$, $\{I_i^L(p,q)|1 \leq p \leq U, 1 \leq q \leq V\}_{i=1}^m$ and $\{I_t^L(p,q)|1 \leq p \leq U, 1 \leq q \leq V\}$, respectively. U and V denote the number of patches in the direction of row and column respectively and (p,q) is the location of the patch. In order to facilitate the representation without causing misunderstanding, we ignore the term of (p,q) below.

2.2 Data-Driven Local Eigentransformation Based Face Hallucination

As mentioned above, PCA is one kind of linear method where training images are described by projections onto a hyperplane embedded in the original image space. However, the face image observed is in a high-dimensional non-linear space, even if we divide it into small patches according to position. The face image(patch) space is not the well known Euclidean space, but should be a non-linear and smooth manifold. Therefore, it is inefficient to globally applying PCA to this distribution [14].

Several extensions to PCA have been proposed to overcome this problem [14, 15]. Local PCA methods are the best choices that it obtains a collection of locally linear descriptions by applying PCA to a local region of the data. Thus, local PCA can be used to produce compact models of non-linear data distributions [15]. As the data of local region is linear, which means it is in the Euclidean space, the data partition by Euclidean distance criterion can well isolate the target region.

To capture the local region of the data manifold, we search the K nearest neighbors in all LR training patches with the same position according to the Euclidean distance:

$$d_i = ||I_t^L - I_i^L||_2^2, 1 \leq i \leq m. \tag{1}$$

The K LR nearest neighbor patches and corresponding HR ones can be represented by

$$I_{(K)}^L = \left\{ I_k^L \,\middle|\, I_k^L \in I^L, k \in C_K\left(I_t^L\right) \right\}$$

and

$$I_{(K)}^H = \left\{ I_k^H \,\middle|\, I_k^H \in I^H, k \in C_K\left(I_t^L\right) \right\},$$

where $C_K\left(I_t^L\right)$ denotes the indexes of K nearest neighbor patches of the input patch in LR training patches. Note that the HR neighbor patches share the same indexes with LR neighbor patches. These neighbors form the local linear distritubiton of the entire patch image space, and its features can be well characterized by PCA.

The LR and HR mean patches can be computed by $\overline{m}_l = \frac{1}{m} \sum\limits_{i=1}^{m} I_i^L$ and $\overline{m}_h = \frac{1}{m} \sum\limits_{i=1}^{m} I_i^H$ respectively. And the demeaned K LR and HR nearest neighbor patches can be represented as $\overline{I}_{(K)}^L$ and $\overline{I}_{(K)}^H$, respectively.

Then covariance matrix of $\overline{I}_{(K)}^L$ can be found as $\overline{I}_{(K)}^L \left(\overline{I}_{(K)}^L\right)^T$. Let E_l be the matrix of eigenvectors of the covariance matrix, which can also be called eigenfaces (based on patches). The eigenvectors are arranged in such a way that respective eigenvalues are in decreasing order.

For a LR image patch I_t^L, a weight vector is computed by projecting it onto these eigenfaces,

$$\mathrm{w} = E_l^T \left(I_t^L - \overline{m}_l\right). \tag{2}$$

The patch can be reconstructed from the $Q(Q < K)$ eigenfaces $E_l = [e_1, \cdots, e_Q]$ that

$$\hat{I}_t^L = E_l w + \bar{m}_l. \tag{3}$$

Given the K LR nearest neighbor patches $\bar{I}_{(K)}^L$, according to singular value decomposition theorem, E_l also can be computed from:

$$E_l = \bar{I}_{(K)}^L V_l \Lambda_l^{-\frac{1}{2}}, \tag{4}$$

where V_l and Λ_l are the eigenvector and eigenvalue matrix for $\left(\bar{I}_{(K)}^L\right)^T \bar{I}_{(K)}^L$.

From Eq.(2)(3)(4), the reconstructed image patch can be represented by

$$
\begin{aligned}
\hat{I}_t^L - \bar{m}_l &= E_l E_l^T \left(I_t^L - \bar{m}_l\right) \\
&= \left(\bar{I}_{(K)}^L V_l \Lambda_l^{-\frac{1}{2}}\right) \left(\Lambda_l^{-\frac{1}{2}} V_l^T \left(\bar{I}_{(K)}^L\right)^T\right) \left(I_t^L - \bar{m}_l\right) \\
&= \bar{I}_{(K)}^L \left[V_l \Lambda^{-1} V_l^T \left(\bar{I}_{(K)}^L\right)^T \left(I_t^L - \bar{m}_l\right)\right] \\
&= \bar{I}_{(K)}^L c.
\end{aligned}
\tag{5}
$$

Thus, we have

$$c = V_l \Lambda^{-1} V_l^T \left(\bar{I}_{(K)}^L\right)^T \left(I_l^L - \bar{m}_l\right). \tag{6}$$

Using this coefficient and replacing LR image patches $\bar{I}_{(K)}^L$ by the corresponding HR ones $\bar{I}_{(K)}^H$, and replacing $\bar{m}_l$ with the HR mean patch $\bar{m}_h$ in Eq.(5), the HR image patch $\hat{I}_h^H$ can be reconstructed. After all the HR patches are reconstructed, integrate them according to the original position. The HR image can be generated by averaging pixel values in the overlapping regions.

3   Experiments Results and Analysis

3.1   Database and Parameter Settings

In the following sections, the effectiveness of our proposed method will be demonstrated by experiments on samples from FEI database [16], which contains 400 images captured from 200 individuals (100 men and 100 women). Among them, 180 individuals are randomly chosen as the training set, and the remaining 20 are regarded as testing set. All the HR face images are aligned by the two eyes and fixed at 120×100 pixels. LR images are generated by blurring of averaging filter and down-sampling with a factor of 4, and we interpolate the LR input image to the HR grid, which can be seen as the input. In our method, we set the size of HR image patch to 16×16 pixels with 8 pixels overlapped, and the variance accumulation contribution rate of PCA is set to 99%. Note that all parameters are set empirically. All the results will be evaluated by the two commonly used criteria PSNR and SSIM [17], which is good at capturing the loss of image structure between reconstructed image and original HR image.

3.2   Influence on the Performance of Different K

In order to assess the influence on the performance of nearest neighbor number K on the result of the proposed method, we conduct experiments on different K with different noise levels and show performance in terms of PSNR and SSIM in Fig. 2. As shown in this figure, the nearest neighbor number K plays an important role in the proposed method. By selecting a relatively small value, e.g., $K = 75$, the proposed method achieves the best performance. When the value of K exceeds 75, the performances decrease sharply. This certifies the importance of introducing local geometry structure. When $K=360$, it reduces to patch eigentransformation learning approach without considering the local manifold geometry. These demonstrate the efficiency of the proposed local patch based eigentransformation learning with neighbors searching.

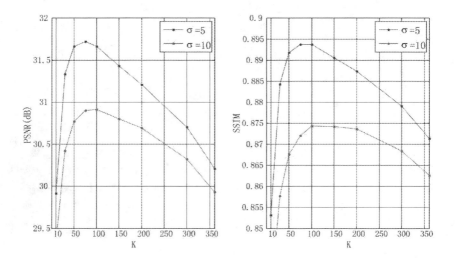

Fig. 2. The average PSNR(dB) and SSIM results of all 40 test images with different nearest neighbor number K

3.3   Performance Compared with State-of-the-Art

In this section, we compare the proposed approach with bicubic interpolation method and four other state-of-the-art learning-based face hallucination algorithms, i.e. Wang *et al.* [4], NE [7], LSR [8] and LcR [12]. Note that in the experiments we carefully tune the parameters of these comparison methods to get the best performance. To demonstrate the robustness to noise of the proposed DLE method, we show the objective and subjective results of different methods with different noise levels, e.g., Additive White Gaussian Noise (AWGN) with standard deviation of 5 and 10 respectively.

Table. 1 tabulates the average objective results in terms of PSNR and SSIM of all 40 test face images. We can observe that our method performs much better

Table 1. The average PSNR(dB) and SSIM results of all 40 test images for different methods

Methods	Criterions	Gaussian Noise	
		$\sigma=5$	$\sigma=10$
Bicubic	PSNR	26.63	24.78
	SSIM	0.7153	0.5212
Wang [4]	PSNR	27.31	26.68
	SSIM	0.7661	0.7495
LSR [8]	PSNR	28.21	23.92
	SSIM	0.7519	0.5361
NE [7]	PSNR	28.25	23.92
	SSIM	0.7560	0.5362
LcR [12]	PSNR	30.29	28.33
	SSIM	0.8600	0.8166
DLE without Pre-processing	PSNR	29.47	25.89
	SSIM	0.8206	0.6700
DLE	PSNR	31.72	30.90
	SSIM	0.8936	0.8720
Improvement	PSNR	**1.43**	**2.57**
	SSIM	**0.0336**	**0.0554**

than all other methods when images are contaminated with noise. The average improvements in term of PSNR over the second best approach (Jiang *et al.*'s LcR) are 1.43 dB and 2.57 dB when σ is set to 5 and 10 respectively.

The subjective reconstruction results with different noise levels are given in Fig. 3. We can observe that bicubic interpolation method produces the most blurry faces with most noise preserved while other learning based methods generate plausible HR images of better quality. The face contours of Wang's method are with obvious "ghost" effects, which mainly because of the poor representation ability of PCA based global face model. Since the least squares solution is not stable, results of LSR arise obvious dark spots and some faces still keep lots of noise, just like bicubic interpolation method. When compared with our proposed method, the results of LcR [12] appear to lack of details and blur the edges especially for mouth and eyes. After looking through Fig. 3, we can observe that the details in areas of mouth, nose and eyes reconstructed by our method are more similar to HR image than other approaches. We attribute this to the powerful representation ability of DLE. In order to extensively verify the effectiveness of our proposed method, we give some reconstruction results of realistic images from the CMU+MIT face database [18]. We can see in Fig. 4 that DLE can produce reasonable results even though the test images are contaminated by strong noise and are very different from the training examples.

Both the objective and subjective results show that the proposed DLE method can be applied to hallucinate the details of facial features from an LR face image especially with strong noise.

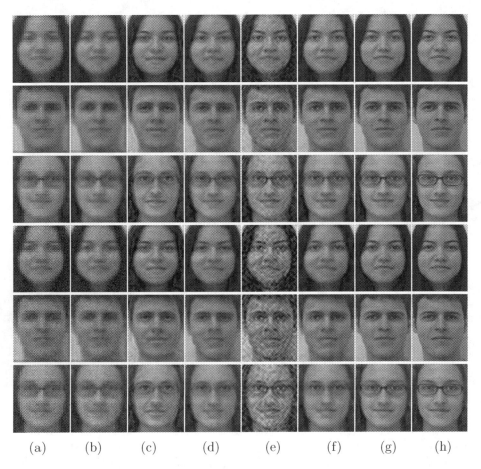

(a)   (b)   (c)   (d)   (e)   (f)   (g)   (h)

Fig. 3. Face hallucination results by different approaches when $\sigma = 5$ (the first three rows) and $\sigma = 10$ (the last three rows). (a) Input LR images; (b) Results of Bicubic; (c) Results of Wang *et al.* [4]; (d) Results of NE [7]; (e) Results of LSR [8]; (f) Results of LcR [12]; (g) Results of our method; (h) Original HR images.

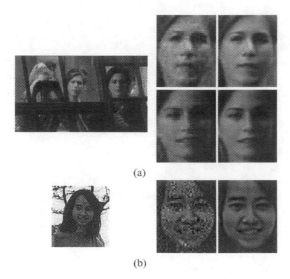

(a)

(b)

Fig. 4. Face hallucination results of realistic images from the CMU+MIT face database. The input pictures are at the left, while the manually aligned LR images are in the middle, and the results at the right.

4   Conclusions

In this paper, we present an effective and noise robust method for face hallucination from a single LR image. Since face images are embedded in high-dimensional non-linear space and linear PCA model cannot capture its non-linear characteristics, we search some samples that are most similar to the input image patch and then use the tailored training set to perform eigentransformation based face hallucination reconstruction. The effectiveness and robustness to noise of the proposed approach have been demonstrated with several experiments.

Acknowledgement. This work was supported in part by the National Key Technologies Research and Development Program under Grant 2013AA014602, in part by the National Natural Science Foundation of China under Grant 61231015, Grant 61172173, Grant 61303114 and U1404618, in part by the Major Science and Technology Innovation Plan of Hubei Province under Grant 2013AAA020, in part by the Guangdong-Hongkong Key Domain Break-through Project of China under Grant 2012A090200007, in part by the China Post-Doctoral Science Foundatio n under Project 2013M530350, in part by the Specialized Research Fund for the Doctoral Program of Higher Education under Grant 20130141120024, and in part by the Key Technology Research and Development Program, Wuhan, China, under Grant 2013030409020109, in part by Guangdong-Hongkong Key Domain Breakthrough Project of China under Grant 2012A090200007, in part by the Fundamental Research Funds for the Central Universities (2042014kf0250).

References

1. Wheeler, F., Weiss, R., Tu, P.: Face recognition at a distance system for surveillance applications. In: Proc. IEEE Conf. Biometrics: Theory Applications and Systems (BTAS), pp. 1–8 (2010)
2. Baker, S., Kanade, T.: Hallucinating faces. In: Proc. IEEE Conf. on Automatic Face and Gesture (FG), pp. 83–88 (2000)
3. Jiang, J., Hu, R., Han, Z., Wang, Z., Lu, T., Chen, J.: Locality-constraint iterative neighbor embedding for face hallucination. In: Proc. IEEE Conf. on Multimedia and Expo (ICME), pp. 1–6 (2013)
4. Wang, X., Tang, X.: Hallucinating face by eigentransformation. IEEE Trans. Systems, Man, and Cybernetics. Part C 35(3), 425–434 (2005)
5. Abdi, H., Williams, L.: Principal component analysis. Wiley Interdisciplinary Reviews: Computational Statistics 2(4), 433–459 (2010)
6. Li, X., Lam, K.M., Qiu, G., Shen, L., Wang, S.: Example-based image super-resolution with class-specific predictors. J. Vis. Commun. Image Represent. 20(5), 312–322 (2009)
7. Chang, H., Yeung, D., Xiong, Y.: Super-resolution through neighbor embedding. In: Proc. IEEE Conf. Comput. Vis. Pattern Recognit. (CVPR), vol. 1, pp. I–275 (2004)
8. Ma, X., Zhang, J.P., Qi, C.: Hallucinating face by position-patch. Pattern Recognition 43(6), 2224–2236 (2010)
9. Yang, J., Wright, J., Huang, T., Ma, Y.: Image super-resolution as sparse representation of raw image patches. In: Proc. IEEE Conf. Comput. Vis. Pattern Recognit. (CVPR), pp. 1–8 (2008)
10. Li, B., Chang, H., Shan, S., Chen, X.: Aligning coupled manifolds for face hallucination. Signal Processing Letters 16(11), 957–960 (2009)
11. Yang, J., Wright, J., Huang, T., Ma, Y.: Image super-resolution via sparse representation. IEEE Trans. Image Process. 19(11), 2861–2873 (2010)
12. Jiang, J., Hu, R., Han, Z., Lu, T., Huang, K.: Position-patch based face hallucination via locality-constrained representation. In: Proc. IEEE Conf. on Multimedia and Expo (ICME), pp. 212–217 (2012)
13. Srivastava, A., Lee, A.B., Simoncelli, E.P., Zhu, S.C.: On advances in statistical modeling of natural images. J. Math. Imag. Vis. 18(1), 17–33 (2003)
14. Kambhatla, N., Leen, T.K.: Dimension reduction by local principal component analysis. Neural Computation 9(7), 1493–1516 (1997)
15. Tipping, M., Bishop, C.: Mixtures of probabilistic principal component analyzers. Neural Computation 11(2), 443–482 (1999)
16. Thomaz, C., Giraldi, G.: A new ranking method for principal components analysis and its application to face image analysis. Image and Vision Computing 28(6), 902–913 (2010)
17. Wang, Z., Bovik, A., Sheikh, H., Simoncelli, E.: Image quality assessment: From error visibility to structural similarity. IEEE Trans. Image Process. 13(4), 600–612 (2004)
18. Rowley, H.A., Baluja, S., Kanade, T.: Neural network-based face detection. IEEE Trans. on Pattern Analysis and Machine Intelligence 20(1), 23–38 (1998)

Lip Segmentation Based on Facial Complexion Template

Chenyang Sun[1], Hong Lu[1,*], Wenqiang Zhang[1], Xiaoxin Qiu[1],
Fufeng Li[2,*], and Hongkai Zhang[2]

[1] Shanghai Key Laboratory of Intelligent Information Processing,
Fudan University, Shanghai, 200433, China
[2] Laboratory of Information Access and Synthesis of TCM Four Diagnosis,
Shanghai University of Traditional Chinese Medicine, 201203, China
honglu@fudan.edu.cn, lff@shutcm.edu.cn

Abstract. In Traditional Chinese Medicine (TCM), lip diagnosis is an important diagnostic method to judge whether a person is healthy or not. Lip images can reflect the physical conditions of organs in the body. Lip diagnosis has a long history in China and the lips are analyzed by experienced doctors with their nude eyes. This method is not objective and efficient especially in the condition of handling many images. Developing an automatic way to split lips from an image is an important and necessary step. What's more, lip segmentation can provide improvement in the areas of speech recognition and speaker authentication. To segment lips and facial complexions, many methods are proposed which are based on color spaces such as RGB, HSV, Lab, etc. Other methods are based on different models such as snake, geometry model, etc. This paper proposes a lip segmentation method based on facial complexion template. A facial complexion template can be constructed when the face is detected. We construct the facial complexion template using Hue channel and Saturation channel of color information. By removing the skin similar to facial complexion template values an initial lip image can be got. Finally, by smoothing the lip contour an optimized lip segmentation result can be obtained.

Keywords: Lip segmentation, facial complexion, smooth lip.

1 Introduction

Lip diagnosis is one of the most active areas in Traditional Chinese Medicine (TCM). Normally lips are analyzed by experienced doctors' nude eyes in traditional analysis way. A person's lip color is considered as a symptom and sign to diagnose whether his spleen or his stomach is healthy or not. The more accurate the lip is segmented, the more reliable diagnosis result can be obtained. Also, handling lip segmentation with digital technology is beneficial for the development of TCM.

Lip color can be classified as deep-red, purple, red, pale, etc. Finding a method which can be applied to different lips is a meaningful thing in recent years. Some methods segment lips based on color information [1]. It is observed that lip pixels and

---

* Corresponding author.

W.T. Ooi et al. (Eds.): PCM 2014, LNCS 8879, pp. 193–202, 2014.

non-lip pixels can be segmented in some color spaces. After the first step, obtain an initial lip contour can be got. To get a complete and better lip, clustering method is performed to find all lip pixels. By clustering, some non-lip pixels can be removed. On the other hand, if the lip pixels and the non-lip pixels have similar color information, it is not easy to obtain the accurate initial lip contour. Especially, when dealing with different complexions, the methods cannot accurately obtain the lip contour.

Another classic way to segment lips, for example geometry model [2], snakes [3]. These methods use the points to approximate the lip contour with spline functions. But these methods needs a priori knowledge about the lip image and cannot work well when the lip pixels and the non-lip pixels have similar color.

[4] proposes a lip segmentation method based on level set model. The lip contour is first obtained the lip contour through Otsu method [5]. And the lip contour is optimized by level set model. However, the results rely on the first step processing. Thus, how to accurately find the lip's initial contour is crucial and important.

Besides been used in TCM, the lip segmentation can be used in speech recognition [6] [7] [8] [9], facial expressions recognition [12], etc. When the pixels around the lip are different from the lip pixels, traditional lip segmentation methods can provide a good result. However, when the difference between the lip pixels and non-lip pixels is not large in feature space, it is not easy to obtain a complete and accurate lip contour by traditional segmentation methods.

To resolve the problems, this paper proposes a facial complexion remove method for lip segmentation. By this method, an accurate initial lip can be obtained. Then the initial lip would be optimized to obtain a smooth lip contour. This paper proposes a facial complexion template method. If the skin value is similar to template's value, the skin can be regarded as non-lip pixels and removed. When an appropriate value is set, the non-lip pixels can be removed and the lip pixels can be kept accurately. The HSV color space is used based on our empirical study. Compared with traditional lip segmentation methods, our method can obtain a more accurate result.

2   Related Work

The test images are from Shanghai University of Tradition China Medicine, and are taken from a collection box. In the center of the box, there exists a hole to put a normal-size face. The images are taken in the condition with appropriate light and by a same camera. The collected images include kinds of facial complexions, i.e. black, red, yellow, cyan, and white. Each kind of facial complexion images' number is 50. In the images, some have the lip pixels similar in color from non-lip pixels. And some are different. Dealing with different complexions, the lip segmentation can obtain different performance. Specifically, since the red facial complexion is similar to lip pixels' color, the lip on red face is hard to be segmented. While the yellow facial complexion is different from lip pixels' color, and the lip on yellow face is easy to be segmented.

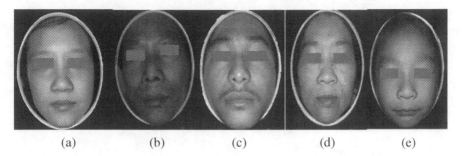

(a)	(b)	(c)	(d)	(e)

Fig. 1. Five kinds of facial complexions images. (a) white facial complexion; (b) black facial complexion; (c) red facial complexion; (d) yellow facial complexion; (e) cyan facial complexion.

3   Lip Segmentation Based on Facial Complexion Difference

Lip segmentation has three steps

(1). Construct a facial complexion template from the detected face
(2). Remove non-lip pixels based on the facial complexion template and obtain an initial lip
(3). Smooth the lip contour and obtain a better lip image

3.1   Construct a Complexion Template

The first step is to find the face and then to extract the skin information. In face detection, boosted cascade of simple features [10] performs well to find human face. After finding the face, the skin pixels are kept and non-skin pixels such as eyes, eyebrows are removed. This method removes impurities. Face skin can be extracted in YCrCb color space [11]. In this method, if the skin value is out of the particular range of Cr or Cb color channel, the skin would be removed.

To construct a facial complexion template, a part of face skin is selected at first. Because the lip is usually at the bottom of face, the skin above mouth would be chosen to build facial complexion template. Then we extract the skin information to construct this template. The facial complexion template values are obtained through a color histogram in which the value represents pixel value from 0 to 255 and its ordinate value represents the corresponding pixel sum value. We find the pixel which has a largest sum on the ordinate and regard this pixel value as face skin value in one color channel.

Based on our empirical study, HSV color space shows a better performance on facial complexion template construction compared with other color spaces. Assume that, k_1 is the largest pixel value in R, G and B color channels, k_2 is the smallest pixel value in R, G and B color channels, from formulas (1) (2) (3) , we can obtain HSV color space transformed from RGB color space. Especially in H and S color channels, face skin color is obviously different from lip color, and an initial lip can be obtained. Assume that H_0 and S_0 construct facial complexion template, H_0 represents facial complexion information in Hue color channel, while S_0 represents facial complexion information in Saturation color channel.

$$V = k_1 / 255 \tag{1}$$

$$S = \begin{cases} 0, (k_1 = 0) \\ (k_1 - k_2) / k_1, (k_2 \neq k_1) \end{cases} \tag{2}$$

$$H = \begin{cases} 60 * (G - B) / (S * V), (S \neq 0, k_1 = R) \\ 60 * (2 + (B - R) / (S * V)), (S \neq 0, k_1 = G) \\ 60 * (4 + (R - G) / (S * V)), (S \neq 0, k_1 = B) \end{cases} \tag{3}$$

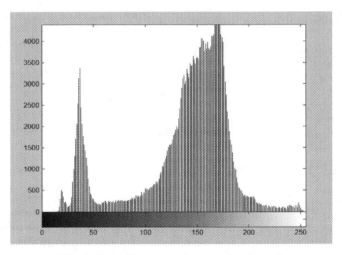

Fig. 2. Color histogram   in one color channel

3.2   Remove Non-lip Pixels to Obtain Initial Lip

Traditional skin removing methods set unalterable value to remove the skin. They set particular value to distinguish lip pixels and non-lip pixels, if the pixel value is out of given scope, it would be removed. But face colors vary a lot. They may work well in some skin scopes, but work badly in some other scopes. To solve this problem, this paper proposes a self-adaption method. This paper's method learns the facial complexion first, then we build a complexion template and remove the non-lip pixels based on the complexion template.

The facial complexion template is constituted by H and S color channel values. Because the difference between lip color and facial complexion is small in H color space, as Equ. (4) shows, if the deviation between template's value H_0 and skin pixel value H is smaller than h_t, the skin pixels can be removed. On the contrast, the difference between lip color and complexion may be obvious in S space, as Equ. (5) shows, if the difference between template's value S_0 and skin pixel value S is smaller than 100, the skin can be removed. After two steps, a majority of skin on the face which is close to facial complexion template is removed and a facial complexion image which contains a raw lip contour can been obtained. Then do some morphologic operations to remove the tiny pixels and fill the tiny holes. Because lip located on bottom of face and lip size can be estimated according to face size, then an initial lip contour can be got through contour-finding method. If this lip satisfies some conditions, then go to 3.3 to do some smooth operations to reduce raw edges and some protrusions, finally, an optimized lip image can be obtained.

The process of skin removing can be shown as follow.

1. Set h_t as 3, do facial complexion removing, if a lip can been found, get lip's value h_l in Hue color channel and get lip's area SS_0, then go to step 2, else h_t adds 1, find lip again, doing this until find a lip and go to step 2.
2. Get the difference value between h_t and h_l and set the difference value as h_d, if $h_d<7$, it means lip color is similar to facial complexion in Hue color channel, else it means they are distinguished. Then go to step 3.
3. h_t adds 1, do facial complexion removing, if a lip can been found, get lip's area SS_1 and get the difference value SS_d between SS_0 and SS_1. If $h_d<7$ and $SS_d<SS_1/4$, return this lip and go to 3.3. If $h_d>=7$ and $SS_d<SS_1/10$, return this lip and go to 3.3. If no lip been found, return the former lip and go to 3.3. If $h_t>8$, return the lip and go to 3.3. Else set $SS_0=SS_1$ and go to step 3.

$$|H - H_0| < h_t \qquad (4)$$

$$|S - S_0| < 100 \qquad (5)$$

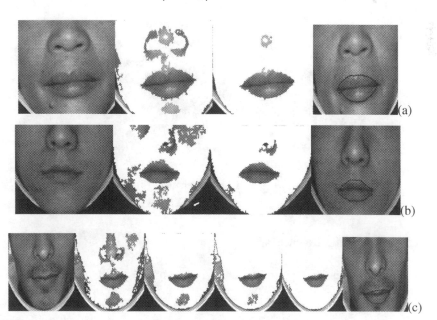

Fig. 3. Facial complexion removing images. (a) (b) (c) are processed images with fewer and fewer non-lip pixels and the last one is the lip contour image.

3.3   Lip Edge Tuning and Obtain an Optimized Lip

The initial lip normally has a rough contour and some protrusions, so it is necessary to find a solution to optimize the lip. The lip's edge is usually raw because the pixel colors between the lip and the non-lip is similar, and the segmentation method regards the non-lip pixels as lip pixels. Then it is necessary to smooth the raw edge and reduce protrusions. Median filtering method can be used to smooth the raw edge. Specifically, median filtering method has a kernel which is a square and its size is usually 3×3, 5×5, and 7×7. After the kernel doing a convolution operation with the raw edge, the square pixel values in the processed image would be set same with the pixel in the square center. After median filtering operation, a smooth edge can be obtained.

[4] proposes a "five points average" method to smooth lip contour of edges. From left lip to right lip, a point is drawn every five pixels. Then the points are linked together. This method can remove protrusions. If the protrusions don't be removed accurately, we can set the step to be ten pixels and do the processing again. Then we do an AND operation with five pixels step image. Finally, the lip edge is smoother and protrusions on the image are removed.

(a)   (b)

Fig. 4. Smoothing lip edge images. (a) A lip with raw edge; (b) A lip with smooth edge after a median filtering operation.

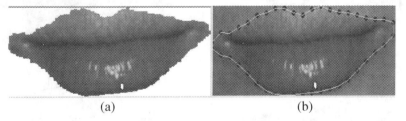

(a)   (b)

Fig. 5. Lip edge tuning results. (a) A lip with protrusions; (b) An optimized lip using "five points average" method.

4   Experimental Results

This paper uses the data set which contains 464 face images and includes five types of facial complexions. To keep each type balance, we sample 50 images in each facial complexion type. Experiments show that lips location can be detected correctly based on facial complexion template method. Furthermore, this method can also accurately segment the lip which is similar to facial complexion.

To evaluate the performance of the lip segmentation, a measure as defined in Equ. (6) is defined. Specifically, we first manually annotate the lip and obtain the lip area S_1. Second, we obtain the lip by the proposed method and obtain the segmented lip area S_2. Then the area overlap (OL) measure is computed as below.

$$OL = 2 * (S_1 \cap S_2) / (S_1 + S_2) * 100\% \tag{6}$$

The larger the OL value, the better the performance. Table 1 shows the performance of lip segmentation on five types of facial complexions.

Table 1. OL in five types of facial complexions

Complexion	OL(%)
White	95.42
Black	96.37
Red	93.19
Yellow	95.02
Cyan	96.71

From Table 1, we can find that, the performance of lip segmentation on five types of facial complexions is promising and robust to different facial complexions. Also, the larger the difference between the lip color and the complexion color, the better segmentation result can be obtained. Also, when dealing with the images where the lip pixels and non-lip pixels have similar color, the proposed method can obtain a good result.

Fig. 6 shows the lip segmentation results of the five facial complexions.

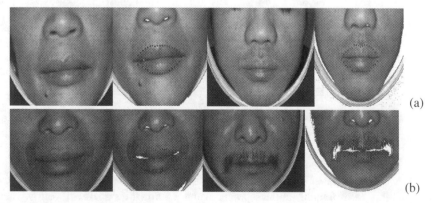

(a)

(b)

Fig. 6. Lip segmentation result of the five types of complexions. (a) (b) (c) (d) (e) are white, black, red, yellow, and cyan facial complexions, respectively. Left image is an input image and the right is a lip segmentation image.

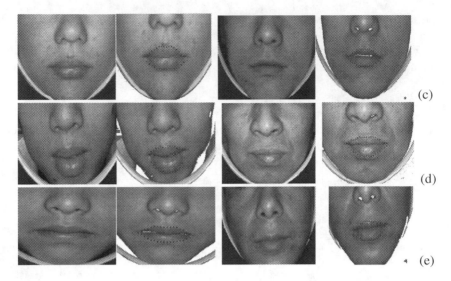

(c)

(d)

(e)

Fig. 6. (*continued*)

Also, we compare the proposed method with fussy c-means (FCM) clustering method and Snake method [3]. SE (segmentation error) is defined to measure the segmentation error as in Equ. (7). The lower the value, the better the performance.

$$SE = (OLE + ILE) / (2 * TL) * 100\%$$ (7)

In Equ. (7), OLE is the number of non-lip pixels being classified as lip pixels. While ILE is the number of lip pixels being classified as non-lip pixels. TL is the number of lip pixels in the real lip region. Table 2 shows the performance of our proposed method and the compared methods.

Table 2. Comparison of SE among FCM, Snake [3] and this paper's method

Methods	FCM	Snake [3]	Ours
SE (%)	14.38	27.43	2.42

It can be observed from Table 2, our method can obtain better result compared with FCM and snake methods.

To prove that this paper's algorithm is applicable to normal pictures which were taken in natural environment, about 100 face images not the person of yellow race were selected form network, experiments show good result. Compared with Yan's [4] and Ghaleh's [1] algorithms, some images are shown as follow.

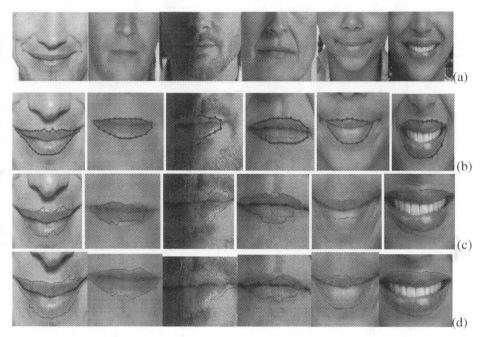

Fig. 7. Compared with other algorithms. (a) input images. (b) this paper's algorithm. (c) Yan's algorithm. (d) Ghaleh's algorithm.

5   Conclusions

This paper proposes a lip segmentation method based on facial complexion template.

Specifically, we first extract face skin information and construct a facial complexion template. Then the non-lip pixels are removed based on the facial complexion template and an initial lip contour is obtained. The proposed method can be used in lip segmentation and can deal with different facial complexions. Also, the method can be used to segment two objects with color difference in a specific color space. Our future work is to segment tongue images and help diagnosis.

Acknowledgements. This work was supported in part by the Natural Science Foundation of China (No. 61170094 and 81373555), Shanghai Committee of Science and Technology (14JC1402202 and 14441904403), and 863 Program 2014AA015101.

References

1. Ghaleh, V., Behrad, A.: Lip contour extraction using RGB color space and fuzzy c-means clustering. In: 2010 IEEE 9th International Conference on Cybernetic Intelligent Systems (CIS). IEEE (2010)

2. Liewa, A.W.-C., Leung, S.H., Lau, W.H.: Lip Contour Extraction Using a Deformable Model. In: 7th IEEE International Conference on Image Processing, pp. 255–258. IEEE Press, Vancouver (2000)
3. Delmas, P., Coulon, P.Y., Fristot, V.: Automatic Snakes for Robust Lip Boundaries Extraction. In: 5th IEEE International Conference on Acoustics, Speech, and Signal Processing, pp. 3069–3072. IEEE Press, Phoenix (1999)
4. Yan, X., Li, X., Zheng, L., Li, F.: Robust Lip Segmentation Method Based on Level Set Model. In: Qiu, G., Lam, K.M., Kiya, H., Xue, X.-Y., Kuo, C.-C.J., Lew, M.S. (eds.) PCM 2010, Part I. LNCS, vol. 6297, pp. 731–739. Springer, Heidelberg (2010)
5. Otsu, N.: A Threshold Selection Method from Gray-level Histograms. IEEE Trans. on Systems, Man, and Cybernetics 9(1), 62–66 (1979)
6. Erber, N.P.: Interaction of audition and vision in the recognition of oral speech stimuli. J. Speech Hearing Res. 12, 423–425 (1969)
7. Zhang, Y., Levinson, S., Huang, T.: Speaker independent audio-visual speech recognition. In: Proc. IEEE Int. Conf. Multimedia and Expo, New York, vol. 2, pp. 1073–1076 (July 2000)
8. Rabi, G., Lu, S.: Visual speech recognition by recurrent neural networks. In: Engineering Innovation: Voyage of Discovery Electrical and Computer Engineering 1997, St. Johns, Nfld., Canada, vol. 1, pp. 55–58 (May 1997)
9. Luettin, J., Thacker, N.A., Beet, S.W.: Visual speech recognition using active shape models and hidden Markov models. In: Proc. IEEE Int. Conf. Acoustics, Speech, and Signal Processing, Atlanta, GA, vol. 2, pp. 817–820 (May 1996)
10. Viola, P., Jones, M.: Rapid object detection using a boosted cascade of simple features. In: Proceedings of the IEEE Computer Society Conference on Computer Vision and Pattern Recognition, CVPR 2001, vol. 1. IEEE (2001)
11. Hsu, R.L., Abdel-Mottaleb, M., Jain, A.K.: Face detection in color images. IEEE Transactions on Pattern Analysis and Machine Intelligence 24(5), 696–706 (2002)
12. Hammal, Z., Couvreur, L., Caplier, A., Rombaut, M.: Facial expression recognition based on the belief theory: Comparison with different classifiers. In: Roli, F., Vitulano, S. (eds.) ICIAP 2005. LNCS, vol. 3617, pp. 743–752. Springer, Heidelberg (2005)

Saliency-Based Deformable Model
for Pedestrian Detection

Xiao Wang[1], Jun Chen[1,2], Wenhua Fang[1], Chao Liang[1,2,*], Chunjie Zhang[3],
Kaimin Sun[4], and Ruimin Hu[1,2]

[1] National Engineering Research Center for Multimedia Software,
School of Computer, Wuhan University, Wuhan, 430072, China
[2] Research Institute of Wuhan University in Shenzhen, China
[3] School of Computer and Control Engineering,
University of Chinese Academy of Sciences, Beijing, 100190, China
[4] School of Remote Sensing Information Engineering,
Wuhan University, 129 Luoyu Road, Wuhan, 430079, China
cliang@whu.edu.cn

Abstract. Pedestrian detection, which is to identify category (pedestrian) of object and give the position information in the image, is an important and yet challenging task due to the intra-class variation of pedestrians in clothing and articulation. Previous researches mainly focus on feature extraction and sliding window, where the former aims to find robust feature representation while the latter seeks to locate the latent position. However, most of sliding windows are based on scale transformation and traverse the entire image. Therefore, it will bring computational complexity and false detection which is not necessary. To conquer the above difficulties, we propose a novel Saliency-Based Deformable Model (SBDM) method for pedestrian detection. In SBDM method we present that, besides the local features, the saliency in the image provides important constraints that are not yet well utilized. And a probabilistic framework is proposed to model the relationship between Saliency detection and the feature (Deformable Model) via a Bayesian rule to detect pedestrians in the still image.

Keywords: Pedestrian detection, Saliency-Based Deformable Model, Saliency Detection, Bayesian rule.

1 Introduction

Recently, pedestrian is an important component part over a large public areas, such as visual surveillance, robotics, and automotive safety. In these scenarios, pedestrian detection is becoming a hot research spot in the computer vision community. During the past three years, a lot of research effort [13, 4, 5, 14, 15, 17] has been devoted to this field. However, the latent scale and the position of the pedestrian are unknown, so they usually resize the sliding window or

---

* Corresponding author.

W.T. Ooi et al. (Eds.): PCM 2014, LNCS 8879, pp. 203–210, 2014.

Fig. 1. Examples of salient regions extracted from original color images in different scenarios. Upper part of the figure shows pedestrian in different scenarios and the lower part shows corresponding salient regions. We can see from the above image that pedestrians have been included in salient map.

/and image many times when detecting by sliding window, and this is so-called multiple scales detection [6]. The selection of a proper scale and the step width of the sliding window will greatly affect the algorithm's precision and efficiency. This is more obvious in the case of high-resolution images. Fortunately, we can get the approximate location of pedestrians by salient regions and avoid the above problems. As we can see from the Fig .1, salient regions contain objects (such as pedestrians) of the image in different scenarios.

Generally speaking, pedestrian detection can be considered as a very important part of visual retrieval problem [11], given a query pedestrian image taken in one image, the algorithm is expected to search the same pedestrian captured by other image. Typically, it consists of two stages feature extraction and scale transformation, where the former aims to find robust feature representation while the latter seeks to give the bounding box precisely. The combination of the histograms of oriented gradients (HOG) features and linear SVM learning machine, proposed by Dalal et al. in [3], has been proven as a competitive method to detect pedestrians. Farenzenaet al. [8] has divided the image of person into 5 regions by exploiting symmetry and asymmetry perceptual principles, and then combine multiple color and texture features to represent the appearance of people. Wanli Ouyang [12] has proposed pedestrian detector which is learned with a mixture of deformable part-based models to effectively capture the unique visual patterns appearing in pedestrians. However, a majority of detectors surveyed in [7] remain complex. Because they detect pedestrian from full image, rather than effective local regions where the object exists.

In this paper, we propose a novel Saliency-Based Deformable Model (SBDM) which combines the salient detection model and the Deformable Model combined by Bayesian rule to detect the latent pedestrians in the still image. It is easy to see from Fig. 2 that we tend to focus on effective local regions of the image rather than the entire. More precisely, we can get saliency degree by histogram based contrast method [2]. Therefore, we can get the effective local regions by combining saliency regions and original image. And deformable model feature [8] is extracted from the effective local regions at the reference scale, and finally a logistic regression classifier is adopted to detect the pedestrian object.

Summarizing, the contribution of our work is two-fold. (1) it takes full advantage of the local regions where the objects exist rather than the whole image which contains a lot of interference information, thus greatly reduces the false alarm rate and the computational complexity. (2) a new probabilistic framework is proposed to model the configuration relationship between results of the salient detection model and the deformable model via Bayesian rule.

2   The Proposed Pedestrian Detection Framework

Our framework has fused the salient region detection model and the traditional deformable model descriptor by Bayesian rule. Rather than summing these votes, we model all the random variables S_R (rectangular window R of the image is salient region) and P_{SR} (we can detect pedestrian from the S_R) in a probabilistic method so that we can determine the final probabilistic values via the Bayesian inference process. Thus, we are interested in modeling the joint posterior of S_R and P_{SR} given local region R, and apply the Bayesian theorem then gives:

$$p(S_R, P_{SR}|R) \propto p(R|S_R, P_{SR})p(S_R, P_{SR}) \tag{1}$$

Our framework is illustrated in Fig. 2. We now focus on the prior and likelihood terms separately.

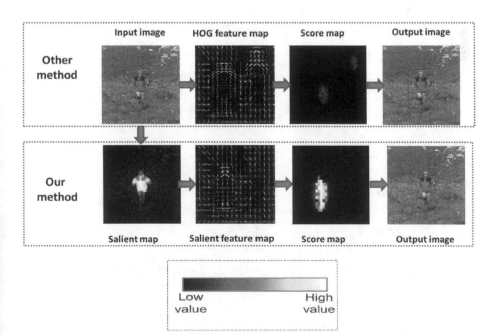

Fig. 2. The framework of our pedestrian detection system

2.1 Salient Region Prior Model

According to a histogram based contrast method [2] which defines saliency values for image pixels using color statistics of the input image. The pixels with the same color value have the same saliency value under this definition, since the measure is oblivious to spatial relations. Therefore, it is so easy to get saliency value for each pixel as,

$$V(k) = \sum_{i=1}^{N} d(k, i), \tag{2}$$

I is the image in Lab color space; $d(k, i)$ is the color distance metric between pixels k and i, and N is the number of pixels in image I. To be more precise, the new saliency degree can be seen as a weight, and the weight is calculated by

$$\omega_k = \frac{1}{N} \sum_{k \subset I} \frac{1}{1 + e^{-V_k}} \tag{3}$$

where ω_k is the weight of k in appearance model, V_k is the salience value of k pixel in image. And the image detected can be written as:

$$D(k) = I(k)\omega_k \tag{4}$$

where $D(k)$ is the new pixel of k in the new image detected. It is so easy to get $p(P_{SR}|S_R)$ from detecting portion, thus we get:

$$p(S_R, P_{SR}) = p(P_{SR}|S_R)p(S_R) \tag{5}$$

2.2 Deformable Likelihood Model

We follow the framework of deformable models [9, 16, 10, 1] and describe an object by a non-rigid constellation of parts location and appearance. In order to explicitly model occlusion, we use a binary part visibility term. Each part is defined by the location of a bounding box $p_i = (p_i^l, p_i^r, p_i^t, p_i^b)$ in the image and the binary visibility state v_i. The score of a model α in the image $\mathbf{I}$, which gives model parts locations $P = (p_0, \ldots, p_n)$ and visibility states $V = (v_1, \ldots, v_n)$, $v_1 \in \{0, 1\}$, is defined as follow:

$$S(I, P, V) = \max_{c \in \{1..C\}} S(I, P, V, \alpha_c) \tag{6}$$

Pedestrian parts are always occluded because of the presence of other pedestrians and self-occlusions. The locations of occluded parts may have consistent appearance, because occlusions often do not happen at random. We use occlusions by learning separate appearance parameters A^o for occluded parts. The bias terms b_i and b_i^o control the balance between occluded and non-occluded appearance terms in S_A. One mixture component of the model has a tree T and edges structure with nodes E corresponding to object parts and relations among parts respectively.

$$S(I, P, V, \alpha_c) = \sum_{i \in T} S_A(I, p_i, v_i, \alpha_i) + \sum_{(i,j) \in E} S_D(p_i, p_j, \alpha_i) \qquad (7)$$

where the unary term S_A provides appearance score using image features $\phi(I, p_i)$,

$$S_A(I, p_i, v_i, \alpha_i) = v_i(A_i.\phi(I, p_i) + b_i) + (1 - v_i)(A_i^0.\phi(I, p_i) + b_i^0) \qquad (8)$$

and the binary term S_D defines a quadratic deformation cost $S_D(p_i, p_j, \alpha_i) = d_i.\psi(p_i - p_j)$ with $\psi(p_i - p_j) = \{dx; dy; dx^2; dy^2\}$ where $dx = p_i^{x_1} - (p_j^{x_1} + \mu_{ij}^x)$ and $dy = p_i^{y_1} - (p_j^{y_1} + \mu_{ij}^y)$). Notably, the score function (6) linearly depends on the model parameters $\alpha_c = \{A_0; \ldots; A_n; A_0^o; \ldots; A_n^o; d_1; \ldots; d_n; B\}$. To represent multiple appearances of an object, our full model combines a mixture of c trees described by parameters $\alpha = \{\alpha_1; \ldots; \alpha_c\}$.

2.3   The Probabilistic Framework

As a result of substituting (5) into (1), we get the posterior:

$$p(S_R, P_{SR}|R) \propto p(R|S_R, P_{SR})p(P_{SR}|S_R)p(S_R) \qquad (9)$$

More specifically, we can also get the the posterior by joint distribution. We model the joint distribution over all the random variables (S_R,P_{SR}), so we can determine probabilistic values as follow:

$$p(S_R, P_{SR}|R) = \frac{p(S_R, P_{SR}, R)}{\sum_R p(R|S_R, P_{SR})p(S_R, P_{SR})} \qquad (10)$$

where $p(S_R, P_{SR}|R)$ is the probability of pedestrian in a rectangular of image, and P_{SR} is results pedestrian detection from effective salient regions.

$$p(S_R, P_{SR}, R) = p(R|S_R, P_{SR})p(S_R, P_{SR}) \qquad (11)$$

As a result of substituting (5) and (11) into (10), we get the final expression for the posterior:

$$p(S_R, P_{SR}|R) = \frac{p(R|S_R, P_{SR})p(P_{SR}|S_R)p(S_R)}{\sum_R p(R|S_R, P_{SR})p(S_R, P_{SR})} \qquad (12)$$

3   Experiment

Different experiments were conducted to evaluate our method. In this section, we used the INRIA still image database to compare our approach (use of feature subsets and mean features) against previous works. The proposed approach is validated by comparing with several state-of-the-art pedestrian detection methods on INRIA datasets. The database contains 1774 pedestrian positive examples

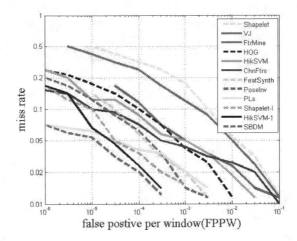

Fig. 3. Results of FPPW on the INRIA persons dataset

and 1671 negative images without pedestrian. The pedestrian annotations were scaled into a series of windows whose size is 64×128 and included a margin of 16 pixels around the pedestrians.

The dataset was divided into two, where 1,000 pedestrian annotations and 1,000 person-free images were selected as the training set, and 774 pedestrian annotations and 671 person-free images were selected as the test set. For each cascade level, the Logitboost algorithm [7] was trained using all the positive examples and $N_n = 10,000$ negative examples generated by boostrapping. Detection on the INRIA pedestrian dataset is challenging since it includes subjects

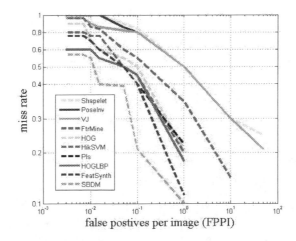

Fig. 4. Results of FPPI on the INRIA persons dataset

with a wide range of variations in pose, clothing, illumination, background, and partial occlusions.

Fig. 3 shows that the performance of our method is comparable to the state-of-the-art approaches. We compare ours experiments results with the state-of-art approaches on INRIA dataset. The x-axis corresponds to false positives per window(FPPW), the y-axis corresponds to the miss rate, and we plot the detection error tradeoff curves on a log-log scale for INRIA dataset.

Fig. 4 compares our approache with other state-of-the-art methods. Our detector is competitive in terms of the detection quality with respect to ChnFtrs [7], provides significant improvement over HOG+SVM and others. The x-axis corresponds to false positives per image (FPPI), the y-axis corresponds to the miss rate.

4   Conclusion

Previous researches mainly focus on feature extraction and sliding window, where the former aims to find robust feature representation while the latter seeks to locate the latent position. However, most of sliding windows are based on scale transformation and traverse the entire image. Therefore, it will bring computational complexity and false detection which is not necessary. To conquer the above difficulties, we propose a novel Saliency-Based Deformable Model (SBDM) method for pedestrian detection. In SBDM method we present that, besides the local features, the saliency in the image provides important constraints that are not yet well utilized. And a probabilistic framework is proposed to model the relationship between Saliency detection and the feature (Deformable Model) via a Bayesian rule to detect pedestrians in the still image.

Acknowledgement. The research was supported by the National Nature Science Foundation of China (61303114, 61231015, 61170023, 61303154, F020503), the Specialized Research Fund for the Doctoral Program of Higher Education (20130141120024), the Technology Research Project of Ministry of Public Security (2014JSYJA016), the Fundamental Research Funds for the Central Universities (2042014kf0250), the China Postdoctoral Science Foundation funded project (2013M530350), the major Science and Technology Innovation Plan of Hubei Province (2013AAA020), the Guangdong Hongkong Key Domain Breakthrough Project of China (2012A090200007), the Key Technology R&D Program of Wuhan (2013030409020109), and the Special Project on the Integration of Industry, Education and Research of Guangdong Province (2011B090400601).

References

[1] Azizpour, H., Laptev, I.: Object detection using strongly-supervised deformable part models. In: Fitzgibbon, A., Lazebnik, S., Perona, P., Sato, Y., Schmid, C. (eds.) ECCV 2012, Part I. LNCS, vol. 7572, pp. 836–849. Springer, Heidelberg (2012)

[2] Cheng, M.M., Zhang, G.X., Mitra, N.J., Huang, X., Hu, S.M.: Global contrast based salient region detection. In: 2011 IEEE Conference on Computer Vision and Pattern Recognition (CVPR), pp. 409–416. IEEE (2011)

[3] Dalal, N., Triggs, B.: Histograms of oriented gradients for human detection. In: IEEE Computer Society Conference on Computer Vision and Pattern Recognition, CVPR 2005, vol. 1, pp. 886–893. IEEE (2005)

[4] Ding, Y., Xiao, J.: Contextual boost for pedestrian detection. In: 2012 IEEE Conference on Computer Vision and Pattern Recognition (CVPR), pp. 2895–2902. IEEE (2012)

[5] Dollár, P., Appel, R., Kienzle, W.: Crosstalk cascades for frame-rate pedestrian detection. In: Fitzgibbon, A., Lazebnik, S., Perona, P., Sato, Y., Schmid, C. (eds.) ECCV 2012, Part II. LNCS, vol. 7573, pp. 645–659. Springer, Heidelberg (2012)

[6] Dollár, P., Belongie, S., Perona, P.: The fastest pedestrian detector in the west. In: BMVC, vol. 2, p. 7 (2010)

[7] Dollar, P., Wojek, C., Schiele, B., Perona, P.: Pedestrian detection: An evaluation of the state of the art. IEEE Transactions on Pattern Analysis and Machine Intelligence 34(4), 743–761 (2012)

[8] Farenzena, M., Bazzani, L., Perina, A., Murino, V., Cristani, M.: Person re-identification by symmetry-driven accumulation of local features. In: 2010 IEEE Conference on Computer Vision and Pattern Recognition (CVPR), pp. 2360–2367. IEEE (2010)

[9] Felzenszwalb, P.F., Girshick, R.B., McAllester, D., Ramanan, D.: Object detection with discriminatively trained part-based models. IEEE Transactions on Pattern Analysis and Machine Intelligence 32(9), 1627–1645 (2010)

[10] Johnson, S., Everingham, M.: Learning effective human pose estimation from inaccurate annotation. In: 2011 IEEE Conference on Computer Vision and Pattern Recognition (CVPR), pp. 1465–1472. IEEE (2011)

[11] Kostinger, M., Hirzer, M., Wohlhart, P., Roth, P.M., Bischof, H.: Large scale metric learning from equivalence constraints. In: 2012 IEEE Conference on Computer Vision and Pattern Recognition (CVPR), pp. 2288–2295. IEEE (2012)

[12] Ouyang, W., Wang, X.: Single-pedestrian detection aided by multi-pedestrian detection. In: 2013 IEEE Conference on Computer Vision and Pattern Recognition (CVPR), pp. 3198–3205. IEEE (2013)

[13] Ouyang, W., Zeng, X., Wang, X.: Modeling mutual visibility relationship in pedestrian detection. In: 2013 IEEE Conference on Computer Vision and Pattern Recognition (CVPR), pp. 3222–3229. IEEE (2013)

[14] Paisitkriangkrai, S., Shen, C., Hengel, A.V.D.: Efficient pedestrian detection by directly optimizing the partial area under the roc curve. In: 2013 IEEE International Conference on Computer Vision (ICCV), pp. 1057–1064. IEEE (2013)

[15] Yan, J., Lei, Z., Yi, D., Li, S.Z.: Multi-pedestrian detection in crowded scenes: A global view. In: 2012 IEEE Conference on Computer Vision and Pattern Recognition (CVPR), pp. 3124–3129. IEEE (2012)

[16] Yang, Y., Ramanan, D.: Articulated pose estimation with flexible mixtures-of-parts. In: 2011 IEEE Conference on Computer Vision and Pattern Recognition (CVPR), pp. 1385–1392. IEEE (2011)

[17] Zeng, X., Ouyang, W., Wang, X.: Multi-stage contextual deep learning for pedestrian detection. In: 2013 IEEE International Conference on Computer Vision (ICCV), pp. 121–128. IEEE (2013)

Age Estimation Based on Convolutional Neural Network

Chenjing Yan[1], Congyan Lang[1], Tao Wang[1], Xuetao Du[2], and Chen Zhang[2]

[1] School of Computer and Information Technology, Beijing Jiaotong University, China
{13120439,cylang,twang}@bjtu.edu.cn
[2] China Mobile Group Design Institute Co., Ltd., Beijing, China
{duxuetao,zhangchen}@cmdi.chinamobile.com

Abstract. In recent years, face recognition technology has become a hot topic in the field of pattern recognition. The human face is one of the most important human biometric characteristics, which contains a lot of important information, such as identity, gender, age, expression, race and so on. Human age is a significant reference for identity discrimination, and age estimation can be potentially applied in human-computer interaction, computer vision and business intelligence. This paper addresses the problem of accurate estimation of human age. An age estimation system is generally composed of aging feature extraction and feature classification. In the feature extraction part, well-known texture descriptors like the Gabor wavelets and the Local Binary Patterns (LBP) have been utilized for the feature extraction. In our method, we use Convolutional Neural Network (CNN) to extract facial features. We gain the convolution activation features through building a multilevel CNN model based-on abundant training data. In the feature classification part, we divide different ages into 13 groups and use the Support Vector Machine (SVM) classifier to perform the classification. The experimental results show that the performance of the proposed method is superior to that of the previous methods when using our aging database.

Keywords: age estimation, convolutional neural network, SVM, feature extraction, classification.

1 Introduction

With the development of human-computer interaction (HCI), the research on facial image has been extensively carried out in many areas including image processing, pattern recognition and computer vision. The human face contains an amount of important information related to personal characteristics. The identity, emotional state, ethnic origin, gender, age, and head orientation of a person are all shown in a face image [1]. Among all these information, age is one of the most significant characteristics which is widely used in many applications, such as human-computer interaction, surveillance monitoring, and video content analysis. For example, an automatic age estimation system can not only improve the human–computer interface, but also prevent under ages from accessing cigarettes, alcohol, and pornographic websites. Therefore, facial age estimation has attracted increasing attentions from scholars in the field of computer vision and pattern recognition.

W.T. Ooi et al. (Eds.): PCM 2014, LNCS 8879, pp. 211–220, 2014.
© Springer International Publishing Switzerland 2014

As an interesting sub-problem of face recognition [2], age estimation is a very challenging problem for some reasons. Firstly, age estimation is a multi-class problem where each year can be classified into a single class. It is more difficult than two-class problem and cannot be directly solved by a binary classifier. This nature leads to an over-fitting problem when the photo database is not abundant enough [3]. Secondly, it's very difficult to collect people's facial image with different ages. Even though the FG-NET (Face and Gesture Recognition Research Network)[4]age face database has solved some problems, in which each person only has photos at some ages. More importantly, facial characteristics of human do not depend only on age, but are affected by many factors. In fact, gender differences, health status, living habits, living geography and weather conditions all have influences on facial characteristics (Figure 1). To solve all these problems, scholars from all over the world put forward a lot of unique opinions and solutions.

Fig. 1. They are all 30 years old. Gender differences, living geography and weather conditions all have influences on facial characteristics.

To realize the age estimation, several age estimation algorithms have been published in the past decade. Lanitis et al. [1] investigated 3 different classifiers for automatic age estimation. The face images in their database are only between 0-30, so it's hard to perform age estimation in the whole age group. Kwon and Lobo [5] classified facial images into three age groups: babies, young adults and senior adults using the distance ratio of facial components and the wrinkle features. But they used a very small database containing only 47 photos in their research. Geng et al. [6] proposed a subspace approach named AGES (Aging pattern Subspace) for automatic age estimation. However, this approach has high computational complexity and is difficult to be applied in real applications. All these approaches suffer from two problems: first, the training databases are relatively small; second, the age categories are not accurate enough. It therefore demands continuous attention on accurate age estimation.

Considering of this, this paper addresses the problem of accurate age estimation. The age estimation system is composed of aging feature extraction and feature classification. In the feature extraction part, we extract the feature using the convolutional neural network. Convolutional neural network uses restructuring and the surface of the same weight shared method incorporating the process of feature extraction in the multilayer perceptron (MLPs). We train the deep convolutional model using the same supervision way as Krizhevsky [7].And the deep convolutional activation features are also extracted through the network. This model shows good performance on the 1000 categories classification. So we assume that the activation features in the hidden layer of the network are also very useful in face recognition and age estimation. It is verified in our experiments. In the classification part, we divide all the ages into 13 groups with different labels, and use SVM classifier to perform the multi-label classification. Gabor filter is also considered in feature extraction and compared with the convolutional neural network in our experiments. The experimental results show that the proposed deep learning method provides better results than traditional methods. We put our algorithm into practice by building a real-time age estimation program using C++ language.

The rest of this paper is organized as follows: In Section 2, we briefly review the previous works on facial age estimation. In Section 3, the methods for facial feature extraction and classification are discussed. In Section 4, several experiments are shown to demonstrate the performance of our approach. Finally, conclusions and future works are outlined in Section 5.

2 Related Work

Facial image processing has been studied under a long term. And age estimation is one of the most interesting topics attracting increasing attention because of its great value. In the last decade, there were several age estimation algorithms published, and the goals of these algorithms can be separated into two categories: One is to estimate the actual age (e.g., 24-year old); the other is to classify a person into an age range, such as baby, teenager, middle-ager, or elder. In the following, we give a brief review of the previous works in age estimation.

In 1999, Kwon and Lobo [5] worked on the age estimation problem. They completed the estimation by classifying ages from facial images into 3 age groups as babies, young adults and senior adults. They calculated different facial ratios based on face features to distinguish babies from young adults and senior adults. Then they used wrinkles to distinguish seniors from babies and young adults. They used 47 pictures in their experiments and achieved good performance. But the categories were too few to put into practical application.

Hayashi et al. [8] divided all the ages into several age groups, with ten years in a group. They mainly used the wrinkle feature and complexion feature to finish age and sex recognition. They extracted the wrinkle feature using Hough Transform. Then they extracted the color feature of the face and performed the histogram equalization. Even though they divided the ages into detailed groups, they still didn't get the good result.

Lanitis et al. [1] proposed the approach firstly using Active Appearance Models (AAMs) [9], which combined shape and intensity variation in facial images. They used quadratic regression function to fit age function model. Because age estimation is a complex nonlinear regression problem and the age range of span is generally large, quadratic function isn't a good solution to estimate the age of all the facial images.

Based on Lanitis's work, Geng et al. proposed the aging pattern subspace (AGES) [6], which further considered the identity information and the ordinal relationship of ages during feature extraction. They proposed a personalized age estimation method that described the long-term aging process of a person and estimated his/her age by minimizing the reconstruction error. However, they still can't solve the problem that person's facial features could still be similar in different age ranges.

Yang and Ai [10] investigated demographic classification using age, gender and ethnicity information with Local Binary Pattern (LBP). Guo et al. [11] introduced an age manifold scheme and combined SVR and SVM to learn and predict human ages.

Choi S.E. et al. [12] used the method of combination of global and local features in the feature extraction part, according to the respective advantages and disadvantages of global and local features. Wei et al.[3] put forward a new method for age estimation, which taking the internal information into account. After feature extraction, they used relevant component analysis (RCA) for adjustment. And in dimension reduction phase they used LLPP dimension and MFA algorithm.

Taking all these into consideration, we use the convolutional neural network to extract facial feature in this paper. And then we use SVM classifier to realize the multi-label classification of age estimation.

3   Age Estimation

3.1   System Framework

The proposed system frame is illustrated in Figure 2. For every image presented to the system, preprocessing is applied to transform the image into the template used in database. Then we extract features with the convolutional neural network. After feature extraction process, the image is classified to one of the age classes based on the trained models. After that, we compute his\her accurate age according to the confidence degree by a linear function.

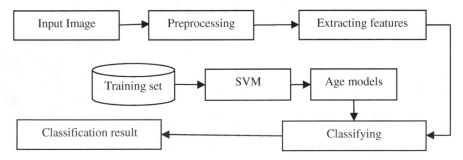

Fig. 2. General system structure

3.2   Feature Extraction

Convolutional neural network is one of the deep learning methods which is widely used in image recognition and speech analysis. In this paper, we use the convolutional neural network to extract features.

Basic Principle of Convolutional Neural Network
Convolutional neural network (CNN) is a multilayer neural network system. In each layer there are lots of two-dimensional planes which are composed of many independent neurons. At a given level, each neuron input comes from the local neighborhood of the previous level, combined with the weight of this level. CNN could have lots of layers and each layer possesses lots of feature maps. The output of one layer is the input of next layer. Figure 3 is a conceptual example of the CNN.

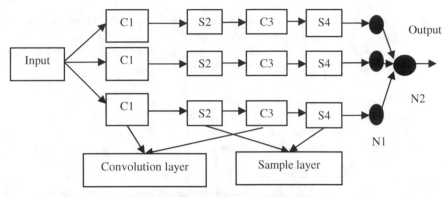

Fig. 3. The conceptual example of the CNN

There is an input layer (input), two convolution layer (C1, C3) and two sample layers (S2, S4). The input image is convolved and biased to produce three feature maps at the C1 level. Each group of four pixels in the feature maps are added, weighted, combined with a bias, and passed through a sigmoid function to produce the C3 level. The hierarchy then produces S4 in a manner analysis to S2. N1, N2 is fully connected layer, like recognizable neurons. Finally these pixel values are rasterized and presented as a single vector input to the convolutional neural network at the output.

Structure of Convolutional Neural Network
In our approach, a deep convolutional model is first trained in a fully supervised setting using a state-of-the-art method proposed by Krizhevsky et al.[7]. The difference between our method and Krizhevsky's method is that we used tight connection between the third, fourth and fifth layer while Krizhevsky used sparse connection.

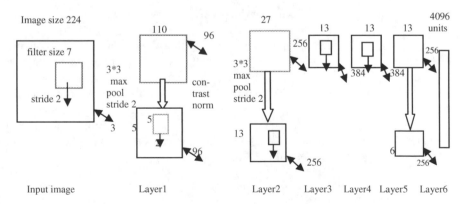

Fig. 4. Convolution neural network model

Figure 4 is structure of the convolutional neural network we used in this paper. There are eight layers in the convolutional neural network. Input layer requested the input images cut into 224×224. We use 96 convolution kernels each size is 7×7 to perform convolution on the input layer, getting the first layer----the convolution layer.

Then we use biases, pooling and contrast normalization to get the feature map of the first layer. And we can get the feature map of the 2,3,4,5 layer with the same operation. We get the sixth layer by connecting all the feature maps above it. By pooling we achieve the useful information with 4096 neurons.

Deep Convolutional Activation Features

In this paper, the deep convolutional activation features are also extracted through the network. We choose this model due to its performance on a difficult 1000-way classification task, hypothesizing that the activations of the neurons in its late hidden layers might serve as very strong features for a variety of object recognition tasks.

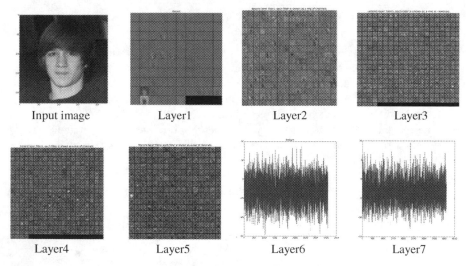

Input image   Layer1   Layer2   Layer3

Layer4   Layer5   Layer6   Layer7

Fig. 5. A Deep Convolutional Activation Feature

Figure 5 shows the feature maps of the first to the seventh layer of the convolutional neural network. The eighth layer of the network also is a fully connected layer and the decision layer. But it is meaningless to this paper, so we don't show its feature map. We describe the activation feature of the nth layer as DeCAFn. According to Jeff Donahue's [13] research, we know that the earlier convolution layers contain less semantic representation than the later layers. And the DeCAF6 feature is better than the other features both in objects recognition and objects classification. So we choose the DeCAF6 feature to perform our experiment.

3.3   Age Classification

Age estimation can be seen as a multi-class classification problem, so traditional classification algorithms such as the nearest centroid classifier [1, 14] and the support vector machine (SVM) can be directly applied for age estimation. After extracting features, we need to perform the classification using multi-label classification. We choose the SVM classifier to classify the facial image into the right age group.

4   Experiments

4.1   Age Databases

In the experiment, we use the WebFace database[15] with facial images ranging from 1 years old to 80 years old. We choose some frontal images of human face in good quality. We divide the people between 0 to 80 years old into 13 groups. In each group we choose 220 pictures (totally 2860 pictures) to form the training set and 56 pictures (totally 728 pictures) to form the testing set. And each group is separated into two parts by gender. If we can collect more pictures in good quality, the result of our experiment will be greatly improved. We extract the deep convolutional activation features of all the images, and use the SVM classifier to estimate age.

We first train the gender model, which can distinguish the gender with the accuracy of 98.1%. Then we train the age models separately for male and female. Based on the gender model and age models, we can obtain the final classification result.

4.2   Experimental Results and Analysis

In this part, we firstly show the result based on convolutional neural network.

Table 1. The confusion matrix of classification

	total	1	2	3	4	5	6	7	8	9	10	11	12	13	Accuracy
1	56	54	2	0	0	0	0	0	0	0	0	0	0	0	0.9643
2	56	1	54	0	0	0	0	1	0	0	0	0	0	0	0.9643
3	56	0	3	46	5	0	0	1	0	0	1	0	0	0	0.8214
4	56	1	0	6	40	8	0	0	0	1	0	0	0	0	0.7143
5	56	0	0	0	7	41	7	1	0	0	0	0	0	0	0.7321
6	56	0	1	0	0	7	39	9	0	0	0	0	0	0	0.6964
7	56	0	1	0	0	0	6	38	9	0	0	1	1	0	0.6786
8	56	0	0	0	0	1	0	4	39	11	0	0	0	1	0.6964
9	56	0	1	0	0	0	0	2	7	42	2	1	1	0	0.7500
10	56	0	0	0	0	1	0	1	0	8	38	5	3	0	0.6786
11	56	0	0	0	0	0	0	0	1	1	4	45	5	0	0.8036
12	56	0	0	0	0	0	0	0	1	0	0	7	38	10	0.6786
13	56	0	0	0	0	0	0	0	0	0	1	0	11	44	0.7857

The confusion matrix shows the result of classification. Bold red numbers stand for the images classified to the right group, while blue italic numbers represent the images classified to the neighboring groups. From Table 1 we can see that the result of the classification in this paper by using the convolutional activation features is good. The accuracy ranges from 67.86% to 96.43% in different groups and the average is 76.64%.

As we all know, age is hardly to estimate, even human can't estimate one's age accurately in many cases. So, we think it is acceptable if the algorithm predicts the age to its neighbor groups. We computed the total number of the right group and its neighbor groups. For example, when we compute the accuracy of 21~25 group, we count numbers which are predicted to 16~20 group, 21~25 group and 26~30 group together. The contrastive result is shown in Table 2 and Figure 6.

Table 2. DeCAF6 feature result and the accuracy considering neighboring groups

	1~4	5~10	11~15	16~20	21~25	26~30
Accuracy	0.9643	0.9643	0.8214	0.7143	0.7321	0.6964
Accuracy	1.0000	0.9821	0.9643	0.9643	0.9821	0.9821

	31~35	36~40	41~45	46~50	51~55	56~60	60~80
Accuracy	0.6786	0.6964	0.75	0.6786	0.8036	0.6786	0.7857
Accuracy	0.9464	0.9643	0.9107	0.9107	0.9643	0.9821	0.9821

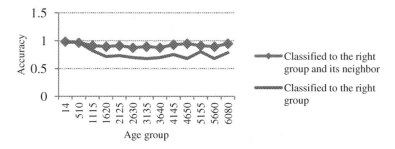

Fig. 6. Accuracy comparison

From Table 2 and Figure 6 we can see that the accuracy is greatly improved and almost all images are well classified in this case. Because the feature difference between the neighboring groups is small, the proposed algorithm may classify the image wrongly to its neighbor groups.

4.3   Experiment Comparison

In order to compare the result, we perform the experiment with the Gabor feature. We extract the Gabor feature of all the pictures and train the same gender and age models using the SVM classifier. Then we realize the age estimation based on the trained models. We compare the results between Gabor feature and DeCAF6 in Figure 7.

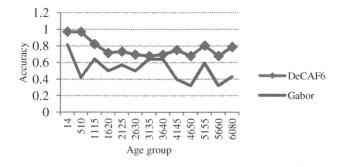

Fig. 7. Comparison of DeCAF6 feature and global Gabor feature

Based on Figure 7, we can see that the estimating accuracy of the DeCAF6 feature is higher than the accuracy of the Gabor feature. So we can conclude that the De-CAF6 feature based on convolutional network performs better on age estimation than the Gabor feature. The result of age estimation is improved by our method.

4.4   Experiment Application

In order to put the algorithm into application, we build an real-time automatic age estimation system using C++ language and OpenCV library. We gain real-time images by a computer's camera and extract the feature of each image based on the method proposed in this paper. After this we use predicting function in the LIBSVM library based on pre-trained models to classify the person in each image to the right age group. Finally, we compute accurate age according to the returned confidence degree by a linear function. Some run-time samples of our system are shown in Figure 8.

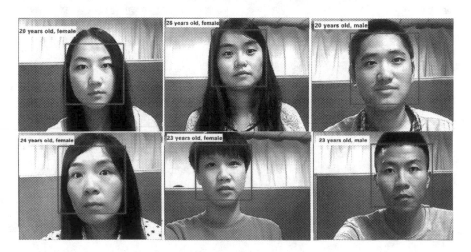

Fig. 8. Some run-time samples of the real-time age estimation system

5   Conclusion

In this paper, a new age estimation approach based on convolution activation features is proposed. We extract the features using the convolutional neural network which are effective for a variety of object recognition tasks. And we complete the age estimation based on gender. Experiments show that the proposed feature achieves better performance than the original Gabor feature. We also build a real-time age estimation system based on the proposed method.

In future, we intend to make further researches on age estimation and try to improve the accuracy by integrating key points of faces. We also hope to put our algorithm into more practical applications.

Acknowledgement. This work is supported by National Nature Science Foundation of China (No.61272352, No.61300071), Beijing Natural Science Foundation (No.4142045), the Fundamental Research Funds for the Central Universities (No.2014JBM025), Beijing Higher Education Young Elite Teacher Project (No.YETP0547), the Open Project Program of the National Laboratory of Pattern Recognition (NLPR) (No.201306286). China Mobile research fund of Ministry of Education (No. MCM20130421).

References

1. Lanitis, A., Draganova, C., Christodoulou, C.: Comparing different classifiers for automatic age estimation. IEEE Transactions on Systems, Man, and Cybernetics, Part B: Cybernetics 34(1), 621–628 (2004)
2. Zhao, W.Y., Chellappa, R., Phillips, P.J., Rosenfeld, A.: Face Recognition: A Literature Survey. ACM Computer Surveys 35(4), 399–459 (2003)
3. Chao, W.L., Liu, J.Z., Ding, J.J.: Facial age estimation based on label-sensitive learning and age-oriented regression. Pattern Recognition 46(3), 628–641 (2013)
4. The FG-NET Aging Database (2002),
   `http://sting.cycollege.ac.cy/~alnantis/fgnetaging.html`
5. Kwon, Y.H., Lobo, N.V.: Age classification from facial images. Computer Vis., Image Understand. 74(1), 1–21 (1999)
6. Geng, X., Zhou, Z.H., Smith-Miles, K.: Automatic age estimation based on facial aging patterns. IEEE Transactions on Pattern Analysis and Machine Intelligence 29(12), 2234–2240 (2007)
7. Krizhevsky, A., Sutskever, I., Hinton, G.E.: Image Net Classification with Deep Convolutional Neural Networks. In: NIPS, vol. 1(2), p. 4 (2012)
8. Hayashi, J., Yasumoto, M., Ito, H., et al.: Age and gender estimation based on wrinkle texture and color of facial images. In: Proceedings of the 16th International Conference on Pattern Recognition, vol. 1 (2002)
9. Cootes, T., Edwards, G., Taylor, C.: Active appearance models. IEEE TPAMI (2001)
10. Yang, Z.G., Ai, H.Z.: Demographic classification with local binary patterns. In: Proc. of International Conference on Biometrics, pp. 464–473 (2007)
11. Guo, G.D., Mu, G.W., Fu, Y., Huang, T.S.: Human age estimation using bio-inspired features. In: Proc. of the IEEE Conf. on Computer Vision and Pattern Recognition (CVPR 2009), pp. 112–119 (2009)
12. Choi, S.E., Lee, Y.J., Lee, S.J., et al.: Age estimation using a hierarchical classifier based on global and local facial features. Pattern Recognition 44(6), 1262–1281 (2011)
13. Donahue, J., Jia, Y.Q., Vinyals, O., et al.: Decaf: A deep convolutional activation feature for generic visual recognition. arXiv preprint arXiv:1310-1531 (2013)
14. Lanitis, A., Taylor, C.J., Cootes, T.F.: Toward automatic simulation of aging effects on face images. IEEE Transactions on Pattern Analysis and Machine Intelligence (PAMI) 24(4), 442–455 (2002)
15. Ni, B., Song, Z., Yan, S.: Web image mining towards universal age estimator. In: ACM MM (2009)

A Curvature Filter and Normal Clustering Based Approach to Detecting Cylinder on 3D Medical Model

Yuan Gao[1], Lifang Wu[1], Yuxin Mao[1], and Jinqiao Wang[2]

[1] School of Electronic Information and Control Engineering,
Beijing University of Technology, Beijing, China, 100124
[2] Chinese Academy of Science Institute of Automation, Bejing, China, 100190

Abstract. In this paper, we propose a cylinder detection approach based on curvature filtering and normal clustering. We first estimate the curvature of the vertexes on each triangle, reserve these triangles which have the characteristics of a cylinder, then the triangles are clustered by the normal. Then all the triangles are transformed onto a new coordinate system, which the Z axis is parallel to the normal. Finally, the cylinders are detected by the Hough transformation in the 2D plane. The experimental results show that our proposed algorithm has good performance.

Keywords: Cylinder Detection, 3D Model, curvature, normal.

1   Introduction

Cylinders are used widely in many kinds of 3D models, such as industrial models, pipeline transport models, and medical models. Therefore, cylinder detection is a key issue in such applications.

However, a 3D model is a set of 3D data, it is not easy to extract the cylinders from only the grid information of 3D models and even the points cloud data. Many researchers proposed some algorithms. Yong-Jin Liu[1]et al. proposed a hierarchical structure detection and decomposition method to detect Cylinders in Large-Scale Point Cloud of Pipeline Plant according to the structural characteristics of point cloud in pipeline plants. Their work reduced the difficult pipeline-plant reconstruction problem in IR^3 into a set of simple circle detection problems in IR^2. Dong-Ming Yan et al. [2] used Lloyd iteration to implement different quadric surface segmentation and it could extract the sides of cylinders.

In this paper, we mainly focus on detecting the pinhole in the shape of a cylinder on the medical 3D model as shown in Fig. 1. This kind of 3D model is small, but the number of triangles reaches the magnitude of $10^4 \sim 10^6$. Some cylinders on the 3D model are incomplete and difficult to detect. We first estimate the curvature of the vertexes on each triangle and filter out almost 60% triangles by curvature distribution of three vertexes. It reduces the much computation cost. Then the preserved triangles are clustered by their normals. The triangles are classified into several sub-clusters. Then each class of triangles is projected onto a plane perpendicular to their normal. Then the circles are detected by Hough transformation in 2D plane. These circles are mapped into the 3D model and the cylinder could be obtained.

W.T. Ooi et al. (Eds.): PCM 2014, LNCS 8879, pp. 221–226, 2014.

Fig. 1. The example 3D model and cylinders

2   The Proposed Algorithm

For a 3D model, we first calculate the discrete Gaussian curvature of each vertex on each triangle. Then all the triangles which include at least two vertices of relatively far from zero curvatures are preserved and others are discarded. then these preserved triangles are clustered into several clusters by their normals. Triangles in Each cluster are mapped into a planes perpendicular to their normal by coordinate transformation. Finally, the 2D Hough transformation is practiced to detect the circle in the plane. Finally, the cylinder is obtained from the circles by the inverse coordinate transformation. The framework of our algorithm is shown in Fig. 2.

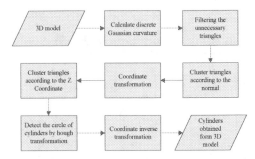

Fig. 2. Framework of the proposed algorithm

2.1   Filtering Out the Triangles by the Curvature of the Vertexes

We first get the topology of a model, which presents the relationship between triangles, vertexes and boundaries $M\{T,V,B\}$, where T is the triangles set, V is the vertexes set and B is the boundary set. There are many approaches to estimate curvature in 3D spaces, the popular one is Meyer's Voronoi algorithm. An improved Voronoi algorithm[3] is utilized to calculate Gaussian discrete curvature as shown in Equation (1)

$$K = \frac{2\pi - \sum \theta_i}{S} \tag{1}$$

Where, K is Gaussian discrete curvature of a vertex, as shown in Figure 3, θ_i is the angle between V_0 and V_{i+1}, S is the sum of triangles belonging V_0.

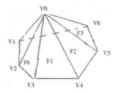

Fig. 3. Illustration of Gaussian discrete curvature calculation

In Fig. 3, if all the triangles related vertex v0 are in a plane, its Gaussian curvature is 0. Therefore, a vertex whose curvature is approximately equal to 0 should be removed. For a triangle, if two of its vertexes are far from zero, we think this triangle possibly is a triangle on the cylinder, they are preserved, and others should be discarded.

2.2   Clustering by Normal

On the 3D model, there are several cylinder, some of cylinders have the same direction. All the triangles on these cylinders have the similar normals. Therefore, all the preserved triangles are clustered by their normal. Let's suppose there is total Tc preserved triangles.

The normal of the first preserved triangle T_1 is set the normal of the first cluster. Suppose, by the ith triangle, there are total M clusters.

We compute the angles between the normal f_i of the ith triangle and all the known cluster normals $f_1, \ldots f_M$.

$$Angle(T_i, Cm) = \arccos(\frac{f_i \bullet f_m}{|f_i| \times |f_m|}) \quad m=1,2,\ldots M \tag{2}$$

If one of angles is smaller than an assigned threshold, this triangle is set to the corresponding cluster. Otherwise, a new cluster ((M+1)th cluster)is generated, and its normal is the normal of the ith triangle. Finally, all the preserved triangles are clustered. Without loss of generality, let's suppose we have total N clusters, the nth cluster includes pn triangles.

2.3   Coordinate Transformation

For the nth cluster, all the triangles in this cluster are transformed into a plane perpendicular to the normal of the nth cluster. In the new coordinate system, only the Z axis is determined by the normal of the cluster. Let's suppose the normal is represented as $\{\alpha, \beta, \gamma\}$, and the new origin is (x_0, y_0, z_0) in the original coordinate. We could get the transformation matrix T_i represented as Equation (3).

$$T_i = \begin{bmatrix} \dfrac{\alpha \bullet \beta}{\sqrt{\alpha^2 + \beta^2}} & \dfrac{\beta \bullet \gamma}{\sqrt{\alpha^2 + \beta^2}} & -\sqrt{\alpha^2 + \beta^2} & 0 \\[3mm] -\dfrac{\beta}{\sqrt{\alpha^2 + \beta^2}} & \dfrac{\alpha}{\sqrt{\alpha^2 + \beta^2}} & 0 & 0 \\[3mm] \alpha & \beta & \gamma & 0 \\[3mm] x_0 & y_0 & z_0 & 1 \end{bmatrix} \tag{3}$$

For a triangle vertex (x_0, y_0, z_0), the new coordinates (x',y',z') could be computed from Equation (4).

$$[x', y', z', 1] = [x, y, z, 1]T_i^{-1} \tag{4}$$

Where T_i^{-1} is the inverse matrix of the T_i. Figure 4 show three example clusters after coordinate transformation. it is obvious that the cluster in Fig 4(a) include some real cylinders but the clusters in Fig 4(b) do not include the cylinders.

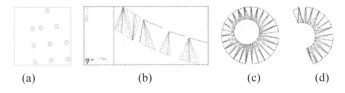

(a)　　　　　　　　(b)　　　　　　　　(c)　　　　(d)

Fig. 4. The projection of coordinate transformation in XOY plane. (a) The cluster with cylinders, (b) the cluster without cylinders. (c) The sub-cluster includes a circle from (a), The sub-cluster includes part of a circle form (a).

Furthermore, a cluster could be divided into several sub-clusters by their z coordinates. Two example sub-clusters are shown in Fig 4(c), (d).

2.4 Hough Transformation and Cylinder Detection

By now, the original cylinder detection in 3D space is transferred into a circle detection in the 2D plane. An improved Hough Transformation[5] is utilized to detect the circle in the plane. From the transformation, we could get the center (cx, cy, cz) of the circle, cz is the z coordinate of the corresponding sub-cluster. The center is transformed in the original 3D space from the Equation (5)

$$[Cx, Cy, Cz, 1] = [cx, cy, cz, 1]T_i \tag{5}$$

3 Experiments

In out experiments, the model includes 378000 triangles and 32 cylinders, as shown in Fig6. The program is run on a desktop PC with Intel(R) Core(TM) I3 CPU, running at 3.30 GHz with 2-GB RAM.

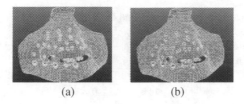

Fig. 5. The cylinder detection results. (a) the input model. (b) The results of cylinder detection.

Our approach detects total 32 cylinders. 29 of them are correct cylinders as shown in Figure 5(b). Two of them are false cylinders as shown in Fig 6(a), (b). Their back part is smoothing. Therefore, the cylinders are detected from the opposite again. Three cylinders are missed as shown in Fig. 6(c), (d), (e). These cylinders are influenced by their neighboring cylinders. Therefore, they are missed.

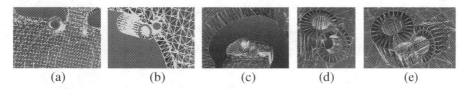

Fig. 6. The missed cylinders

4   Conclusion and Future Work

In this paper, we propose a cylinder detection algorithm based on curvature filtering and normal clustering. Vertexes curvature in a triangle is used to filter out most of unrelated triangles. Then the preserved triangles are clustered by their normal. All the triangles in a cluster are transformed into a plane perpendicular to the normal, and the circle in the plane is detected from the Hough transformation. Finally, the cylinders' coordinates are obtained by transforming the circle coordinate into the 3D space. The experimental results demonstrate that our approach could correctly detect 29 of 32 cylinders. But there are also two false cylinders and three missed cylinders. The future work will focus on the cylinder detection in complex 3D model, in which some cylinders are influenced by others.

References

1. Liu, Y.-J., Zhang, J.-B., Hou, J.-C., Ren, J.-C., Tang, W.-Q.: Cylinder Detection in Large-Scale Point Cloud of Pipeline Plant. IEEE Transactions on Visualization and Computer Graphics 19, 1700–1707 (2013)
2. Yana, D.-M., Wang, W., Liu, Y., Yang, Z.: Variational Mesh Segmentation via Quadric Surface Fitting. Computer-Aided Design 44(11), 1072–1082 (2012)

3. Xu, S., Mei, X., Geng, J.: Matching method of 3D model based on curvature characteristics of teeth. Computer Engineering and Design 34(8) (2013)
4. Xu, D.: 3D graphics coordinate transformation between the two kinds of coordinate systems. Journal of XiDian University 23(3) (1996)
5. Xiao, L., Cai, C., Zhou, C., Ding New, M.: fast algorithm of Hough transform detection of circles. Application Research of Computer 24(10) (2007)

Image Compositing Based on Hierarchical Weighted Blending

Huihui Wei and Qimin Cheng*

The Department of Electronics and Information Engineering/Wuhan National
Laboratory for Optoelectronics, Huazhong University of Science and Technology,
Wuhan 430074, China
chengqm@hust.edu.cn

Abstract. Recent image compositing methods mainly focus on the compositing for normal images, while for shadow ones, these methods may be less effective due to the special structure in the shadow area. Besides, many of these methods may generate problems of serious color distortion or cannot realize seamless blending. In order to improve these problems, we propose a new hierarchical weighted method based on an alpha matte for image composition, especially for those with shadows. In our method, we divide the blending area into different layers according to the alpha matte, and implement a hybrid method combining gradient based method with transformed alpha blending as well as weights in these layers respectively. By conducting a series of experiments, we demonstrate the superiority of our proposed method.

Keywords: image compositing, hierarchical blending, weight, shadow.

1 Introduction

Image compositing has been a hot research topic in the field of digital image processing. It is widely used in image editing, animation, special film effects, and also an important technique which can assist image retrieval, object detection, visual tracking, etc [1].

Existing image compositing methods are mainly divided into two categories: gradient based and alpha blending. Before compositing, foregrounds need to be extracted from source images. According to the value of the extracted mask, the extracting form can be divided into hard extraction (mask with value 0 or 1) and soft extraction (alpha matte with values 0, 1 and between both).

Gradient based methods mainly use hard extraction [8–11, 23] to obtain the foreground. In Gradient based methods [8, 16, 17, 20], the blending problem was transformed into solving the Poisson equation for blending image seamlessly. However, this kind of methods may lead to the serious color distortion problem. To solve this problem, [19] provided an edit tool by integrating Poisson image editing for the user to conveniently manipulate the appearance of the source image. [24] proposed a hybrid method by combining MVC(Mean-Value Coordinates) image cloning method with alpha blending.

---

* Corresponding author.

W.T. Ooi et al. (Eds.): PCM 2014, LNCS 8879, pp. 227–237, 2014.

While alpha blending methods use soft extraction technology (matting) to get the alpha matte, and these methods usually need users to provide some constraints first. Usually, the constraints may be a trimap [7,12–14] or scribbles [6,15]. Once the alpha matte is ready, the blending result I can be obtained by following

$$I = \alpha * f_F + (1 - \alpha) * f_B \tag{1}$$

Where f_F and f_B stand for the foreground and background of the image respectively, and α represents the opacity of the foreground. Alpha blending fits for the situation that the new background shares very similar features with that of the source images, while obvious blending artifacts will occur if not.

Most of the above methods mainly focus on the compositing problems for normal images, while for shadow ones, the problem has seldom been mentioned how to solve efficiently, which has been also proposed as a future work in [5]. Inspired by the recent compositing methods, we propose a hybrid hierarchical weighted method based on an accurate alpha matte. However, unlike the previous work of this kind, we concentrate our research on how to realize a seamless blending as well as decrease the color distortion problem. Besides, we focus mainly on the shadow image compositing problem rarely involved in recent works. This method can also be further used in many image processing tasks, such as image retrieval [2] and classification [3,4].

In our work, because of the special structure of the shadow images, accurate masks are essential to be obtained first. By hierarchical, we mean the blending area is divided into different layers including outer, border and inner region according to the alpha matte. As for weighted, we modify the gradient based method by adding a weight in the inner region. Compared with previous methods, the benefits of our proposed method outstand in two aspects. Firstly, by implementing a hybrid method in the border region, the source image and the new background can be blended seamlessly for their good continuity in both color and texture. Secondly, weight adjustment for the inner region makes it possible that the source image preserves its original color as much as possible to avoid the serious color distortion.

2   Hierarchical Weighted Image Compositing

In out method, an accurate alpha matte is needed in the first step, how to get the accurate alpha matte will be discussed later. The main procedure is as $Fig.1$ shows, after we input the source image and obtain its precise alpha matte, we can conduct our hierarchical weighted method following the main scheme shown as the dashed rectangle. Firstly, we divide the blended area into outer, border and inner region respectively according to the alpha matte. Then, in the border region, we conduct a hybrid method of combining transformed alpha blending and gradient based method. For the inner region, we first calculate the illumination contrast coefficient of the source and target image around the border region, and use this coefficient to define the weight for the weighed blending. Finally, we can get a new satisfying composited result.

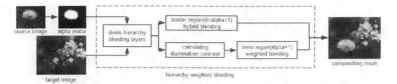

Fig. 1. Procedure and main scheme of our proposed method

2.1   Obtaining Accurate Blending Area

Since extracted masks by hard extraction are binary, when the source is a shadow image, if the shadow area is merely marked as background, the composited results will lose features in shadow area completely; otherwise, serious mismatch may occur in the shadow area of composited results. So, we choose the soft extraction method to obtain the accurate extracted mask (alpha matte) for the next compositing.

Shahrian E et al. [7] proposed an excellent matting method, however, all the experiments were conducted on normal images, so, this method may not be so accurate for shadow ones. However, considering robust matting [6] can further refine the alpha matte, we choose [7] to extract the preliminary alpha matte first, and then adjust the details by robust matting [6] to obtain a more precise one for shadow images. $Fig.2$(c-f) shows an example that we use hard extraction(graph cut) and soft extraction methods including Bayesian matting [12], sampling [7] and detailed adjusting by robust matting [6] to get the alpha matte for the shadow image ($Fig.2$(a)) respectively.

2.2   Hierarchical Image Compositing Solution

According to the value α of the alpha matte, we divide the blending area into three regions: outer ($\alpha = 0$), border ($0 < \alpha < 1$) and inner ($\alpha = 1$) region respectively, as Fig.3 shows. In our method, the gradient based method derives from Poisson editing theory [8], in which the compositing process was modeled as the interpolation in a directed vector field. Supposing the function of the result image is f_p, the membrane interpolation in the blending progress is equivalent to a minimization problem.

$$\min_{f_p} \iint_{\Omega} |\nabla f_p - v|^2 \quad \text{with} \quad f_p\,|_{\partial\Omega} = f^*\,|_{\partial\Omega} \tag{2}$$

Where, v is the guidance vector field for the source image, Ω and $\partial\Omega$ correspond to the blended interior and border area respectively, and f^* is defined as the function of the target image. Because the function definition in the border of the composited image is equal to that of the target image, the composited results can be seamlessly blended.

However, method following this theory also led to the color distortion problem. We improved this problem by adding different weights to keep the color fidelity.

(a)　　　　　　　　(b)

(c)　　　　　(d)　　　　　(e)　　　　　(f)

Fig. 2. (a) shadow image *pillar*. (b) trimap of (a). (c) hard extraction by graph cut, all shadow area has been labeled as foreground. (d) alpha matte by Bayesian matting [12], too much noise in shadow area and boundary. (e) comprehensive sampling method [7], no description of shadow area. (f) our result by adjusting, more smooth and refined both in the shadow area and boundary.

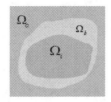

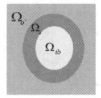

Fig. 3. Different regions in blending area. **Fig. 4.** Luminance contrast sampling. Ω_{ib} $\Omega_o(\alpha = 0)$, $\Omega_b(0 < \alpha < 1)$, and $\Omega_i(\alpha = 1)$ corresponds to regions ($\alpha > 0$) including represents for the outer, border and inner both border and inner region in source and region respectively. target image, Ω_s is the luminance contrast sampling area.

In the border region, we use a transformed alpha blending method based on the original alpha blending equation in formula (1), we substitute f_F for Poisson editing result f_p. Thus, the improved alpha blending can be defined as

$$I_b = \alpha * f_p + (1 - \alpha) * f_B \tag{3}$$

Because the composited result in the border region I_b is the sum of Poisson editing result and background following different weights, the source and target image can be blended seamlessly in the boundary as well as keep good continuity in both color and texture. While for the inner region, to solve the problem of color distortion as well as realize a smooth blending transition between the border and inner region, we control the composited result I_i by adding weights to the Poisson editing result f_p and the foreground f_F respectively. The weighted blending solution can be expressed as

$$I_i = (1 - w) * f_p + w * f_F \tag{4}$$

w stands for the color fidelity weight, if we set w to

$$w = (1 - \beta) \tag{5}$$

Then, the final compositing result f can be given by

$$f = \begin{cases} f_B & (\alpha = 0) \\ \alpha * f_p + (1 - \alpha) * f_B & (0 < \alpha < 1) \\ \beta * f_p + (1 - \beta) * f_F & (\alpha = 1) \end{cases} \tag{6}$$

2.3 Blending Weight by Illumination Contrast Solution

Now, we discuss how to define the weight coefficient w in (4). To obtain a proper weight, we first introduce the luminance contrast coefficient, and use it to adjust the weight for an optimized value. As $Fig.4$ shows, Ω_{ib} is the blending area ($\alpha > 0$) consisting of both border and inner region, Ω_s is the luminance sampling region, which is also the outer region band with equally wide and closest to Ω_{ib}. The progress for calculating luminance contrast coefficient is as the following steps: firstly, transform the source and target image into IHS space; secondly, extract the luminance component in the corresponding Ω_s region of both the source and target image and mark them as I_s and I_t respectively. Then, the luminance contrast coefficient c can be defined as

$$c = average(I_s)/average(I_t) \tag{7}$$

According to (7), if the source is brighter than the target, $c > 1$, otherwise, $c < 1$. In order to cast the weight in the proper range between 0 and 1, we use:

$$w = \begin{cases} min(1, max(\sigma 1 * c + \sigma 2, \sigma 1 * 1/c + \sigma 2)) & (c > \tau \text{ or } 1/c > \tau) \\ \delta & (otherwise) \end{cases} \tag{8}$$

Through experiments, we find that $\tau = 2, \delta = 0.6, \sigma 1 = 0.25, \sigma 2 = 0.1$ works well in practice. For example, in $Fig.5$, the proper w computed according to formula (7,8) is $0.72(Fig.5(f))$, it shows that too small weight($Fig.5(b - e)$) cannot decrease the color penetration, while too big weight($Fig.5(g)$) may cause unnatural transition between the border and inner region, and our solution can make a better balance and achieve a more optimized result.

(a)   (b)

(c) w=0.25   (d) w=0.45   (e) w=0.6   (f) w=0.72   (g) w=0.9

Fig. 5. Weight influence on compositing results. When the source image (a) is to be pasted onto the target image (b), different w(weight) may influence the compositing result in foreground color and natural blending between the border and inner region (b-g), and our result gives a better balance (f).

Fig. 6. Water bird. a shadow image compositing example under the similar color condition. (a) source image. (b) alpha matte of the source image by [6, 7]. (c) target image. (d) direct pasting. (e) Poisson editing [8]. (f) drag-and-drop pasting [16]. (g) IHS transform [18]. (h) alpha blending using alpha matte by [7]. (i) our method(using alpha matte in (b)). (j) enlarged view.

3   Experimental Results and Discussion

In this section, we show the experiments on a series of different source and target images, we compare our proposed method with direct pasting, Poisson editing [8], drag-and-drop pasting [16], IHS transform [18] and alpha blending to do the contrast analysis. Because color and texture are two main factors that will influence the final compositing results, we also conduct our experiments analysis on image pairs from the two factors respectively.

Color Factor Analysis. $Fig.6$ shows an example that the source shadow image ($Fig.6(a)$) and the target image ($Fig.6(c)$) share similar color tone in the background, the enlarged part ($Fig.6(j)$) shows blending artifacts around the boundary by Poisson editing [8] ($Fig.6(e)$). By drag-and-drop pasting ($Fig.6(f)$) [16], the feather of the bird changed. Result by IHS transform ($Fig.6(g)$) [18] shows even greater changes in the color. In ($Fig.6(h)$) (alpha blending result), the bird's shadow in the water almost vanished. Result by our method is in ($Fig.6(i)$), compared with Poisson editing (as enlarged part in ($Fig.6(j)$) shows), less artifacts occur and the composited result looks more real.

When the color between the background of the source and target image differs a lot, as $Fig.7$ shows. $Fig.7$(e-i) are the composited results by different methods mentioned above respectively. It shows Poisson editing ($Fig.7(e)$) produces obvious artifacts at the blending boundary, inconsistent color results appear both in Poisson editing ($Fig.7(e)$) and drag-and-drop pasting ($Fig.7(f)$). color

Fig. 7. Land bird. a shadow image compositing example with the condition of great color differences. (a) source image. (b) alpha matte by [6,7]. (c) target image. (d) direct pasting. (e) Poisson editing [8]. (f) drag-and-drop pasting [16]. (g) IHS transform [18]. (h) alpha blending using alpha matte by [7]. (i) our method(with alpha matte in (b)).

Fig. 8. Umbrella. a shadow image compositing example under the different texture condition. (a) source image. (b) alpha matte using [6,7]. (c) target image. (d) direct pasting. (e) Poisson editing [8]. (f) drag-and-drop pasting [16]. (g) IHS transform [18]. (h) alpha blending using alpha matte by [7]. (i) our method(with alpha matte in (b)).

of composited result distorted seriously by IHS transform ($Fig.7(g)$), and alpha blending ($Fig.7(h)$) shows color mismatch between the composited result and the grass land. However, our proposed method ($Fig.7(i)$) not only reserves the color of the bird better, but also keeps the characteristics of the shadow area in good continuity with that of the grass land.

Texture Factor Analysis. $Fig.6$ and $Fig.7$ show the example that the textures between the image pairs are similar. While, $Fig.8$ shows the different condition, the shadow of the source image($Fig.8(a)$) has great texture difference with the desert($Fig.8(c)$). Results by Poisson editing($Fig.8(e)$) [8] and drag-and-drop pasting($Fig.8(f)$) [16] show color distortion as well as texture mismatch. IHS transform method ($Fig.8(g)$) [18] produces obvious changes in color. Composited result

Fig. 9. Pillar. another shadow image compositing example under the same condition with $Fig.8$. (a) source image. (b) alpha matte by [6,7]. (c) target image. (d) direct pasting. (e) Poisson editing [8]. (f) drag-and-drop pasting [16]. (g) IHS transform [18]. (h) alpha blending using alpha matte by [7]. (i) our method(alpha matte for compositing is as (b) shows).

by alpha blending is in $Fig.8(h)$, artifacts in the shadow area are too obvious. In $Fig.8(i)$, our proposed method gives a more acceptable result that looks more real. $Fig.9$ gives another compositing example with the same condition.

Discussion. Our method is also useful for normal image compositing, as $Fig.10$ shows. $Fig.10(d-i)$ are the results by different methods including ours, it shows that by our proposed method, the color tone of the tower can keep a better match with the surroundings, and also the color continuity of the tower itself is reserved well.

Evaluations. It's hard to give accurate assessment on the compositing results depending highly on the human visual perception. However, in [18], the objective index of AG (average gradient) in the field of image fusion is first used for image composition. While some works, for example, [22] chose the subjective assessment by joining the strength of volunteers for the quality assessment which depends much on human factor. Impressed by [18,22], we take both of the two evaluations to assess our methods on the experimented images above. The objective index AG in [21] is defined as

$$\bar{G} = \frac{1}{m*n} \sum_{i=1}^{m-1} \sum_{j=1}^{n-1} \sqrt{[f(i,j) - f(i+1,j)]^2 + [f(i,j) - f(i,j+1)]^2} \quad (9)$$

According to [21], the greater $\bar{G}$ is, the clearer an image is, the average gradient contrast results of the above experimental images are shown in $Tab.1$.

Fig. 10. Tower. a normal image compositing example. (a) source image. (b) alpha matte by [7]. (c) target image. (b) direct pasting. (e) Poisson editing [8]. (d) drag-and-drop pasting [16]. (e) IHS transform [18]. (f) alpha blending with alpha matte by [7]. (g) our method.

Table 1. Average gradient(AG) contrast among different methods

Methods Images	Poisson editing	Drag-and-drop	IHS transform	Alpha blending	Ours
Water bird	11.156	11.153	11.421	11.197	11.433
Land bird	8.757	8.6271	9.6542	9.2671	9.9385
Umbrella	4.7652	5.6059	5.9165	5.0355	5.9491
Pillar	2,8703	2.8785	2.9207	2.8726	3.0313
Tower	7.4915	7.5506	7.9238	8.5509	8.4969

Table 1 shows the values of AG in the last column (our method) are bigger than other columns in most cases, in other words, our proposed method can give a clearer new composited image to some extent.

Simultaneously, for subjective evaluation, we invited some volunteers to take part in our study. For each group of composited results, we do not indicate which result is obtained by which method and order the results randomly. The participants can give scores ranging from 1(the worst) to 5(the best) to each result, then, we gather these scores and compute the average score for each method, as $Fig.11$ shows.

The diagram above shows the average score of our method is higher than that of other methods for each example, which means the composited results by our method look more real than other methods referred in this paper, this further demonstrates that results by our proposed method have a better accessibility for human perception.

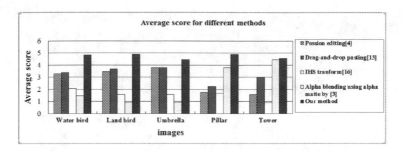

Fig. 11. Contrast of average score for different methods

4   Conclusion

Many of the recent methods for image compositing may lead to color distortion or artifacts, also seldom involved shadow image compositing. In this work, we propose a hierarchical image compositing method based on alpha matte to mark different blending (outer, border and inner) areas repectively, and use a hierarchical weighted method by combining Poisson editing and transformed alpha blending as well as weights based on the luminance contrast ratio in different layers. We also implement a series of experiments to compare our proposed method with other state-of-art methods. The experiment results show that our proposed method can realize a successful image compositing for both shadow images and normal ones, and the objective and subjective evaluations further demonstrate the effectiveness of our proposed method.

References

1. Gao, Y., Wang, M., Zha, Z., Shen, J., Li, X., Wu, X.: Visual-Textual Joint Relevance Learning for Tag-Based Social Image Search. IEEE Transactions on Image Processing 22(1), 363–376 (2013)
2. Gao, Y., Ji, R., Liu, W., Dai, Q., Hua, G.: Weakly Supervised Dictionary Learning with Attributes. IEEE Transactions on Image Processing (2014)
3. Gao, Y., Wang, M., Tao, D., Ji, R., Dai, Q.: 3D Object Retrieval and Recognition with Hypergraph Analysis. IEEE Transactions on Image Processing 21(9), 4290–4303 (2012)
4. Yang, Y., Wang, X., Guan, T., Shen, J., Yu, L.: A Multi-dimensional Image Preference Prediction Model for User Generated Images in Social Networks. Information Sciences 281, 601–610 (2014)
5. Sunkavalli, K., Johnson, M.K., Matusik, W., et al.: Multi-scale image harmonization. ACM Transactions on Graphics (TOG) 29(4), 125 (2010)
6. Wang, J., Cohen, M.F.: Optimized color sampling for robust matting. In: IEEE Conference on Computer Vision and Pattern Recognition, CVPR 2007, pp. 1–8. IEEE (2007)
7. Shahrian, E., Rajan, D., Price, B., et al.: Improving Image Matting Using Comprehensive Sampling Sets. In: 2013 IEEE Conference on Computer Vision and Pattern Recognition (CVPR), pp. 636–643. IEEE (2013)

8. Prez, P., Gangnet, M., Blake, A.: Poisson image editing. ACM Transactions on Graphics (TOG) 22(3), 313–318 (2003)
9. Boykov, Y., Veksler, O., Zabih, R.: Fast approximate energy minimization via graph cuts. IEEE Transactions on Pattern Analysis and Machine Intelligence 23(11), 1222–1239 (2001)
10. Li, Y., Sun, J., Tang, C.K., et al.: Lazy snapping. ACM Transactions on Graphics (ToG) 23(3), 303–308 (2004)
11. Rother, C., Kolmogorov, V., Blake, A.: Grabcut: Interactive foreground extraction using iterated graph cuts. ACM Transactions on Graphics (TOG) 23(3), 309–314 (2004)
12. Chuang, Y.Y., Curless, B., Salesin, D.H., et al.: A bayesian approach to digital matting. In: Proceedings of the IEEE Computer Society Conference on Computer Vision and Pattern Recognition, CVPR 2001, vol. 2, pp. II-264–II-271. IEEE (2001)
13. Zheng, Y., Kambhamettu, C.: Learning based digital matting. In: 2009 IEEE 12th International Conference on Computer Vision, pp. 889–896. IEEE (2009)
14. Chen, Q., Li, D., Tang, C.K.: KNN matting. In: 2012 IEEE Conference on Computer Vision and Pattern Recognition (CVPR), pp. 869–876. IEEE (2012)
15. Levin, A., Rav Acha, A., Lischinski, D.: Spectral matting. IEEE Transactions on Pattern Analysis and Machine Intelligence 30(10), 1699–1712 (2008)
16. Jia, J., Sun, J., Tang, C.K., et al.: Drag-and-drop pasting. ACM Transactions on Graphics (TOG) 25(3), 631–637 (2006)
17. Farbman, Z., Hoffer, G., Lipman, Y., et al.: Coordinates for instant image cloning. ACM Transactions on Graphics (TOG) 28(3), 67 (2009)
18. Ding, Y., Wei, X., Pang, H., et al.: A new framework for the fusion of object and scene based on IHS transform. In: 3rd International Conference on Digital Image Processing, pp. 80091W–80091W-6. International Society for Optics and Photonics (2011)
19. Bie, X.H., Huang, H.D., Wang, W.C.: Free appearance-editing with improved poisson image cloning. Journal of Computer Science and Technology 26(6), 1011–1016 (2011)
20. Wang, D., Jia, W., Li, G., Xiong, Y.: Natural image composition with inhomogeneous boundaries. In: Ho, Y.-S., et al. (eds.) PSIVT 2011, Part II. LNCS, vol. 7088, pp. 92–103. Springer, Heidelberg (2011)
21. Qu, G., Zhang, D., Yan, P.: Information measure for performance of image fusion. Electronics Letters 38(7), 313–315 (2002)
22. Hays, J., Efros, A.A.: Scene completion using millions of photographs. ACM Transactions on Graphics (TOG) 26(3), 4 (2007)
23. Price, B.L., Morse, B., Cohen, S.: Geodesic graph cut for interactive image segmentation. In: 2010 IEEE Conference on Computer Vision and Pattern Recognition (CVPR), pp. 3161–3168. IEEE (2010)
24. Ding, M., Tong, R.F.: Content-aware copying and pasting in images. The Visual Computer 26(6-8), 721–729 (2010)

Multifold Concept Relationships Metrics

Wenyi Cao[1], Richang Hong[1], Meng Wang[1], and Xiansheng Hua[2]

[1] Hefei University of Technology, HeFei, China
{wenyicao2012,hongrc.hfut,eric.mengwang}@gmail.com
[2] Microsoft Research, Redmond
xshua@microsoft.com

Abstract. How to establish the relationship between concepts based on the large scale real-world click data from commercial engine is a challenging topic due to that the click data suffers from the noise such as typos, the same concept with different queries etc.

In this paper, we propose an approach for automatically establishing the concept relationship. We first define five specific relationships between concepts and leverage them to annotate the images collected from commercial search engine. We then extract some conceptual features in textual and visual domain to train the concept model. The relationship of each pairwise concept will thus be classified into one of the five special relationships. Experimental results demonstrate our proposed approach is more effective than Google Distance.

Keywords: Auto visual concept net, visual conceptual feature, concept relationship, similarity measurement.

1 Introduction

To explore semantic relationship between concepts has been a hot research topic, since it has been widely applied on natural language processing, object detection, and multimedia retrieval. However, important as it can be, a rapid and robust approach that can discover the complex concept relationship and apply to a large-scale real-world network environment has not been proposed in the industry and the academia.

Some artificial attempts, such as the Cyc project[5] and the WordNet project[7] , have been carried out. However, To design such network for distinguishing the complex concept relationship and to fill up the network by linguistic experts are some long term and labor-intensive works. Although great efforts have been cost to enlarge the scale of concept network, the concept database is still limited and difficult to update comparing to the overall information on the web.

Some approaches to automatically explore the relationship by means of measuring the difference of two concepts, such as Normalized Google distance[1] and Flickr distance[9], have been proposed in recent years. These approaches cost little human effort and can cover far more concepts in the web. However, they only measure the distance of pairwise concepts and can't recognize the kind of concept relationship, so this type of approach is limitedly applied in a variety

W.T. Ooi et al. (Eds.): PCM 2014, LNCS 8879, pp. 238–247, 2014.
© Springer International Publishing Switzerland 2014

of application environment. So it is promising to put forward an approach to recognise the different relationship like hyponymy and meronymy.

In this paper, we propose the Multifold Concept Relationships Metrics (MCRM) approach to explore multifold concept relationships based on the Clickture[3], which collects the user's click data from the commercial search engine. Since Clickture is a large-scale image-query-click database, it can effectively simulate the real-world application environment to explore the concept relationship. In this approach, we will define five specific concept relationships and extract 8-dimensional conceptual features to distinguish the different concept relationships. The 8-dimensional conceptual features are defined with three levels according to Clickture, including textual information, visual information, the co-occurrence of images. Once extracting the conceptual features is accomplished, we will learn a multiple classifier to classify the concept relationship between two concepts. The contributions of the paper are summarized as follows:

* To the best of our knowledge, we are the first to give multifold definitions of the concept relationship in visual domain and to perform a comprehensive study on the role of conceptual features in classifying the concept relationship.
* We propose a rapid and effective approach to establish the large-scale multifold visual concept relationship network.

The rest of the paper is organized as follows. Section 2 presents some related works. Section 3 elaborates these conceptual features separately. In section 4 we will introduce our multiple classifier for classifying the different concept relationships. Section 5 presents the experimental results. Section 6 concludes the paper.

2   Related Work

2.1   Google Distance

Google distance is proposed to compute the distance between two concepts derived from the number of hits returned by Google search engine when querying both concepts. The concept relationship with the same or similar meanings in a natural language sense tend to be "close" in units of Google distance, while concepts with dissimilar meanings tend to be farther apart.

The normalized Google distance derives from the earlier normalized compression distance (Cilibrasi & Vitanyi 2003)[2]. NGD can ideally measures the textual conceptual relationship, i.e. the same and similarity, when the pairwise concepts frequently occur in the same web page. However, NGD is insufficient to measure the concept relationship among daily life, since the knowledge of concept relationship stems from human perception and 80% of human cognition comes from visual information.

2.2   Flickr Distance

In order to make up for the lack of Normalized Google Distance, Lei Wu et .al. propose the Flickr Distance to explore the relationship between two concepts in visual domain by measuring the square root of Jensen-Shannon divergence between the visual language models corresponding to these two concepts. If two concepts more frequently appear in the same image, the relationship is more similarity and the Flickr distance is small, otherwise dissimilarity and large.

Since the Flickr distance measures the similarity between two concepts based on the images in visual domain, it is in accord with the human cognition. However, the Flickr distance only measures the similarity but doesn't consider the type of relationship between two concepts. Thus it is not quite reasonable to directly apply the Flickr distance to the real-world environment. In addition that the Flickr distance is established on the Flickr data, However, Flickr data is very noisy. The computational cost of Flickr distance is quite expensive, so the Flickr distance is inadequate in the large-scale environment.

3   Concept Relationships and Features

In this section, we will elaborate conceptual relationships and features,include five specific relationships, CS,TS,HH,PR and UR, separately and 8-dimensional conceptual features, ED,EMD,JK,CK,JH,CH,ES and NGD, separately.

Clickture-Full. A large-scale data set, named Clickture, is sampled from one year click log of a commercial image search engine. It consists of a big table with 212.3 million triads:

$$Clickture = < K, Q, C > \qquad (1)$$

A triad $< K, Q, C >$ means that the image "K" is clicked "C" times in the search results of query "Q" in one year. It has 40 million unique image keys and 73.6 million unique queries.

Clickture-Lite a data set randomly samples from Clickture-Full and contains 1 million images, 11.7 million queries and 23.1 million image-query pairs.

Conceptual Feature. Different pairwise concepts have different concept relationships and distances, such as the Normalized Google distance. In order to distinguish these concept relationships, we will extract 8-dimensional conceptual features and each dimension of conceptual features is one kind of the distances between two concepts.

Complete Similarity(CS). It is the relationship between the same two concepts, such as Justin Bieber and J.B. or BeiJing and Peking.

Type Similarity(TS). It is the relationship that these two concepts have different types of the same object, such as Husky and Bulldog(the same object=dog) or Goldfish and Angelfish(the same object=fish).

Hypernym Hyponym(HH). It is super-subordinate relation that A is one kind of B, such as Husky and Domestic dog or The Great Wall and BeiJing.

Parallel Relationship(PR). It is the relationship that two concepts share the same super base class, such as Husky and Parrot (super base class=animal).

Unknown Relationship(UR). For the completeness of concept relationships, we add the unknown relation that we don't mention above.

Visual Word(VW). We assume that the number of visual words is limited and different concept relationships have different visual word distribution. We are given a database of pairs (x_i, y_i), i=1,...,m. In each pair, x_i is the image feature, such as sift, and y_i is the concept. We assume that visual word vw_i is generated by k-means.

$$X = (x_1, x_2, ..., x_m)$$
$$g(x) = argmin_{i=1,...,k}||X - vw_i|| \tag{2}$$

The conceptual feature can be measured by the divergence between the distributions of visual words.

Edit Distance(ED). It is a way of quantifying how dissimilar two strings (e.g., query) are to one another by counting the minimum number of operations,such as insert, delete and substitute, required to transform one string into the other. Edit Distance is regarded as one dimensional conceptual feature to measure the conceptual relationship between the queries of two concepts.

Earth Mover's Distance(EMD). The EMD[6] is widely used in content-based image retrieval to compute distances between the color histograms of two digital images. The EMD between concepts q_i and q_j is defined as follows:

$$EMD(q_i, q_j) = \frac{1}{N * N} \sum_{x \in \Gamma_{q_i}, y \in \Gamma_{q_j}} emd_{xy} \tag{3}$$

where $emd_{i,j}$ is the distance between the color histograms of images i and j. Γ_q the top n image set of concept q order by click count. Since the computational cost of EMD is super-cubic, the N is set to 50. $EMD(q_i, q_j)$is the one dimensional conceptual feature.

Jensen Kmeans(JK). This conceptual feature is computed by measuring the Jensen-Shannon divergence between two visual word probability distributions generated by kmeans[4]. For each visual word probability distribution $P_{c_{i1}}$ can be computed:

$$if \quad g(x) = argmin_{i=1,...,k}||X - vw_i||$$
$$then \quad num\_of\_vw_i = num\_of\_vw_i + 1$$
$$P_{c_{i1}} = \frac{num\_of\_vw_i}{N} \tag{4}$$

where N is the total number of visual word in the concept c_1. $num\_of\_vw_i$ represents the number of vw_i in the concept c_1. JK can be computed as follow:

$$JK(C_1|C_2) = \sum_{i=1}^{n} [\frac{p_{i_1}logp_{i_1} + p_{i_2}logp_{i_2}}{2} - (\frac{p_{i_1} + p_{i_2}}{2})log(\frac{p_{i_1} + p_{i_2}}{2})] \tag{5}$$

where JK represents the the Jensen-Shannon divergence between concepts c_1 and c_2.

Cosine Kmeans(CK). This conceptual feature is computed by measuring the cosine similarity between two visual word probability distributions generated by kmeans. For each visual word probability distribution $P_{c_{i1}}$ can be computed as above. The cosine similarity can be computed as follow:

$$CS(c_1, c_2) = \frac{\sum_{i=1}^n p_{i_1} p_{i_2}}{\sqrt{\sum_{i=1}^n p_{i_1}^2 \sum_{i=1}^n p_{i_2}^2}} \tag{6}$$

$$CS_{dist}(c_1, c_2) = 1 - CS(c_1, c_2)$$

where CS_{dist} represents the the cosine distance between concepts c_1 and c_2.

Jensen Hash(JH). This conceptual feature is computed the same as Jensen Kmeans(JK), except for the way of visual word generated. The Visual Word(VW) is generated by Bin's Hash coding[8] as follow:

$$VW_{ik} = \begin{cases} 0, if & v_{ik} > mean_k; \\ 1, & otherwise \end{cases} \tag{7}$$

Where VW_{ik} represents the k_{th} dimensional feature of visual word of image i. VW_i is the value of bin vector VW_{ik}.

Cosine Hash(CH). This conceptual feature is computed the same as Cosine Similarity(CS), except for the way of visual word generated. The Visual Word(VW) is generated by Bin's Hash coding as follow:

$$VW_{ik} = \begin{cases} 0, if & v_{ik} > mean_k; \\ 1, & otherwise \end{cases} \tag{8}$$

Where VW_{ik} represents the k_{th} dimensional feature of visual word of image i. VW_i is the value of bin vector VW_{ik}.

Exemplar Similarity(ES). This conceptual feature is computed by measuring the similarity between the exemplars of two concepts. Exemplar can be generated by performing kmeans on the image feature set of this concept and one centroid is one exemplar($k << N$).

Denote by Γ_c the representative exemplar set of concept c. Then the exemplar similarity between concepts c_1 and c_2 is defined as follows:

$$Sim_e(C_1|C_2) = exp(-\frac{1}{k * k} \sum_{x \in \Gamma_{c_1}, y \in \Gamma_{c_2}} \frac{\|x - y\|^2}{\delta^2}) \tag{9}$$

Note that δ is set to the median value of all pair-wise Euclidean distances between exemplars of different concepts.

Normalize Google Distance(NGD). This conceptual feature is proposed to calculate the distance between two concepts in the search results from Google search engine. So we make some slight changes to let it more suitable for our problem. The Google distance is defined as follow:

$$NGD(C_1, C_2) = \frac{max(log f(C_1), log f(C_2)) - log f(C_1, C_2)}{log N - min(log f(C_1), log f(C_2))} \tag{10}$$

Where $NGD(C_1, C_2)$ represents the Normalized Google distance between concepts C_1 and C_2. $f(C_1)$, $f(C_2)$ and $f(C_1, C_2)$ denote the number of images labeled C_1, C_2, both C_1 and C_2, separately. N is the total number of images in the click data.

4   Relationship Learning

In this section, we will elaborate the structure of the training set and the machine learning method to classify the relationship between two concepts.

4.1   Training Data

For exploring the relationship between these concepts, we will extract 8-dimensional conceptual features and manually annotate the type of relationship between two concepts. If we have n concepts, the size of data set is (n*(n-1)/2). The data set can express as follow:

$$x = (ED, EMD, NGD, ES, CK, JK, CH, JH)$$
$$y = \{r | r \in \{CS, TS, HH, PR, UR\}\}$$
$$training\_data = (x_i, y_i) | i = 1, ..., \frac{n(n-1)}{2}$$

$$(11)$$

The conceptual relationship y represents the five specific relationships between any two concepts ascending order by the distance between two concepts. The 8-dimensional conceptual features x are extracted from different aspects of Clickture data set. The concept feature ED describes the difference of five conceptual relationships in textual queries. ED of CS is close to 0, such as hyena and hyaena, but ED of PR is close to 1, such as weasel and miniature poodle. These conceptual features EMD,JK,CK,JH,CH,ES derive from basic image features,such as sift and LLC. We use these conceptual features to capture the difference of concept relationship in visual domain. EMD and ES stem from basic image feature, so these two conceptual features can easily capture these differences of image content between two concepts. JK,CK,JH and CH are from visual word distribution. Since different latent topic concepts have different visual word distributions, these conceptual features can easily find the difference of latent topic of different concepts. The conceptual feature NGD is used to describe the concurrence of image file. The NGD is close to 0 between hyena and hyaena, because of a mass of the same image files in these two concepts. In according to the difference of 8-dimensional conceptual features, we can recognise the type of concept relationship.

4.2   Relationship Learning Approach

We pick up a portion of the common concepts and construct the labeled data set as above. We apply the multiple classifier SVM to train the conceptual-semantic model to classify the relationship of two concepts.

5   Experiments on Concept Relationships

In this section, we will demonstrate that the conceptual features are effective to explore the five specific relationships. We will apply the conceptual model to rapidly and effectively explore the relationship between a portion of Clickture concepts.

5.1   Experiment I

Experiment Setting. In experiment I, we use Clickture-Lite to filter out queries with no less than 50 clicked images and form a candidate concept list. We select 423 common concepts from the candidate list as concept set. Eventually, the concept database have 423 different concepts, 0.231 million unique images and 3.21 million repeatable images. We adopt this approach mentioned in the section 4.1 to construct a training data set, which has 423*422/2=89253 labeled data. The data set is evenly split into nine folds, each of which contains the five specific relationships. We choose one fold to tune the parameters of multiple classifier SVM. Among the remaining eight folds, we select four folds to train the model and the rest to evaluate the performance. We switch the train folds and the test folds to compute the mean performance. All the learning and testing are carried out on a stand-alone machines (Intel-Xeon 2.00GHz CPU, 32G RAM, Windows Server 2012 Standard).

Experiment Result. Table1-3 illustrate the performance of the different conceptual features, separately. The following observations can be obtained:

* The overall conceptual features obtains about 92.78% mean precision and 43.44% mean recall relative improvements compared to the best in eight concept features. It shows that the comprehensive conceptual features are necessary to measure the complex conceptual-semantic relationships.
* The complete similarity(CS) relationship obtains 99.22% accuracy, 91.14% precision and 92.27% recall. It shows that CS can be captured very easily, because these two concepts share a mass of the same images.
* The Earth Mover's distance obtains the lowest performance. The EMD in computer science is to compare two grayscale images that may differ due to dithering, blurring, or local deformations . It shows that these reasons are not the major factor of distinguishing different concepts.

5.2   Experiment II

Experiment Setting. In experiment II, we use Clickture-Lite to filter out queries with no less than 100 clicked images and form a candidate concept list with 7370 concepts. These 7370 concepts and their images are the first candidate list. We choose 20 concepts and their images from 423 concepts in the experiment I as the second candidate list. We generate the data set to measure the relationship between these two candidate lists. The size of data set is 20*7370

Table 1. Accuracy of The five Specific Relationships

RelationShip	ED	EMD	NGD	ES	CK	JK	CH	JH	ALL
CS	0.9277	0.6018	0.8834	0.6974	0.9535	0.9408	0.9400	0.9436	0.9922
TS	0.7462	0.7672	0.7544	0.7672	0.7721	0.7745	0.7359	0.7531	0.8714
HH	0.6262	0.5885	0.8851	0.8298	0.8583	0.8583	0.8583	0.8583	0.9247
PR	0.6415	0.7156	0.7092	0.6990	0.7278	0.6832	0.6785	0.6838	0.8184
UR	0.5782	0.6184	0.6195	0.4373	0.8390	0.7829	0.7065	0.7012	0.8534
Mean	0.7039	0.6583	0.7703	0.6861	0.8301	0.8079	0.7839	0.7880	0.8920

ED: Edit Distance; EMD: Earth Mover's Distance; NGD: Normalized Google Distance; ES: Exemplar Similarity; CK: Cosine K-means; JK: Jensen K-means; CP: Cosine Hash; JH: Jensen Hash; ALL: overall features listed above; CS: complete-similarity relationship; TS: type similarity relationship; HH: hypernym hyponym relationship; PR: parallel relation; UR: unknown relation; Mean: the mean performance of five specific relationships.

Table 2. Precision of The five Specific Relationships

RelationShip	ED	EMD	NGD	ES	CK	JK	CH	JH	ALL
CS	0.3612	0.0264	0.2793	0.0233	0	0.4216	0.4112	0.4379	0.9114
TS	0.1834	0	0.1682	0	0.5071	0.5146	0.4494	0.4784	0.7214
HH	0.0759	0.0728	0.5839	0.0777	0	0	0	0	0.7301
PR	0.3149	0	0.4861	0.4422	0.5144	0.4406	0.3721	0.3662	0.6712
UR	0.1585	0.3067	0.3965	0.2548	0.7881	0.5973	0.5012	0.4957	0.7713
Mean	0.2188	0.0812	0.3828	0.1596	0.3619	0.3948	0.3468	0.3556	0.7611

Abbreviations are the same as above

Table 3. Recall of The five Specific Relationships

RelationShip	ED	EMD	NGD	ES	CK	JK	CH	JH	ALL
CS	0.6413	0.4456	0.9518	0.1356	0	0.7262	0.6692	0.7503	0.9227
TS	0.0263	0	0.0565	0	0.7499	0.5523	0.5938	0.6520	0.7294
HH	0.3573	0.3914	0.7543	0.0453	0	0	0	0	0.7434
PR	0.2219	0	0.3420	0.2223	0.7661	0.4212	0.1875	0.1438	0.7089
UR	0.3740	0.2360	0.5576	0.4732	0.6208	0.8070	0.8044	0.8228	0.7142
Mean	0.3242	0.2146	0.5324	0.1753	0.4274	0.5013	0.4510	0.4738	0.7637

Abbreviations are the same as above

elements. We will judge whether 20 queries can exactly find the other queries in 7370 queries with the five specific relationships. We will show this experiment result in Experiment II.

Experiment Results. This is an application that use the conceptual-semantic model to classify the concept relationship in a real-world environment.

Table 4. Top2 concept list of CS concept relationship

query	Top1	Top2
Alaskan Husky	husky	Husky puppies
Barack obama	Pictures of obama	obama
Black cat	Black cats	Cat drawings
bulldog	Bull dog	Bulldog pictures
coast	pics of beaches	Images of the beach
justin	Justinbieber	Justin bieber
Gray wolf	Wolf pictures	Pictures of wolfs
laptop	computers	Computer pics

Table 5. precision@10 of the five specific relationships

CS.P@10	TS.P@10	HH.P@10	PR.P@10	UR.P@10
0.9873	0.2073	0.5478	0.3298	0.9765

Table4-5 illustrate the precision top 10 of the specific five relationships of 20 concepts. In the manual evaluation, for each specific relationship of 20 concepts, we select the top10 concept pairs of SVM scores. The size of concept pairs is less than 20*5*10. We present them to 6 independent users. For each concept pair, the user is required to give a score 0 or 1. 1 indicates that the concept pair has the specific relationship,otherwise 0. By averaging these scores form all the users, we get the final score for each of concept pairs. If the score is greater than 0.5, the concept pair has the specific relationship. In the end, we compute the precision@10 for each specific relationship, which is shown in Table 5. We can find

* Complete similarity relationship (CS) obtains 98.73% precision. For such high precision, a mass of complete similarity concept pairs exist in the concept set.
* Type similarity relationship (TS) obtains only 28.73% precision. For such low precision, the number of concept pairs with this relationship is far less than this of CS concept pairs. For example, Barack Obama doesn't have a concept with TS. For this reason, It has a bigger possible to find a wrong concept.
* We obtain a more consistent and robust performance in this limited concept set(about 7370 test concepts). It demonstrates that our approach is effective to find these relationships between two concepts.

In summary, our approach is effective and robust to rapidly explore the specific relationship between two concepts. It also easily establishes a large-scale relationship network for these simple conceptual features.

6 Conclusions and Future Works

In this paper, we propose a multifold concept relationships metrics approach to exploit the specific relationship between two concepts. Different from the Google Distance and Flickr Distance, the proposed approach is able to find the different relationships between two concepts. Comparing with WordNet, the proposed approach can easily enlarge the scale of concepts with a low cost.

We will focus our future works in two directions. First, we will study the comprehensive concept relationship in the internet and expand to the specific conceptual-semantic relationships. Second, we will explore more conceptual features to improve the performance of our approach.

Acknowledgement. This work is partially supported by National 863 project 2014AA015204, National Science Foundation of China 61472116.

References

1. Cilibrasi, R., Vitányi, P.M.B.: The google similarity distance. IEEE Trans. Knowl. Data Eng. 19(3), 370–383 (2007)
2. Cilibrasi, R., Vitányi, P.M.B.: Clustering by compression. CoRR, cs.CV/0312044 (2003)
3. Hua, X.-S., Yang, L., Wang, J., Wang, J., Ye, M., Wang, K., Rui, Y., Li, J.: Clickage: towards bridging semantic and intent gaps via mining click logs of search engines. In: ACM Multimedia, pp. 243–252 (2013)
4. Huang, H., Cheng, Y., Zhao, R.: A semi-supervised clustering algorithm based on must-link set. In: Tang, C., Ling, C.X., Zhou, X., Cercone, N.J., Li, X. (eds.) ADMA 2008. LNCS (LNAI), vol. 5139, pp. 492–499. Springer, Heidelberg (2008)
5. Lenat, D.B.: Cyc: A large-scale investment in knowledge infrastructure. Commun. ACM 38(11), 32–38 (1995)
6. Levina, E., Bickel, P.J.: The earth mover's distance is the mallows distance: Some insights from statistics. In: ICCV, pp. 251–256 (2001)
7. Miller, G.A.: Wordnet, a lexical database for the english language, Cognition Science Lab. Princeton University (1995)
8. Wang, B., Li, Z., Li, M., Ma, W.-Y.: Large-scale duplicate detection for web image search. In: ICME, pp. 353–356 (2006)
9. Wu, L., Hua, X.-S., Yu, N., Ma, W.-Y., Li, S.: Flickr distance. In: ACM Multimedia, pp. 31–40 (2008)

A Comparison between Artificial Neural Network and Cascade-Correlation Neural Network in Concept Classification

Yanming Guo[1,2], Liang Bai[2], Songyang Lao[2], Song Wu[1], and Michael S. Lew[1]

[1] LIACS Media Lab, Leiden University, Niels Bohrweg 1, 2333 CA Leiden, The Netherlands
[2] College of Information Systems and Management,
National University of Defense Technology, Changsha, China
{y.guo,s.wu,m.s.lew}@liacs.leidenuniv.nl, xabpz@163.com,
laosongyang@vip.sina.com

Abstract. Deep learning has achieved significant attention recently due to promising results in representing and classifying concepts most prominently in the form of convolutional neural networks (CNN). While CNN has been widely studied and evaluated in computer vision, there are other forms of deep learning algorithms which may be promising. One interesting deep learning approach which has received relatively little attention in visual concept classification is Cascade-Correlation Neural Networks (CCNN). In this paper, we create a visual concept retrieval system which is based on CCNN. Experimental results on the CalTech101 dataset indicate that CCNN outperforms ANN.

Keywords: artificial neural network, cascade-correlation neural network, deep learning, concept classification.

1 Introduction

Machine learning is a branch of artificial intelligence. It is focused on learning the hidden patterns from data automatically, and has long been an active area. There are two major development waves in machine learning [1]: shallow learning and deep learning.

Shallow learning arose in the 1980s, when back propagation (BP) [2] was introduced to artificial neural networks (ANN) [3]. The usage of the BP algorithm enabled ANN to learn the statistical regularities from the known training set. Roughly speaking, the BP algorithm worked well for one hidden layer but deteriorated for deep networks (multiple hidden layers) owing to the problem of vanishing or exploding gradients [4]. Nevertheless, it resulted in a significant step forward in machine learning. In the 1990s, a variety of other shallow learning methods appeared, such as support vector machines, boosting, logistic regression, etc. These methods have been successful in practical applications and are still receiving attention today.

Deep learning (for a good example, see Hinton and Salakhutdinov [5]) represents another development wave in machine learning, and has experienced rapid

W.T. Ooi et al. (Eds.): PCM 2014, LNCS 8879, pp. 248–253, 2014.

development recently [6, 7]. Deep learning is a set of algorithms that attempt to model high-level abstractions in data by using architectures composed of multiple non-linear transformations [8]. Compared to shallow learning, deep learning emphasizes the depth of the modal structure and highlights the importance of feature study. The purpose of deep learning is to grasp the inner nature of the data, but not just the statistical regularity. In actual operation, it would build a model which contains multiple hidden layers.

Deep learning originates from ANN. There are two major deep learning algorithms that derive from ANN: cascade-correlation neural network (CCNN) and convolutional neural network (CNN). As the effectiveness of CNN has been widely studied in recent years [9, 10], the advantages of CCNN in computer vision area has received relatively little attention. The goal of this paper is to examine how CCNN performs relative to ANN.

2   Basic Theory

In this section we introduce some fundamental notions underlying ANN and CCNN.

2.1   Artificial Neural Network

Artificial neural networks (ANN) are an information processing paradigm which is inspired by biological nervous systems. The traditional ANN usually has a fixed topology, in which the connections go forward from the input layer to the output layer, via one or more hidden layers.

For a given input, its value is propagated along the connections, reaching the neurons in the next layer. Each of the connections in ANN has an associated weight, the input value is firstly multiplied with the weights of the connections and summed up, forming a new input for the neuron, and then the new input is executed through an activation function, as Equation 1 shows,

$$y(x) = g\left(\sum_{i=0}^{n} w_i x_i\right) \qquad (1)$$

x_i is the ith neuron in the previous layer, w_i is the weight determining how much should this neuron be weighted, g is the activation function deciding how to achieve the output based on the sum of the input. The output of the activation function is the output of the neuron. This will eventually give the final output the network, if it differs from the desired value, the weights of the connections will be adjusted to minimize this difference, until the error is acceptably small.

Although ANN can get relatively good results in some applications, it is difficult to determine the number of hidden layers and hidden neurons in each hidden layer; it is also hard to determine the optimal activation function and may be time-consuming to train a variety of parameters that can reach a global optimum.

The drawbacks of ANN give rise to a variety of dynamic algorithms which do not need to depend on many parameters. One of the most important approaches to dynamically adjust the network is CCNN, which will be described below.

2.2   Cascade-Correlation Neural Network

Cascade correlation neural network (CCNN) are "self-organizing" networks, which was first proposed by Fahlman and Lebiere[11]. Instead of just adjusting the weights in a network of fixed topology, the algorithm can dynamically change the architecture of an ANN by adding connections and neurons.

CCNN starts with a simple ANN structure with no hidden neurons. During the training process, each hidden unit receives connections from the original inputs as well as the pre-existing hidden units, the connections are initialized with small random weights, creating a pool of candidate units. To add a single hidden unit at a time, the candidate which has the best performance is chosen as the final hidden unit, the others are discarded, once the hidden unit is added to the net, its weight is frozen. The cascade architecture is shown in Fig 1.

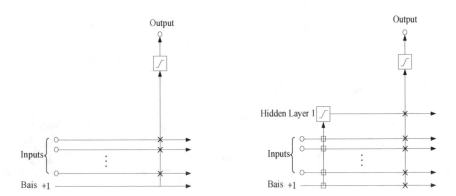

Fig. 1. The cascade architecture

For CCNN, each new unit adds a new one-unit "layer" to the network, which leads to very deep networks and the creation of very powerful high-order feature detectors.

3   Concept Classification System

In this paper, we use the following concept classification process to compare the performance of ANN and CCNN, which is shown in Fig.2.

Generally, the concept classification process can be divided into two phases: training phase and testing phase. The goal of training phase is to get a classifier which has the ability to distinguish differences between different categories. We further divide this phase into 4 steps. First, we extract the dense-sift feature [12] from dataset as step 1, and then generate the vocabulary utilizing Approximate Nearest Neighbors

[13] in step 2. Next, we assign the images with visual words in the vocabulary, generating a histogram of visual words for each image in step 3. Finally, we train the classifier from those histograms as step 4, in which we will generate the classifiers from ANN and CCNN respectively. The testing phase reflected in step 5 mainly tests the performance of the classifiers which have been modeled. It should be noted that the training data never overlaps with the testing data.

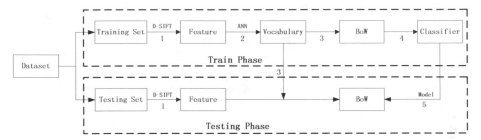

Fig. 2. Concept classification system

4   Experiments

Dataset: To make a comparison of the concept classification between ANN and CCNN, we select 20 categories from CalTech101 which have a large number of images (more than 80 images), namely airplanes, bonsai, brain, Buddha, butterfly, chandelier, ewer, Faces, grand_piano, hawksbill, kangaroo, ketch, laptop, Leopards, menorah, Motorbikes, scorpion, starfish, sunflower and watch. In each category, we randomly select 50 images as the training set, and other 30 images as the testing set. Therefore, we can get 1000 images for training, 600 images for testing.

Experimental Setups: Due to space limitations, we only briefly summarize the details of the experiment. It can be divided into three steps: 1) local feature extraction; 2) vocabulary generation and word assignment; 3) classifier training and testing.

In the first step, we extract the Dense-Sift feature with the extraction step as 8 pixels, finally we get a 128-dimension descriptor for each image.

In the next step, we formulate two vocabulary sizes (5000, 10000) of the visual words from the training set, taking advantage of Approximate Nearest Neighbors method, and assign each image with the visual words.

In the last step, classifiers are trained using the histogram of visual words. For the ANN algorithm, we use a symmetric sigmoid as the activation function. To test the overall concept classification performance of ANN, we set the hidden layer number to 1,2,3,4 respectively, and for each layer, we set the number of hidden neuron nodes to 20, 50 and 100. We further set the desired error as 0.0001 for both ANN and CCNN.

Results and Discussions: To compare the performance of concept classification between ANN and CCNN, we select the commonly used evaluation criteria, precision (number of correct matches / number of number of matches) and recall (number of correct matches / number of correspondences). The final concept classification precision is shown in table 1, while the recall in table 2.

Table 1. Precision of concept classification

Vocabulary Size	Neuron nodes	1-hidden layer	2-hidden layer	3-hidden layer	4-hidden layer	CCNN
5000	20	0.2997	0.2606	0.1958	0.2642	0.4138
5000	50	0.2942	0.2702	0.2225	0.2328	0.4138
5000	100	0.3618	0.2733	0.2445	0.1449	0.4138
10000	20	0.2704	0.1734	0.2032	0.2076	0.3999
10000	50	0.2839	0.2233	0.2033	0.2354	0.3999
10000	100	0.2916	0.2247	0.1996	0.1435	0.3999

Table 2. Recall of concept classification

Vocabulary Size	Neuron nodes	1-hidden layer	2-hidden layer	3-hidden layer	4-hidden layer	CCNN
5000	20	0.2617	0.2700	0.1767	0.2450	0.3983
5000	50	0.2733	0.2617	0.2483	0.2550	0.3983
5000	100	0.3583	0.2700	0.2400	0.1417	0.3983
10000	20	0.2583	0.1750	0.2250	0.1800	0.3983
10000	50	0.2733	0.2100	0.1950	0.2067	0.3983
10000	100	0.2867	0.1983	0.1933	0.1267	0.3983

From the table, we can conclude that, 1) for ANN, the performance does not improve with the increase of hidden layers and the hidden neurons in each layer, demonstrating that the abstraction ability for ANN would not necessarily improve when we increase the depth of the network, this is possibly owning to the fact that ANN suffers from the problem of vanishing or exploding gradients when more hidden layers are used. 2) The method of CCNN significantly outperforms ANN in concept classification, no matter how we set the hidden layers and hidden nodes for ANN, which demonstrated that CCNN could retain most useful hidden units, or features, for the later use of classification.

We further compare the classification precision between the best result of ANN (vocabulary size=5000, hidden layer number = 1, hidden neuron number =100) and CCNN for each of the categories, which shows that for most categories, CCNN performs better than the best result of ANN, as is displayed in Fig 3.

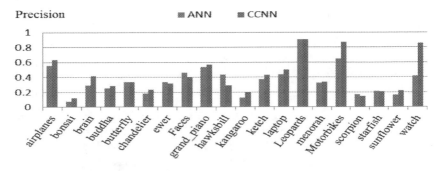

Fig. 3. Comparison of precision for ANN and CCNN

5   Conclusion

Deep learning is a promising paradigm in representing high-level abstractions. Because deep convolutional neural networks have been widely studied, we mainly focus on the comparison of cascade correlation neural networks and classic artificial neural networks in visual concept classification. We designed a visual concept detection system utilizing SIFT-based bag of words and performed tests on the well-known CalTech101 dataset. The results from our experiments show that the deep cascade correlation network significantly outperforms the classic neural network and provides a promising research direction regarding deep learning.

Acknowledgments. We are grateful to the China Scholarship Council (CSC), Leiden University and National University of Defense Technology for their support.

References

1. Weston, J., Ratle, F., Mobahi, H., Collobert, R.: Deep learning via semi-supervised embedding. In: Montavon, G., Orr, G.B., Müller, K.-R., et al. (eds.) NN: Tricks of the Trade, 2nd edn. LNCS, vol. 7700, pp. 639–655. Springer, Heidelberg (2012)
2. Rummelhart, D.E.: Learning representations by back-propagating errors. J. Nature 323(9), 533–536 (1986)
3. Yegnanarayana, B.: Artificial neural networks. M. PHI Learning Pvt. Ltd. (2009)
4. Mikolov, T., Kombrink, S., Deoras, A., et al.: RNNLM-Recurrent neural network language modeling toolkit. In: Proc. of the 2011 ASRU Workshop, pp. 196–201 (2011)
5. Hinton, G.E., Osindero, S., Teh, Y.W.: A fast learning algorithm for deep belief nets. J. Neural Computation 18(7), 1527–1554 (2006)
6. Ranzato, M., Susskind, J., Mnih, V., et al.: On deep generative models with applications to recognition. In: Proc. CVPR, pp. 2857–2864 (2011)
7. Le, Q.V.: Building high-level features using large scale unsupervised learning. In: Proc. Acoustics, Speech and Signal Processing (ICASSP), pp. 8595–8598 (2013)
8. Bengio, Y., Courville, A., Vincent, P.: Representation learning: A review and new perspectives. J. IEEE Trans. PAMI, 1798–1828 (2013)
9. Cireşan, D.C., Meier, U., Masci, J., et al.: High-performance neural networks for visual object classification. J. arXiv preprint arXiv:1102.0183 (2011)
10. Krizhevsky, A., Sutskever, I., Hinton, G.E.: ImageNet Classification with Deep Convolutional Neural Networks. J. NIPS, 1106–1114 (2012)
11. Fahlman, S.E., Lebiere, C.: The cascade-correlation learning architecture. J. Advances in Neural Information Processing Systems, 424–532 (1989)
12. Lazebnik, S., Schmid, C., Ponce, J.: Beyond bags of features: Spatial pyramid matching for recognizing natural scene categories. In: Proc. CVPR, vol. 2, pp. 2169–2178 (2006)
13. Indyk, P., Motwani, R.: Approximate nearest neighbors: towards removing the curse of dimensionality. In: Proc. of the Thirtieth Annual ACM Symposium on Theory of Computing, pp. 604–613. ACM (1998)

Location-Based Hierarchical Event Summary for Social Media Photos

Weipeng Zhang, Jia Chen, Jie Shen, and Yong Yu

Dept. of Computer Science & Engineering, Shanghai Jiao Tong University, China
{weipengzhang,jiachen,jieshen,yyu}@apex.sjtu.edu.cn

Abstract. In this paper, we propose a system named "LHES" to detect location-based social events on flexible time scales and generate a hierarchical summary for the event. Particularly, we focus on social events that happened at landmarks. Flexible time scales include month, day, hour and minute. For each landmark, our LHES system generates a hierarchical (tree style) summary, in which the root node gives a snapshot of the entire event and child nodes span different time periods (beginning, ending, etc.) of the parent event. To generate such a summary, we use both visual cues (e.g., color, texture) and metadata (e.g., time stamp, image tags, titles and description). Our online demo is available at **http://hed.apexlab.org**.

Keywords: event detection, hierarchical, social media, clustering.

1 Introduction

In this paper, we propose the idea of event detection on flexible time scales and build a system to generate a hierarchical event summary based on such detection. With the increasing popularity of the smart phones equipped with GPS, more and more location-based applications have been proposed. Among the multimedia community, researchers have proposed applications such as [2]. In our work, we also integrate the geographical information into the system. Another field related to our work is event detection [2], [1], [4]. They use supervised or unsupervised methods to detect event at a fixed time scale such as one day or one hour. Different from them, we detect event at flexible time scales and generate a hierarchical summary for the event based on such detection.

For a social event, especially for a grandiose one, it is possible that the entire event was consisting of several subsidiary events. An example is shown in Figure 1. Assuming that the Olympic Games is taken as the entire event, we can easily see that it can be split into subsidiary events (e.g. swimming, tennis) that happened at different time. Recursively, a subsidiary event could also be split into more events. This encourages us to detect events at flexible time scales and organize them into a hierarchical summary along time axis.

We give a formalized description of the above example. Events detected at different time scales can be organized into a tree structure. In this structure, child node events are detected at finer time scale than parent node events. The

W.T. Ooi et al. (Eds.): PCM 2014, LNCS 8879, pp. 254–257, 2014.

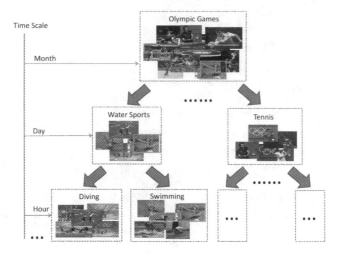

Fig. 1. An event example with hierarchical structure

child node events span different time period of the parent node event. For each node, we generate a summary including both image and text like the method in work [3].

The contribution of our paper is two-fold: 1) we detect events at flexible time scales; 2) we generate a hierarchical event summary based on contextual and visual evidence. To the best of our knowledge, our work is the first approach to detect location-based event at flexible time scales.

Fig. 2. The Main Web UI of Our LHES System

2   System Overview

2.1   User Interface

We aim to design a concise and friendly user interface as shown in Figure 2. In the top panel, one can select a landmark to browse its associated events. On the bottom panel, there is a list of sample landmark-centric events for quick examination.

The realtime effect for hierarchical detection is illustrated in Figure 3. We arrange all of the events chronologically by our system, where the date that the event happened is displayed on the left. And the event-associated photos, as well as the contextual summary are assembled on the right. One may notice that there exist some events with a hierarchical structure, for which the photo ensemble of the entire event is placed on the top of the right panel, and the one for its child event is placed below. In this way, we provide a fundamental representation for each event: time period, visual and contextual description of the underlying event, as well as its possible hierarchical structure. These are the basic elements of an event, which answer the question: when, where and what happened.

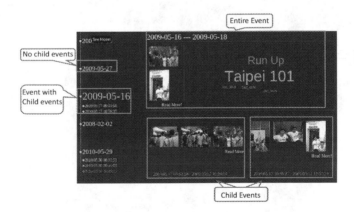

Fig. 3. Hierarchical event summary by our LHES system

2.2   Technical Details

LHES has indexed over 1.1 million Flickr landmark images. We focus on landmark rather than other arbitrary location. The LHES system consists of two principle components: 1) event detection on flexible time scales; 2) hierarchical event summary generation. In this section, we will introduce each component in details.

Event Detection on Flexible Time Scales. This component is the main contribution of our work. As one landmark is selected by the user, our system will automatically extract those associated photo streams from our database. Within an unsupervised learning framework, social events can be detected on a coarse time scale. To explore its possible child events, we subsequently pool visual (i.e. color, texture) and contextual (i.e. bag of tags) features into our algorithm over the entire photo collections on a finer time scale. This unsupervised learning framework uses two scores on different time scale, including visual and contextual feature score. The visual feature score is measured by the similarity of each two landmark images based on the mixture of image features (i.e. color, texture). The contextual feature score is measured by the number of the shared tags between two image's tag set.

Hierarchical Event Summary Generation. This component aims to provide a hierarchical event summary based on the flexible time scale by using visual and contextual evidence. The visual and contextual summary are generated separately in our work. We mainly consider the brightness, contrast, RGB color channels as the feature descriptors [5] for the visual reproduction as the work of [3]. To abstract the contextual information of one event, we utilize the metadata of its associated photos such as image tags and descriptions. By a threshold of the frequency of each word, our system presents the key words as the summary.

3   Conclusion

In this paper, we propose a novel location-based hierarchical event detection system. Encouraged by the daily instances, we explore the intrinsic connection over events from coarse temporal duration to fine time period, which could be straightforwardly represented as a tree-based structure. To provide an elegant user experience, we devise a concise and precise user interface that is based on a diversified summary from both visual and contextual evidence. Preliminary experiments on several landmarks demonstrate the efficiency and effectiveness of our system.

References

1. Claudiu, S., Mihai, G., Wolfgang, N., Raluca, P.: Bringing order to your photos: event-driven classification of flickr images based on social knowledge. In: CIKM, pp. 189–198 (2010)
2. Hila, B., Mor, N., Luis, G.: Learning similarity metrics for event identification in social media. In: WSDM, pp. 291–300 (2010)
3. Junfeng, Y., Jia, C., Zejia, C., Yihe, Z., Shenghua, B., Zhong, S., Yong, Y.: Searching for diversified landmarks by photo. In: ACM Multimedia, pp. 1337–1338 (2012)
4. Liu, X., Huet, B.: EventEnricher: a novel way to collect media illustrating events. In: ICMR, pp. 303–304 (2013)
5. Songhao, Z., Junchi, Y., Yuncai, L.: Improving Semantic Scene Categorization by Exploiting Audio-Visual Features. In: ICIG, pp. 449–450 (2009)

Towards Natural Gestures for Presentation Control Using Microsoft Kinect

Boon Yaik Ooi[1], Chee Siang Wong[1], Ian KT Tan[2], and Chee Keong Lee[3]

[1] Universiti Tunku Abdul Rahman, Kampar, Malaysia
[2] Multimedia University, Cyberjaya, Malaysia
[3] iEnterprise Online Dot Com, Petaling Jaya, Malaysia
{ooiby,wongcs}@utar.edu.my, ian@mmu.edu.my, cheelee@ieol.com.my

Abstract. Microsoft Kinect was initially developed for gaming and has since been used in a variety of applications. The work here addressed some of the challenges for it to be used as a gesture driven presentation application. The proposed use of Hidden Markov Model to provide a smoother, and hence a step towards a more natural movement, is demonstrated.

1 Introduction

The work described (and demonstrated) here is the development of a gesture recognition solution for users to control presentation slides remotely using the Microsoft Kinect sensor. Conventionally, the challenges of automatic capture and analysis of human motion can be divided into four stages [1], (1) Initialization, (2) Tracking, (3) Pose estimation and (4) Recognition.

The introduction of depth sensors such as Microsoft Kinects skeletal tracking application programming interface (API) [2] has solved the first two stages of the aforementioned stages to the extent that developers can now focus on the 3rd and 4th stages.

In terms of gesture recognition, the problem can be further divided into recognizing static pose and dynamic gesture. There are many attempts to view the problem of dynamic gesture recognition as a series of static pose recognition problems [3] [4] [5]. In general, such approach has 2 stages. The first stage is usually to recognize the static pose, and the second stage is to analyze the relationship between the static poses to recognize the given gesture. These stages are necessary for system to recognize a wide range of actions which are not necessary for controlling presentation slides.

Although the gesture recognition for controlling presentation slides has far less gestures to recognize, the gesture recognition solution has to be fast and accurate to avoid unnecessary delay and interruption on the user presentation. Therefore, the 2-stage approach might not be suitable. In additional to that, developing this gesture recognition solution is not without its own problem, notably to identify the gestures from a sequence of continuous stream of gestures. During a presentation, the user might have presentation gestures that are not meant to

W.T. Ooi et al. (Eds.): PCM 2014, LNCS 8879, pp. 258–261, 2014.
© Springer International Publishing Switzerland 2014

control the slides. Therefore, the system must be able to identify control gestures and ignore presentation gestures.

2   Methodology

The proposed solution is a skeletal-region based (via Kinect API) technique combine with Hidden Markov Model (HMM) to perform fast and accurate gesture recognition to control presentation slides. As the Kinect sensor return us the skeletal information of our user, the skeletal is broken into several regions as shown in Figure 1.

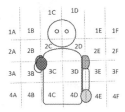

Fig. 1. Regions based on skeletal information

From Figure 1, When a user moves his hand from region 2D to 2F the system will generate a series consists of 2D, 2E and 2F sequence. The system can then use this series to infer that the user has done a left to right swipe. The same applies if the series return by the system is from 2F to 2D then it would infer it a left swipe. This skeletal-region based technique has eliminated the need for a pose classifier and the computation complexity is reduced compare to existing 2-stage approaches.

However, gesture recognition process is not complete with skeletal-region based technique as it has yet to handle the temporal information. Users may move from one region to another at a different pace and might be using different path. Typical gesture recognition systems use a fixed number of frames to capture users gestures where such fixed number of frames approach could not handle a wide variety of users pace. This solution solves the pace issue by capturing user movement based on previous locations. The number of captures is arbitrarily set at 50. This means that the system only captures the user hand location if there are changes based on the previous location. Thus, the system can tolerate a wide range of different gesture speed. Unfortunately, this technique still generates slightly different sequence for the same gesture. For instance, a left swipe, the sequence could be a sequence with different number of 2D, 2E and 2F. This is the point where a simple HMM is used to distinguish and classify these sequences.

Besides gesture speed, the system also has to distinguish one gesture from another. The system has to identify the beginning and the ending of a gesture.

This is solved by having a dedicated start region. The start gesture is when the hand overlaps with the shoulder points as indicated by the red area in Figure 1. At this starting point, the number of captures is reset to 0. The gesture recognition process is initiated every time when the number of captures reaches 50.

3   Functionality

Based on the technique described in previous section, this work developed a complete system that works in the following four modes: Desktop, Microsoft PowerPoint, Browser, and Magnifier. These modes are engaged based on the current context of the presentation computer.

Before the user starts executing any gesture, the user must start with the **Ready** action. This **Ready** action is used in all modes. Figure 2 shows the right hand **Ready** action when the user needs to perform any gesture which starts with Right hand. The height of user palm should be aligned with his/her shoulder. This applies to the left hand as well. Table 1 lists the gestures that users can execute.

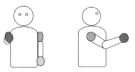

Fig. 2. Ready action for right hand: front view and side view

As an example, when user is in PowerPoint Mode and the user wants to go to the next slide, the user should start with a right hand **Ready** action and perform a right swipe action with the right hand as illustrated in Figure 3.

Fig. 3. Ready Action, (2) Right Hand: Right Swipe

If the user wants to engage Magnifier Mode, he or she starts with both the left and right hand in **Ready** and moves the hands apart. The demonstration video can be found at http://www.youtube.com/watch?v=-LssTzQHjn4

Table 1. List of gestures that users can execute in the system

Modes	Functions	Gestures
General	Go to Desktop	Left Hand: Left Swipe
	Reset	Repeat Left Hand: Left Swipe until reach Desktop
Desktop Mode	Navigate icons Up/Down /Left/Right direction	Right Hand: Up/Down/Left/Right
	Double click the selected icon	Right Hand: Push
	Go to Desktop	Left Hand: Left Swipe
PowerPoint Mode	Start PowerPoint Slide Show	Right Hand: Push
	Next slide	Right Hand: Right Swipe
	Previous slide	Right Hand: Left Swipe
	Tabbing selection to trigger Multi-media content	Left Hand: Up/Down Swipe
	Blank projector screen	Left Hand: Right Swipe
	Quit PowerPoint Slide Show (when in slide show mode)	Left Hand: Left Swipe
	Quit PowerPoint Application (when in edit mode)	Left Hand: Left Swipe
Browser Mode	Scroll page up / down (corresponding with gesture speed)	Right Hand: Up / Down Swipe
	Scroll page left / right	Right Hand: Left/ Right Swipe
	Left Click	Right Hand: Push
	Quit Browser	Left Hand: Left Swipe
Magnifier Mode	Start Magnifier	Hold Left Hand and Right Hand at **Ready** mode
	Zoom In	Move Left Hand and Right Hand further apart
	Zoom Out	Move Left Hand and Right Hand closer together
	Quit Magnifier	Left Hand: Left Swipe

4   Conclusion

In this paper, we briefly described a system that utilizes the Microsoft Kinect hardware and API for the use of presentation using Microsoft PowerPoint. This project is funded by iEnterprise Online Dot Com with support from the MSC Malaysia Innovation Voucher.

References

1. Moeslund, T.B., Hilton, A., Krger, V.: A survey of advances in vision-based human motion capture and analysis. J. Comp. Vision and Image. 104(2), 90–126 (2006)
2. Tracking Users with Kinect Skeletal Tracking,
   http://msdn.microsoft.com/en-us/library/jj131025.aspx
3. Hachaj, T., Ogiela, M.R.: Rule-based approach to recognizing human body poses and gestures in real time. Multimedia Syst. 20(1), 81–99 (2014)
4. Thurau, C., Hlaváč, V.: Recognizing Human Actions by Their Pose. In: Cremers, D., Rosenhahn, B., Yuille, A.L., Schmidt, F.R. (eds.) Visual Motion Analysis. LNCS, vol. 5604, pp. 169–192. Springer, Heidelberg (2009)
5. Otasevic, N.: Recognizing simple human actions by exploiting regularities in pose sequences. Masters thesis. MIT (2013)

3D-Spatial-Texture Bilateral Filter
for Depth-Based 3D Video

Xin Wang[1], Ce Zhu[1], Jianhua Zheng[2], Yongbing Lin[2], and Yuhua Zhang[1]

[1] University of Electronic Science and Technology of China, Sichuan, China
wangxin_chinaman@163.com, eczhu@uestc.edu.cn,
yuhua_zhang@126.com
[2] HiSilicon Technologies Co. Ltd., Beijing, China
{zhengjianhua,linyongbing}@hisilicon.com

Abstract. In the depth-based 3D video system, filters used in texture and depth images denoising, such as bilateral filter and trilateral filter, are generally designed based on calculating the weighted average of reference pixels located around the pixel to be filtered. In this paper, we propose a 3D-spatial-texture bilateral filter by considering the relationship of two pixels in the 3D space, including geometric closeness in the 3D world coordinate, as well as their corresponding texture/color similarity. Accordingly, the weight is defined with two kernels describing two abovementioned factors respectively, namely, a spatial kernel and a range kernel. The experimental results show that better performance can be achieved by using the proposed filter for both texture and depth image denoising, compared with conventional bilateral filter and trilateral filter.

Keywords: 3D-Spatial-Texture Bilateral Filter, 3D World Coordinate, Geometric Closeness, Texture Similarity, Depth-based 3D Video.

1   Introduction

Depth-based 3D video content is comprised of conventional 2D texture/color video and its associated depth images which reflect the distance between an object and a camera in the 3D space. With the content, virtual view images can be synthesized by using depth-image-based rendering (DIBR) [1-2], so that people can flexibly select a viewpoint they like to enjoy the video. It is known that the quality of texture and depth images is the key factor impacting the view synthesis result. However, the content is often contaminated by different kinds of noise during acquisition, coding and transmission, especially depth images are very difficult to be obtained accurately if not impossible. Throughout virtual view synthesis, the noise in depth and texture images will produce geometric distortion and texture/color distortion in synthesized images, respectively. Therefore, it is highly required or desirable to perform noise reduction in texture and depth images in the depth-based 3D video system.

In the past decades, a variety of solutions have been developed for image denoising, which replace a to-be-filtered noisy pixel by a weighted average of neighboring pixels (reference pixels). Each weight reflects the correlation between a reference pixel and the to-be-filtered pixel. Bilateral filter proposed in [3] employs both

W.T. Ooi et al. (Eds.): PCM 2014, LNCS 8879, pp. 262–267, 2014.
© Springer International Publishing Switzerland 2014

geometric closeness and pixel value similarity between pixels to determine a weight. In other words, the weight is defined by two kernels: a spatial kernel and a range kernel. By considering the correlation above, such a bilateral filter is widely used as an edge-preserving method in texture denoising. As an extension of bilateral filter, trilateral filter proposed in [4] is mainly used in depth image denoising. As similarity structure can be observed between depth image and its corresponding texture information, another range kernel measuring the similarity between corresponding texture pixels is added to refine the weight. With this new kernel, the trilateral filter has the capability to partly remove the depth-texture misalignment in boundary regions [5] in the depth image. Considering the factors of conventional methods mentioned above, other special filters with different kinds of kernel functions, such as [6-8], were proposed to further improve the filtering performance.

However, the weight calculation methods of aforementioned filters have a drawback. On the geometric closeness, they tend to consider the distance only in the 2D image plane, which always lead to inaccurate results of describing the relationship between pixels. As a solution, in this paper, we propose a 3D-spatial-texture bilateral filter. The proposed filter is designed with consideration of the relationship of two pixels in the 3D space, including geometric closeness in the 3D world coordinate and texture/color similarity. Accordingly, two kernels are designed to describe the two factors, namely, a spatial kernel and a range kernel. The experiment results show that better rendering quality can be achieved by using the proposed filter in both texture and depth images denosing, compared with conventional methods.

The remainder of the paper is organized as follows. In section 2, the 3D-spatial-texture bilateral filter is presented. Then, experimental results will be provided in section 3, and section 4 concludes this paper.

2   Proposed 3D-Spatial-Texture Bilateral Filter

Pixels in an image are expressions of points on the object surface in the real scene. In other words, the relationship between pixels is reflected by the correlation of their corresponding points in the 3D space. The correlation of two points in the 3D space can be determined by two factors: 1) geometric closeness in the 3D world coordinate; 2) texture/color similarity. Accordingly, the weight is determined by the two factors.

1)   Geometric closeness in the 3D world coordinate

In conventional methods, geometric closeness is defined as the 2D location closeness in the image plane. Actually, in the trilateral filter, depth similarity can be regarded as another kind of geometric closeness which can be named as depth closeness. However, in geometric closeness calculation, conventional methods consider the two kinds of closeness as two independent components, which always make mistakes in some conditions.

As illustrated in Fig. 1, A, B, C and D are pixels in the image plane obtained from points A', B', C' and D' in the 3D space by a camera. In the image plane, the distance between A and B is equal to A and D. In the corresponding depth image, B and C have a same depth value. Through observation, in the 3D space, comparing with B', C' and D' are located closer with A'. In other words, C and D are more relevant with A. But if we calculate the weight by conventional methods, opposite conclusions will be obtained. Therefore, the geometric closeness needs to be determined by jointly considering the 2D location closeness and the depth closeness, that is to say, considering the geometric closeness in the 3D world coordinate.

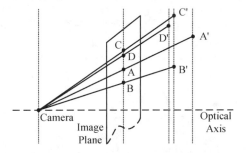

Fig. 1. Illustration of conditions that conventional methods make mistakes in geometric closeness calculation

Based on the analysis above, as in Fig. 2, we firstly project each pixel to the 3D world coordinate by 3D warping technique with Eq. (1) [9], which jointly using the pixel location in the image plane and its depth value.

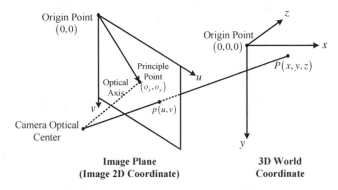

Fig. 2. Illustration of projecting a pixel from the 2D coordinates to the 3D world coordinate

$$P = R^{-1}\left(dA^{-1}p - t\right), A = \begin{pmatrix} f_x & r & o_x \\ 0 & f_y & o_y \\ 0 & 0 & 1 \end{pmatrix}, p = \begin{pmatrix} u \\ v \\ 1 \end{pmatrix}, P = \begin{pmatrix} x \\ y \\ z \end{pmatrix} \quad (1)$$

where $P(x,y,z)$ is the projecting result of pixel $p(u,v)$; R^{-1} and t denote the rotation matrix and translation vector that transform pixels in the image 2D coordinate to the world coordinate; d is the depth value of p; A specifies the intrinsic parameters of the camera, f_x and f_y are the focal lengths in horizontal and vertical directions, r is the radial distortion coefficient, (o_x,o_y) denotes the position of principle (center) point in the image plane.

Then, the geometric closeness of two pixels can be defined by the spatial distance between two corresponding points in the 3D world coordinate.

2)   Texture/color similarity

Like conventional bilateral or trilateral filtering methods, the texture/color similarity can be reflected by the difference between texture pixel values of them.

Consequently, the weight is defined by two kernels describing two abovementioned factors respectively, namely, a spatial kernel and a range kernel. The filtering process of depth and texture image can be formulized in Eq. (2), and L_p is the normalization factor.

$$D_p' = \frac{1}{L_p}\sum_{q\in K} w_q D_q, \quad T_p' = \frac{1}{L_p}\sum_{q\in K} w_q T_q, \quad L_p = \sum_{q\in K} w_q \tag{2}$$

$$w_q = f_S(P,Q) f_R(T_p,T_q)$$

$$f_S(P,Q) = \exp\left(-\left(\left|x_P - x_Q\right| + \left|y_P - y_Q\right| + \left|z_P - z_Q\right|\right)\big/2\sigma_S^2\right) \tag{3}$$

$$f_R(T_p,T_q) = \exp\left(-\left|T_p - T_q\right|\big/2\sigma_R^2\right) \tag{4}$$

where p is the pixel to be filtered, q is one of reference pixels in the reference pixel set K; D_q and T_q denote the depth and texture pixel values of q, D_p' and T_p' are the corresponding filtering results; f_S is the spatial kernel with pixel locations in the world coordinate ($P(x_P,y_P,z_P),Q(x_Q,y_Q,z_Q)$) as inputs, and f_R is the range kernel with texture pixel values (T_p,T_q) as inputs. The kernels can be defined by different kinds of functions, such as Eq. (3)-(4), which increase as the difference between the inputs decrease. σ_S and σ_R are used to balance the results.

3   Experimental Results

The proposed filter is tested on four test sequences, "Ballet", "Breakdancer", "Balloons" and "Kendo". The left and right views are to be filtered, while the intermediate view is to be synthesized. Original left and right views are encoded using HM 13.0 [10] with ALL-Intra mode. QPs are set as 33 for texture and 40 for depth. The loopfilter and SAO (Sample Adaptive Offset) [11] are turned off. After filtering the decoded results with the proposed and conventional methods [3-4], virtual intermediate views are synthesized with VSRS 3.5 (View Synthesis Reference Software). The SSIM [12] quality of the synthesized result is measured with respect to the original intermediate view. For the proposed filter, a series of uniform parameters for all the test sequences are set as follows. The size of reference pixels set is 7×7, σ_S and σ_R in Eq. (3)-(4) are set as 3.5 and 4 for depth, as well as 0.78 and 0.9 for texture.

Fig. 3 shows the SSIM quality of synthesized results by conventional and proposed methods, which demonstrates that the proposed method achieved better performance than the conventional methods. At the same time, corresponding example images of "Ballet" are shown in Fig. 4. For depth image, as the artifacts were removed by proposed methods, the boundaries become sharper, which was helpful to reduce the geometric distortion in synthesized view. For the texture, it can be found that the boundaries of the objects are slightly sharper.

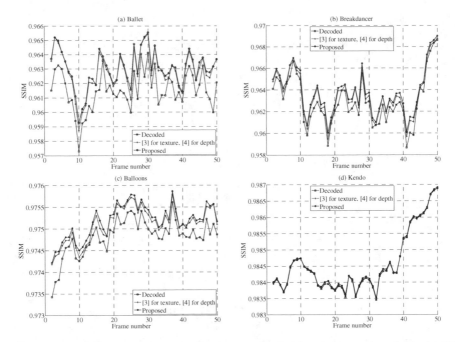

Fig. 3. SSIM quality of synthesized results by conventional and proposed methods: (a) "Ballet", (b) "Breakdancer", (c) "Balloons" and (d) "Kendo"

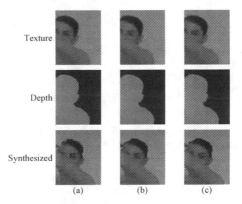

Fig. 4. Example images of the denoising results: (a) decoded, (b) [3] for texture, [4] for depth and (c) proposed

4   Conclusion

Recently, depth-based 3D video has received increasing interest. As different kinds of noise always exist in 3D video content, the denoising technique is playing an important role in 3D video system. In this paper, we proposed a 3D-spatial-texture bilateral filter by considering the relationship of pixels in the 3D space. In this filter, pixels in the 2D image coordinate are projected to the 3D world coordinate firstly. Then, each weight is defined with two kernels: a spatial kernel describing the geometric closeness in the 3D world coordinate and a range kernel describing the texture similarity. Considering the two factors, the relationship between pixels can be determined more accurate. Through the experimental results, we confirmed that the proposed method can achieve better performance in depth-based 3D video denoising, compared with conventional methods.

Acknowledgement. This research is supported by National Natural Science Foundation of China (NSFC, No. 61228102) and a grant from the Ph.D. Programs Foundation of Ministry of Education of China (No.20130182110010).

References

1. Fehn, C.: A 3D-TV approach using depth-image-based rendering (DIBR). In: Proc. Visualization, Imaging and Image Processing (VIIP), pp. 482–487 (2003)
2. Zhu, C., Li, S.: A new perspective on hole generation and filling in DIBR based view synthesis. In: Proc. International Conference on Intelligent Information Hiding and Multimedia Signal Processing (IIH-MSP), Beijing, China, pp. 607–610 (October 2013)
3. Tomasi, C., Manduchi, R.: Bilateral filtering for gray and color images. In: Proc. IEEE International Conference on Computer Vision (ICCV), Bombay, pp. 839–846 (January 1998)
4. Liu, S.J., Lai, P.L., Tian, D., Chen, C.W.: New depth coding techniques with utilization of corresponding video. IEEE Transactions on Broadcasting 57(2), 551–561 (2011)
5. Zhao, Y., Zhu, C., Chen, Z.Z., Tian, D., Yu, L.: Boundary artifact reduction in view synthesis of 3D video: from perspective of texture-depth alignment. IEEE Transactions on Broadcasting 57(2), 510–522 (2011)
6. Oh, K.J., Yea, S., Vetro, A., Ho, Y.S.: Depth reconstruction filter and down/up sampling for depth coding in 3-D video. IEEE Signal Processing Letters 16(9), 747–750 (2009)
7. Min, D.B., Lu, J.B., Do, M.N.: Depth video enhancement based on weighted mode filtering. IEEE Transactions on Image Processing 21(3), 1176–1190 (2012)
8. Jung, S.W.: Enhancement of image and depth map using adaptive joint trilateral filter. IEEE Transactions on Circuits and Systems for Video Technology 23(2), 258–269 (2013)
9. Zhu, C., Zhao, Y., Yu, L., Tanimoto, M. (eds.): 3D-TV System with Depth-Image-Based Rendering. Springer (2013) ISBN 978-1-4419-9963-4
10. Bossen, F., Flynn, D., Suehring, K.: JCT-VC AHG report: HEVC HM software development and software technical evaluation (AHG3). In: JCT-VC of ITU-T SG 16 WP 3 and ISO/IEC JTC 1/SC 29/WG 11 16th Meeting, San Jose CA, US (January 2014)
11. Sullivan, G.J., Ohm, J., Han, W.J., Wiegand, T.: Overview of the high efficiency video coding (HEVC) standard. IEEE Transactions on Circuits and Systems for Video Technology 22(12), 1649–1668 (2012)
12. Wang, Z., Bovik, A.C., Sheikh, H.R., Simoncelli, E.P.: Image quality assessment: from error visibility to structural similarity. IEEE Transactions on Image Processing 13(4), 1057–7149 (2004)

A System for Parking Lot Marking Detection

Bolan Su and Shijian Lu

Institute for Infocomm Research
1 Fusionopolis Way , Singapore 138632
{subl,slu}@i2r.a-star.edu.sg

Abstract. In this paper, we proposed a robust parking lot marking detection technique that is one important component for intelligent transportation systems and assisted/autonomous driving. Our system learns features of parking lot markings from training data and matches these templates to detected features in the test video during runtime. In the proposed system, maximally stable extremal regions (MSER) are used to detect a set of parking lot marking candidates. Features are then extracted from the detected candidates and Support Vector Machine (SVM) is applied to classify the parking lot marking in an efficient manner. With the detected parking lot markings, a parking lot is estimated by fitting two adjacent parking lot markings. The proposed technique is tested on real world street-view videos captured with an in-car camera. The experimental results show that the proposed technique is robust and capable of detecting parking lots under different lighting, marking sizes, and marking poses.

1   Introduction

Autonomous driving research aims to develop cars without human operation, which can lead to a revolution of our modern industry. The autonomous drivings will effectively provide more accurate environment information with different sensors, reduce the operation complexity for human driver, and increase the safeness in driving. These techniques will contribute in building a smart transportation system, for a more safety modern city and lower-energy cost environment.

One of the important techniques for autonomous driving is automatically detection of the traffic markings, which is very helpful for the analysis of the road information with vision sensors such as cameras. Compare with the traffic signs that appear at the side or on top of roads, road markings refer to the signs on the road surface that divide the road surface into different lanes and also provide certain vehicle parking policy information. Although many information provides by road markings is redundant to traffic signs, some of the information are only provided by road markings, such as parking lot markings.

In literature, the research related to road marking detection is quite limited. Most of the research focuses on traffic sign recognition [7,12,4]. A few techniques [3,13] make use of the lane marking for road marking detection, as the

W.T. Ooi et al. (Eds.): PCM 2014, LNCS 8879, pp. 268–273, 2014.
© Springer International Publishing Switzerland 2014

lane marking detection techniques is well-developed [10]. However, these methods will fail when the lane marking does not exists, which often happens in the videos for parking lot detection.

A comparison on road marking detected methods along with a benchmarking dataset are provided in [14]. Burrow et. al. [2] proposed a thresholding method to detect the road markings, but their method does not recognize the type of road markings. In addition, a few feature based approaches [6,8,11,16] are proposed to detect and classify road markings. Li et. al. [8] developed a shape matching technique to detect and recognize different type of road markings by representing the road marking shapes with a binarized bitmap. An eigensapce based method is proposed by Noda et. al. [11] to retain a complete road marking shape under different illumination, and blur conditions. On the other hand, Wu and Ranganathan [16] makes use of MSERs [9] to extract good road marking shapes for feature learning. However, these methods might not be invariance to perspective and rotation.

Fig. 1. Some parking lot marking examples

In this paper, we focus on parking lot markings detection, especially the 'T' and 'L' shapes, as illustrated in Figure 1. Compared with other type of road markings, parking lot markings are more difficult to classify, as they usually have a larger perspective and occluded by parked vehicles. We proposed a robust system that can efficiently detect the parking lot markers. The proposed system first extracts MSERs to detect possible connected component candidates. Several features are then extracted from those candidates and classified using a pre-trained model. Finally, the parking lot area is determined by fitting two adjacent parking lot markings. Note that the system can be easily extended for the detection and recognition of different types of road markings with a little adaptation. Experiments on real world street-view videos captured with an in-car camera show its robustness and effectiveness.

2 Proposed System

The overall flowchart of our proposed system is shown in Figure 2. To reduce the process time, the algorithm processes only on the pre-defined Regions of Interest (ROI), which is highlighted in yellow rectangle as shown in Figure 2. In the following part of this section, we will first discuss the generation of the road lane marking candidates, followed by extracting features from these candidate regions. The training and classification procedure will be illustrated at the end of this section.

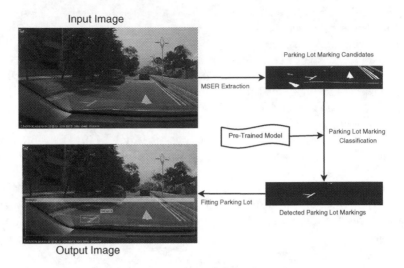

Fig. 2. The flowchart of our proposed parking lot marking detection system

2.1 Road Lane Marking Candidates Generation

It can be observed that the parking lot markings are usually painted on the road surface with the same color. Furthermore, the parking lot markings need to be brighter than the surrounding road surface to make them easily be recognized by the human drivers. So Maximally Stable Extremal Regions (MSERs) [9] can be applied to detect regions of interest, which are connected regions that the pixels inside the regions are always brighter than the pixels on the boundary.

MSERs are effective in detecting parking lot marking regions while a lot of unwanted regions are filtering out, such as vehicles, peoples, and trees. However, there are still lots of road markings are kept in the resulting image, which is illustrated in Figure 2. These connected components can be viewed as parking lot marking candidates, and will be pruned out by the following feature learning step.

2.2 Feature Extraction

Good features are important for object recognition. In our proposed systems, we make use of two well-known features for object recognition: Histogram of Oriented Gradient (HOG) [9] and Local Binary Pattern (LBP) [1] features. It has been shown that combining the LBP descriptor and HOG descriptor can improve the detection performance [15].

HOG feature descriptor decomposes an image block into small non-overlapped squared cells, and computes an histogram of oriented gradients in each cell. These histogram is then normalized and connected together to form a feature vector. Since the parking lot marking candidates are irregular connected components,

Fig. 3. One example of the parking lot detection result of our proposed system

we take the bounding rectangles of the connected components as the input of the HOG feature extractor. Furthermore, the bounding rectangles need to be normalized into the same size beforehand.

LBP feature descriptor is a simple texture operator that labels the pixels of an image by comparing the value with neighbouring pixels to construct a binary map. One of most important property of the LBP operator is its robustness to monotonic gray-scale changes. The extraction of LBP descriptor also follows the same manner as HOG descriptor.

In our proposed system, the image block size is set to $[32, 64]$, and the cell size is set to 16 for both HOG and LBP features empirically. The orientation histogram of HOG feature is set to eight bins. The neighbourhood window of LBP features is set to $[3, 3]$.

2.3   Road Lane Marking Regions Classification

The two features described in last subsection is concatenated to from a feature vector to train a classification model for parking lot marking recognition. To obtain the positive training samples, we first manually label lots of parking lot markings, then apply MSERs on those annotated images to look for connected components within the labelling regions. Those connected components outside the labelling regions are treated as negative training samples.

Linear SVM [5] is then applied to construct the classification models. The model is then applied to identify the parking lot marking regions. Finally, the adjacent parking lot marking regions are fitted to locate the parking lot area. One output sample is shown in Figure 3, where the parking lot markings are highlighted in yellow rectangles, and the parking lot area is estimated using green lines.

3   Experiments

To collect data for verification of our proposed system, we first capture videos contained road side parking lot markings in Singapore with an in-car camera.

Then the parking lot marking regions are manually labelled for training, while the negative samples are generated automatically. In total, there are 1396 annotated sample regions with 698 parking lot regions. We take 80% of them (1117) as training samples, the rest are used as testing samples.

We compare the classification performance by using HOG feature or LBP feature only with using both of the two features. A ten folds cross validation is applied. One confusion matrix of the classification result is shown in Table 1, where 1 denotes the positive sample, 0 denotes the negative sample. The average recognition accuracy of the ten folds cross validation for HOG feature, LBP feature and combined feature are 91.85%, 84.75% and 93.43%. The experimental results show that our proposed method can correctly detect most of the parking lot markings. The performance is further improved by combining HOG feature and LBP feature together.

Table 1. The confusion matrices of different features with 80% samples for training

Confusion Matrices		Predicted Results					
		HOG feature		LBP feature		Combine feature	
		1	0	1	0	1	0
Ground Truth	1	127	12	118	21	127	12
	0	9	131	23	117	7	133

4 Conclusion

In this paper, we propose a novel system for parking lot marking detection. The proposed system makes use of MSERs to generate parking lot marking candidates, and extracts HOG and LBP features of these candidates to prune out the non parking lot regions. The experimental results show good performance of our proposed method. In addition, the proposed method can be easily extent to detection of other types of road markings. In future study, we will test our algorithm on recognition of different road markings, such as arrows, lane marking, etc.. The proposed technique could be a good contribute to autonomous driving research as well as driver-assisted system development.

References

1. Ahonen, T., Hadid, A., Pietikainen, M.: Face description with local binary patterns: Application to face recognition. IEEE Transactions on Pattern Analysis and Machine Intelligence 28(12), 2037–2041 (2006)
2. Burrow, M.P.N., Evdorides, H.T., Snaith, M.S.: Segmentation algorithms for road marking digital image analysis. In: Proceedings of the Institution of Civil Engineers, Transport, vol. 156, pp. 17–28 (2003)
3. Charbonnier, P., Diebolt, F., Guillard, Y., Peyret, F.: Road markings recognition using image processing. In: IEEE Conference on Intelligent Transportation System, ITSC 1997, pp. 912–917 (November 1997)

4. Chen, L., Li, Q., Li, M., Mao, Q.: Traffic sign detection and recognition for intelligent vehicle. In: 2011 IEEE Intelligent Vehicles Symposium (IV), pp. 908–913 (June 2011)
5. Fan, R.E., Chang, K.W., Hsieh, C.J., Wang, X.R., Lin, C.J.: Liblinear: A library for large linear classification. Journal of Machine Learning Research 9, 1871–1874 (2008)
6. Kheyrollahi, A., Breckon, T.: Automatic real-time road marking recognition using a feature driven approach. Machine Vision and Applications 23(1), 123–133 (2012)
7. Kiran, C., Prabhu, L., Abdu, R., Rajeev, K.: Traffic sign detection and pattern recognition using support vector machine. In: Seventh International Conference on Advances in Pattern Recognition, ICAPR 2009, pp. 87–90 (February 2009)
8. Li, Y., He, K., Jia, P.: Road markers recognition based on shape information. In: 2007 IEEE Intelligent Vehicles Symposium, pp. 117–122 (June 2007)
9. Matas, J., Chum, O., Urban, M., Pajdla, T.: Robust wide-baseline stereo from maximally stable extremal regions. Image and Vision Computing 22(10), 761–767 (2004)
10. McCall, J., Trivedi, M.: Video-based lane estimation and tracking for driver assistance: survey, system, and evaluation. IEEE Transactions on Intelligent Transportation Systems 7(1), 20–37 (2006)
11. Noda, M., Takahashi, T., Deguchi, D., Ide, I., Murase, H., Kojima, Y., Naito, T.: Recognition of road markings from in-vehicle camera images by a generative learning method. In: Proceedings of IAPR Conference on Machine Vision Applications (2009)
12. Ruta, A., Li, Y., Liu, X.: Real-time traffic sign recognition from video by class-specific discriminative features. Pattern Recognition 43(1), 416–430 (2010)
13. Vacek, S., Schimmel, C., Dillmann, R.: Road-marking analysis for autonomous vehicle guidance. In: Proceedings of the 3rd European Conference on Mobile Robots (2007)
14. Veit, T., Tarel, J.P., Nicolle, P., Charbonnier, P.: Evaluation of road marking feature extraction. In: 11th International IEEE Conference on Intelligent Transportation Systems, ITSC 2008, pp. 174–181 (October 2008)
15. Wang, X., Han, T., Yan, S.: An hog-lbp human detector with partial occlusion handling. In: 2009 IEEE 12th International Conference on Computer Vision, pp. 32–39 (September 2009)
16. Wu, T., Ranganathan, A.: A practical system for road marking detection and recognition. In: 2012 IEEE Intelligent Vehicles Symposium (IV), pp. 25–30 (June 2012)

Fast Search of Binary Codes with Distinctive Bits

Yanping Ma[1], Hongtao Xie[2], Zhineng Chen[3], Qiong Dai[2],
Yinfei Huang[4], and Guangrong Ji[1]

[1] School of Information Science and Engineering,
Ocean University of China, Qingdao, China, 266100
[2] Institute of Information Engineering, Chinese Academy of Sciences,
National Engineering Laboratory for Information Security Technologies,
Beijing, China, 100093
[3] Interactive Digital Media Technology Research Center, Institute of Automation,
Chinese Academy of Sciences, Beijing, China, 100190
[4] Shanghai Stock Exchange, Shanghai, China, 200120
xiehongtao@iie.ac.cn

Abstract. Although distance between binary codes can be computed fast in hamming space, linear search is not practical for large scale dataset. Therefore attention has been paid to the efficiency of performing approximate nearest neighbor search, in which Hierarchical Clustering Trees (HCT) is the state-of-the-art method. However, HCT builds index with the whole binary codes, which degrades search performance. In this paper, we first propose an algorithm to compress binary codes by extracting distinctive bits according to the standard deviation of each bit. Then, a new index is proposed using com-pressed binary codes based on hierarchical decomposition of binary spaces. Experiments conducted on reference datasets and a dataset of one billion binary codes demonstrate the effectiveness and efficiency of our method.

Keywords: binary codes, approximate nearest neighbor search, binary indexing.

1   Introduction

Nearest neighbor search in large-scale dataset is a core problem in many applications, including image retrieval [1][2], object recognition [3] and computer vision [4][5]. Recently, there has been growing interest in representing visual content and building index in terms of compact binary codes [6][7][8], as they are fast to compute and storage efficient. What's more, binary codes are compared using the Hamming distance [9], which can be computed by executing the XOR operation followed by a few bitwise instructions that can be performed quickly, especially on modern central processing units. Usually millions of binary codes are compared only in less than a second [9].

Even though the distance between binary codes can be computed efficiently, using linear search for exact matching is practical only for small datasets [10][11]. For large-scale datasets, exact linear search will lose its time performance. This is because that the computing capability of the processors is limited, while the size of dataset is infinite. To solve this problem, hashing techniques-based approximate nearest

W.T. Ooi et al. (Eds.): PCM 2014, LNCS 8879, pp. 274–283, 2014.

neighbor (ANN) search algorithms for binary codes have been proposed [6][12][13]. These methods can offer speedups of several orders of magnitude over linear search, at the cost that a fraction of the returned results are approximate neighbors, but usually close in distance to the exact neighbors. So they can improve the search efficiency while sacrificing little precision. However, as pointed out in the literatures [1][9][10], these methods cannot deal with long codes and have limited scalability.

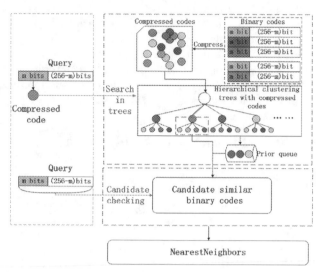

Fig. 1. The framework of our index with compressed binary codes. In index building, binary codes in the dataset are compressed into short distinctive codes. Then, the index is built with compressed codes by constructing multiple hierarchical clustering trees. In searching, compressed code of the query is searched in the index, getting similar candidates. Then, the whole codes of the query and the candidates are used for validity checking.

To improve the ability of indexing binary codes with long codes and large-scale datasets, Muja *et al.* [10] propose an index structure, namely hierarchical clustering trees (HCT) based on hierarchical decomposition of the search space. HCT is the state-of-the-art approach of ANN search for binary codes. In index building, HCT builds multiple random hierarchical clustering trees [20], in which each non-leaf node contains a cluster center and each leaf node contains the dataset items. In searching, HCT starts from the root of the tree using the whole binary code for comparisons, finds the closest cluster until reaching the leaf node and returns the binary codes in the leaf node. While favorable for simplicity and scalability, HCT has one shortcoming. It uses whole binary code for index building and searching. As many bits of binary code are not distinctive [21], it has unnecessary computational overhead. Besides, ANN search in HCT by the whole binary code results in too much candidates for filtering out false positives [9], which dramatically decreases the search efficiency.

To overcome the above problem, in this paper we first put forward an algorithm to compress binary codes by extracting distinctive bits according to the standard deviation of each bit. Then, we build a new index using compressed binary codes based on hierarchical decomposition of binary spaces, as shown in Figure 1. Figure 1 describes the index building and searching processes of our index. In index building, binary codes in the dataset are compressed into short distinctive codes. Then, the index is built with compressed codes by constructing hierarchical clustering trees. In searching, distinctive code of the query is searched in the hierarchical trees, getting candidate similar items. Then, the whole codes of the query and the candidates are used for validity checking, to remove any non nearest neighbors.

There are two advantages of our method. Firstly, with binary code compression, the search efficiency can be obviously improved. Secondly, it is simple and easy to implement. Extensive experiments are conducted on benchmark datasets and we further employ 10^9 binary codes stemmed from ANN_SIFT1B set [14] to assess the large scale search performance. Our method demonstrates excellent improvement in efficiency and the search speed of our index can be significantly improved compared to HCT index [10], while keeping high accuracy. It also leads to consistent performance gains on the large scale database.

2   Fast Matching of Binary Codes with Distinctive Bits

In this section, we first propose an algorithm to compress binary codes by extracting distinctive bits, according to the standard deviation of each bit. Then, a new index using compressed binary codes based on hierarchical decomposition of binary spaces is explained.

2.1   Extracting Distinctive Bits of Binary Code

As there are many indistinctive bits in the whole binary code [21], ANN search with whole code in HCT index [10] leads to unnecessary computation and lots of candidate items [9], which also decrease efficiency. In order to solve this problem, we propose a method using standard deviation to compress binary code.

In statistics theory, the standard deviation shows how much variation or dispersion of data points from the average exists. A low standard deviation indicates that the data points tend to be very close to the mean; a high standard deviation indicates that the data points are spread out over a large range of values [22]. We just use the above properties to compute standard deviation of each bit. The bits with higher standard deviations are more distinctive in the whole binary code. So, we extract short distinctive bits in each binary code to represent the former long one.

In more detail, suppose each binary code has b bits

$$h = (x_1, x_2, ..., x_i, ..., x_b),$$ (1)

where h is a binary code with b bits and x_i is i-th bit. In case of ORB binary codes [8], each one has 256 bits. To learn distinctive bits, we build a training set T with n binary codes and compute the standard deviation of each bit. The function of computing standard deviation is defined as follows:

$$SDValue(x_i) = \sqrt{\frac{1}{n}(\sum_{k=1}^{n}(h_k(x_i) - \sum_{k=1}^{n}h_k(x_i)/n))^2},$$ (2)

where $h_k(x_i)$ is the *i-th* bit of the *k-th* binary code and *n* is the number of binary codes in the training set.

Figure 2 illustrates the standard deviation of each bit with 50000 ORB codes. We can see that the standard deviation values of same bits are relative larger than the other bits. The standard deviation values of a majority of the bits are between 0.3 and 0.5. But the standard deviation values of some bits are below 0.3, even less than 0.2. It indicates these bits have different degrees of distinctiveness.

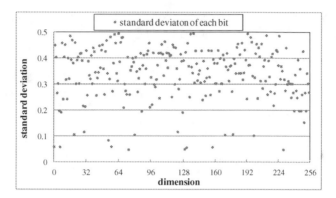

Fig. 2. Training result of standard deviation with 50000 binary codes. *X*-axis is the dimension of binary code. *Y*-axis is standard deviation of each bit.

With equation (2), the steps of learning distinctive bits are as follows:

1. Build a training set *T* for learning, and *T* has no elements in common with the base set;
2. Compute the standard deviation of each bit;
3. Pick *m* bits according to the top *m* $SDValue(x_i)$ values, as the distinctive bits;
4. Record the dimension values of these *m* bits for index building and searching.

2.2   Fast Matching of Binary Codes with Distinctive Bits

With the method proposed in subsection 2.1, we can compress binary code by extracting distinctive bits. In this subsection, we introduce the procedures of index building and searching with short distinctive codes.

2.2.1   Index Building

In index building, we first extract distinctive bits to compress the binary codes of the dataset. Then, the compressed binary codes are clustered into multiple hierarchical clustering trees, as explained in Algorithm 1.

In algorithm 1, $db(d_1, d_2, ..., d_m)$ is the dimension value vector of the distinctive bits and d_m is the dimension value of *m-th* bit. We extract distinctive bits according to *db* for each binary code in the dataset *D* to obtain short distinctive codes D'. Then multiple hierarchical clustering trees are constructed with D', to build the index of distinctive codes, which ensures fast ANN search.

Algorithm1. Building index of distinctive codes

Input: D, dataset; TD, training dataset
Output: index of distinctive codes
Parameters: D', compressed dataset; db, a vector of the dimension value of each distinctive bit; m, the size of db; $Index$, index of distinctive codes.
1: $db \leftarrow$ Train distinctive bits(TD, m)
2: $D' \leftarrow$ Compress binary codes(D, db)
3: $Index \leftarrow$ Build multiple hierarchical clustering trees (D')

2.2.2 Index Searching

The searching process is presented in Algorithm 2. It starts with extracting distinctive bits of the query. Then it searches multiple hierarchical clustering trees to obtain candidate results CR. Finally, exhaustive linear search is executed in CR to filter the false matches with the whole binary codes.

Algorithm 2. Search in index of distinctive codes

Input: q, a query; $Index$, index of distinctive codes
Output: nearest neighbors
Parameters: q', compressed query; db, a vector contains the dimension value of each distinctive bit; CR, candidate results; cn, size of CR.
1: **for** each q **do**
2:      $q' \leftarrow$ Compress binary code (q, db)
3:      $CR \leftarrow$ Search multiple hierarchical clustering trees ($Index$, q', cn)
4:      nearest neighbors$\leftarrow$ Exhaustive search (CR)
5: **end for**

In searching multiple hierarchical trees, it starts with a depth-first traversal of each tree, during which it visits the node closest to the query and recursively explores it, while adding the unexplored neighboring nodes to a priority queue [10]. When reaching the leaf node all the points contained are selected as candidates. After all the trees have been explored once, the searching process is continued by getting the closest node to the query from the priority queue, and resuming depth-first traversal from that node. It ends when the number of candidates exceeds a maximum threshold cn, which specifies the degree of approximation desired from the algorithm.

By building index and searching with distinctive codes, it can not only reduce the computational overhead of tree traversal, but also reduce the number of candidates for validity checking. So it can obviously improve the efficiency of the index, as illustrated in the next section.

3   Experiments and Evaluations

In this section, experiments are conducted to evaluate the proposed index structure. First, we investigate the influences of various parameters, and then compare the overall performance with the state-of-the-art method on famous benchmark dataset.

3.1 Experimental Setup

The experiments are conducted on following three datasets, as shown in Table 1. To obtain binary codes, we crawl 25000 images from Flickr and extract ORB descriptors [8]. The training set has 500,000 ORB codes and is used for learning the distinctive bits. The query set has 10,000 ORB codes, and the base set for building index has 2500,000 ORB codes. Note that the training set, query set and base set have no intersection with each other.

Table 1. Description of datasets. The item is ORB feature of 256 bits.

Dataset	Size
training set	500,000
query set	10,000
base set	2500,000

To evaluate large scale search performance, we use the famous ANN_SIFT1B dataset [14] and the item is 128-D SIFT descriptor. It contains a query set of 10^4 descriptors and a base set of 10^9 descriptors. We use the minimal loss hashing [9] to create datasets of binary codes and each code has 256 bits. The experimental environment is Intel Xeon E5-2620*2(2.00 GHz, 7.2GT/s, 15M cache, 6cores) and with 64 GB memory.

For experimental evaluation, we use speed up to compare our index with HCT [10]. The speed up is defined as follows:

$$speedup = \frac{Linear\_time}{Ann\_time},$$ (3)

where $Linear\_time$ is time cost of linear search and Ann_time is time cost of our method or HCT. Precision is another principle to measure the performance of an index and is defined as follows [10]:

$$precision = \frac{count(AR \cap LR)}{count(LR)},$$ (4)

where AR is the ANN search result of our method or HCT, and LR is the linear search result. For each query, we calculate its precision, then we take the mean value over all queries. In experiments, we evaluate the performance of the algorithm by analyzing the speedup over linear search with respect to the search precision for different settings. Speedup over linear search is used both because it is an intuitive measure of the performance of algorithm and is relatively independent of the hardware [10].

3.2 Influences of Parameters

We implement our method on top of the open source FLANN library [11] to take advantage of the parameter auto-tuning framework provided by the library. The performance of our method is influenced by the following parameters: number of hierarchical clustering trees, branching factor, maximum leaf size, the number of distinctive bits m and the number of candidates. For number of hierarchical clustering trees, branching factor and maximum leaf size, they are auto-confirmed by the FLANN library. Specifically, for our method there are 4 clustering trees, the branching factor is 16 and the maximum leaf size is 200. For considerations of space, we only study the influences of the number of distinctive bits and the number of candidates.

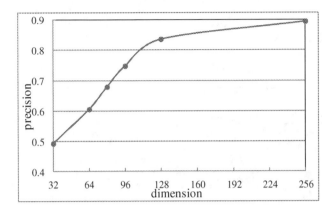

Fig. 3. The precision of our method with respect to different number of distinctive bits

Figure 3 shows the accuracy of our method with different number of distinctive bits. We can see that the more bits we use in building index and the higher search accuracy can be obtained. When we use the whole binary codes, the search accuracy is about 89%. However, when the number grows from 128 to 256, the growth rate of accuracy begins to decline. If we extract 128 distinctive bits for each code, the search accuracy is about 83%. As the speedup of 128bits is much higher than that of 256 bits, as shown in Figure 4, we extract 128 distinctive bits for each binary code, in order to keep an optimal balance of search precision and efficiency.

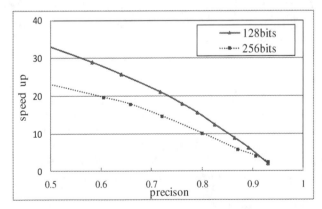

Fig. 4. The precision and speed up of our method over linear search with respect to different number of distinctive bits

Figure 4 shows the speedup and precision with 128 distinctive bits and 256 distinctive bits (non-compression), respectively. We can see that when the search precision is below 90%, code compression has better search efficiency than non-compression.

When the precision is higher than 90%, the speedup of 128 bits is about the same as that of 256 bits. The reason is that compression sacrifices little precision and we need return more candidates when using 128 bits than using 256 bits, to ensure search precision. For many applications, such as image retrieval and filtering [23], the single point searching precision of 0.8-0.9 is sufficient for the entire accuracy. So our method is applicable for these scenarios.

Figure 5 illustrates the impact of the number of candidate results on the performance of our method. As elaborated in the last paragraph, we have to return enough candidates to ensure high accuracy. From Figure 5, we can observe that the accuracy improves when the number of candidates increases. However, when the number of candidates increases from 2000 to 4000, the accuracy nearly stays stable. The more candidates the more exact near neighbors are returned, but the search is more expensive. Therefore, we set the number of candidates to be 2000, which is still less than the number of candidates in HCT [10]. This is because that with short distinctive codes, the dimension of Hamming space and radius of the searching Hamming ball are only a fraction of that for the long codes [9], which results in less candidates.

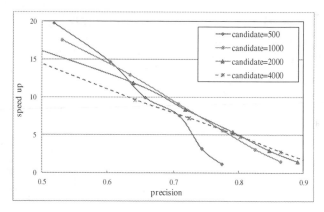

Fig. 5. The precision and speed up of our method over linear search for different number of candidates

3.3   Comparison of Overall Performance

Figure 6 compares the performance of our index with HCT [10]. The experiment is conducted on the dataset with 2500,000 binary codes. The parameters of our method are set as described in the previous section, and the parameters in HCT are auto-tuned by the FLANN library [11]. From Figure 6, we can see that when the search precision is lower than 90%, our index has much better efficiency than HCT. The search speed can be averagely increased by about 55% compared to HCT. When the search precision exceeds 93%, our index and HCT index have similar speed up values, but our method is consistently more efficient than HCT index. This reveals that when we use short distinctive codes, we can get higher speed up. However, when we need the maximal precision, the speed up is not too obvious. So, we should take into account of both precision and speed up to obtain good performance with our index. Fortunately, the precision is still high enough for most scenarios.

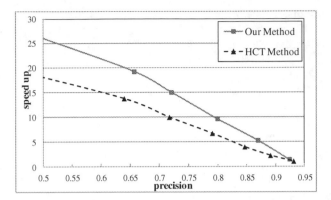

Fig. 6. Comparison between our index and HCT index for dataset with 2500,000 binary codes

3.4   Scalability

We use the ANN_SIFT1B dataset (10^9 items) [14] to demonstrate the scalability of our method, as shown in Figure 7. It can be seen that the search performance of our method scales well with the dataset size and it always has better performance than HCT. The search speed can be averagely increased by about 40% compared to HCT, in dealing with large scale dataset. According to the experiment, our method has higher efficiency, while keeping accuracy.

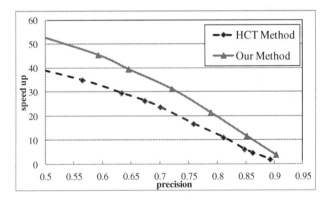

Fig. 7. Comparison between our index and HCT for dataset with 10^9 binary codes

4   Conclusions

In this paper, we introduce an index structure for fast approximate search of binary codes with distinctive bits. The goal is to improve the state-of-the-art HCT index. We first propose an algorithm to compress binary codes by extracting distinctive bits according to the standard deviation of each bit. Then, a new index is proposed using compressed binary codes based on hierarchical decomposition of binary spaces. In brief, the proposed index is simple and cheap, and has been proven by exhaustive experiments.

Acknowledgements. This work is supported by National Nature Science Foundation of China (61303171, 61303175, 61271406), National Key Technology R&D Program (2012BAH13F04), the "Strategic Priority Research Program" of the Chinese Academy of Sciences (XDA06031000), National High Technology Research and Development Program (2011AA01A103).

References

1. Zhang, W., Gao, K., Zhang, Y., Li, J.: Efficient Approximate Nearest Neighbor Search with Integrated Binary Codes. In: ACM MM, pp. 1189–1192 (2011)
2. Chu, W.-T., Li, C.-J., Tseng, S.-C.: Travelmedia: an intelligent management system for media captured in travel. JVCI 22, 93–104 (2011)
3. Torralba, A., Fergus, R., Weiss, Y.: Small codes and large image databases for recognition. In: IEEE CVPR (2008)
4. Zhang, L., Zhang, Y., Tang, J., Gu, X., Li, J., Tian, Q.: Topology Preserving Hashing for Similarity Search. In: ACM MM (2013)
5. Xie, H., Zhang, Y., Tan, J., Guo, L., Li, J.: Contextual Query Expansion for Image Retrieval. IEEE Trans. on Multimedia 16(4) (2014)
6. Salakhutdinov, R., Hinton, G.: Semantic Hashing. International Journal of Approximate Reasoning (2009)
7. Strecha, C., Bronstein, A., Bronstein, M., Fua, P.: LDAHash: improved matching with smaller descriptors. IEEE Transactions on PAMI 34(1), 66–78 (2012)
8. Rublee, E., Rabaud, V., Konolige, K., Bradski, G.: ORB: an efficient alternative to SIFT or SURF. In: IEEE ICCV, pp. 2564–2571 (2011)
9. Norouzi, M., Punjani, A., Fleet, D.J.: Fast search in hamming space with multi-index hashing. In: IEEE CVPR (2012)
10. Muja, M., Lowe, D.G.: Fast matching of binary features. In: CRV (2012)
11. Muja, M., Lowe, D.G.:
    http://people.cs.ubc.ca/~mariusm/index.php/FLANN/FLANN
12. Zitnick, C.L.: Binary coherent edge descriptors. In: Daniilidis, K., Maragos, P., Paragios, N. (eds.) ECCV 2010, Part II. LNCS, vol. 6312, pp. 170–182. Springer, Heidelberg (2010)
13. Weiss, Y., Torralba, A., Fergus, R.: Spectral Hashing. In: Advances in Neural Information Processing Systems (2008)
14. Jegou, H., Douze, M., et al.: Product Quantization for Nearest Neighbor Search. IEEE Transactions on PAMI 33(1), 117–128 (2011)
15. Aly, M., Munich, M., Perona, P.: Distributed kd-trees for retrieval from very large image collections. In: BMVC (2011)
16. Babenko, A., Lempitsky, V.: The inverted multi-index. In: IEEE CVPR (2012)
17. SilpaAnan, C., Hartley, R.: Optimized KD-trees for fast image descriptor matching. In: CVPR (2008)
18. Gionis, A., Indyk, P., Motwani, R.: Similarity search in high dimensions via hashing. In: Proceedings of the International Conference on Very Large Data Bases (1999)
19. Broder, A.Z.: On the resemblance and containment of documents. In: IEEE Compression and Complexity of Sequences, pp. 21–29 (1997)
20. Park, H.S., Jun, C.H.: A simple and fast algorithm for K-medoids clustering. Expert Systems with Applications 36(2), 3336–3341 (2009)
21. Zhang, L., Zhang, Y., Tang, J., Lu, K., Tian, Q.: Binary Code Ranking with Weighted Hamming Distance. In: IEEE CVPR (2013)
22. Bland, J.M., Altman, D.G.: Statistics notes: measurement error (1996)
23. Jegou, H., Douze, M., Schmid, C.: Improving bag-of-features for large scale image search. Int. J. Comput. Vis. 87(3), 316–336 (2010)

Data-Dependent Locality Sensitive Hashing

Hongtao Xie[1], Zhineng Chen[2], Yizhi Liu[3], Jianlong Tan[1], and Li Guo[1]

[1] Institute of Information Engineering, Chinese Academy of Sciences,
National Engineering Laboratory for Information Security Technologies,
Beijing, China, 100093
[2] Interactive Digital Media Technology Research Center, Institute of Automation,
Chinese Academy of Sciences, Beijing, China, 100190
[3] School of Computer Science and Engineering, Hunan University of Science and Technology,
Xiangtan, China, 411201
{xiehongtao,tanjianlong,guoli}@iie.ac.cn,
zhineng.chen@ia.ac.cn, liuyizhi928@gmail.com

Abstract. Locality sensitive hashing (LSH) is the most popular algorithm for approximate nearest neighbor (ANN) search. As LSH partitions vector space uniformly and the distribution of vectors is usually non-uniform, it poorly fits real dataset and has limited performance. In this paper, we propose a new data-dependent LSH algorithm, which has two-level structures to perform ANN search in high dimensional spaces. In the first level, we first train a number of cluster centers, then use the cluster centers to divide the dataset into many clusters and the vectors in each cluster has near uniform distribution. In the second level, we construct LSH tables for each cluster. Given a query, we first determine a few clusters that it belongs to with high probability, and then perform ANN search in the corresponding LSH tables. Experimental results on the reference datasets show that the search speed can be increased by 48 times compared to E2LSH, while keeping high search precision.

Keywords: Locality sensitive hashing, approximate nearest neighbor.

1 Introduction

Nearest neighbor search in high-dimensional space is the core problem in database management, data mining and computer vision [1]. Traditional tree-based indexing methods can return accurate results, but they are time consuming for data with high dimensionalities. It has been shown that when the dimensionality exceeds 10, existing tree-based indexing structures are slower than the linear-scan [2], which is known as "dimensionality curse". To solve this problem, hash-based methods are proposed [3] for approximate nearest neighbor (ANN) search.

Among all the hash-based algorithms, locality sensitive hashing (LSH) [4] is one of the most widely used ANN search methods. It first takes a number of random projection functions to group or collect items close to each other into the same buckets with a high probability. In order to perform similarity query, LSH hashes the query item into some buckets and uses the data items within those buckets as potential candidates for the final results. Moreover, the items in the buckets are ranked according

W.T. Ooi et al. (Eds.): PCM 2014, LNCS 8879, pp. 284–293, 2014.
© Springer International Publishing Switzerland 2014

to the exact distance to the query item to compute the nearest neighbor. The final ranking computation among the candidates is called *short-list* search. LSH is an algorithm based on random projection, so it partitions the vector space uniformly. In contrast, the distribution of the items in vector space is usually non-uniform. Consequently, the distribution of randomly projected values is far from being uniform. In such a situation, querying will cost a lot of time for many disturbance points are probed and it severely limits its search performance.

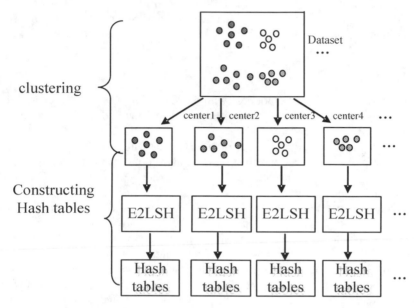

Fig. 1. The framework of our two-level index. In the first level, we group the dataset into clusters though the centers that are trained beforehand. During the second level, we apply E2LSH [4] to each cluster.

To solve aforementioned problem, improved hash algorithms have been proposed [5][6][7]. These methods use more candidates, estimate optimal parameters or use improved hash functions. As they only consider the average runtime or quality of the search process over randomly sampled hash functions, which may result in large deviations in runtime cost or the quality of k-nearest neighbor results [8], they still cannot solve the problem well. In this paper, we present a novel data-dependent LSH algorithm, which is composed of two-level structures. During the first level, we first train a number of centers through K-means algorithm [9] on the training set and then divide the dataset into a number of clusters which are corresponding to these centers. This step gets the similar points together. Within each cluster, the distribution of items is more uniform than that of the items in the whole dataset. During the second level, we apply E2LSH technique to each cluster to construct LSH tables. The uniform distribution of points within clusters results in the uniform distribution of point indices in the corresponding hash tables. The framework of proposed data-dependent LSH algorithm is illustrated in Figure 1.

To further improve the search speed, we propose a novel search pruning method, based on a new distance metric which utilizes information about the relative positions and sizes of all clusters. Given a query, we first compute its distances to all cluster centers and take a few clusters as the candidates. Then, we apply E2LSH search algorithm to the candidate clusters and get the nearest neighbors to the query. The experiments conducted on benchmark datasets show that our algorithm can obviously improve the search speed while keeping high precision.

The rest of this paper is organized as follows. Section 2 gives a review of previous works. Section 3 presents the proposed data-dependent LSH method. Section 4 shows the experimental comparisons. In section 5, we give the conclusion and future work.

2   Literature Review

As the basic LSH algorithm cannot deal with non-uniformly distributed dataset, many techniques have been proposed for improvement. These methods are classed into data independent and data dependent.

Data independent methods do not use the base dataset for building index. Lv *et al.* propose multi-probe LSH [12], which systematically probes the buckets near the query points in a query-dependent manner, instead of only probing the bucket that contains the query point. It can obtain a higher recall ratio with fewer hash tables, but may result in larger selectivity from additional probes. Bawa *et al.* propose LSH-forest [13], which avoids tuning of the parameter by representing the hash table as a prefix tree and the parameter is computed based on the depth of the corresponding prefix-tree leaf node. Dong *et al.* [14] construct a statistical quality and runtime model using a small sample dataset, and then compute parameters that can result in a good balance between high recall ratio and low selectivity. Pan *et al.* propose Bi-level LSH [8]. In the first level, it uses a RP-tree to partition the dataset into sub-groups with bounded aspect ratios and is used to compute well-separated clusters. During the second level, it builds a single LSH hash table for each sub-group along with a hierarchical structure based on space-filling curves. Bi-level LSH is similar to our method in this paper, but as it uses RP-tree to partition the dataset randomly, it has limited performance.

Data dependent methods use the base dataset for building index. Babenko et al. propose inverted multi-index [15], which generalizes the inverted index by replacing the standard quantization within inverted indices with product quantization. In some methods, data are represented by binary codes to reduce memory usage and computation times, as the calculations in Hamming space can be executed by efficient bitwise XOR operation. Among these schemes, spectral hashing [6] is a representative method. It defines a hard criterion for a good code that is related to graph partitioning and uses a spectral relaxation to obtain an eigenvector solution. Using binary codes can improve the efficiency of indexing [16]. However, we only focus on using pure floating-point vector.

3   Data-Dependent Locality Sensitive Hashing

In this section, we first elaborate the details of our data-dependent LSH framework and propose a distance metric. Then we present a new search algorithm for data-dependent LSH.

3.1   Constructing Data-Dependent LSH Index

As LSH cannot deal with non-uniform distribution dataset, it has limited search efficiency. K-means is a popular clustering algorithm and can partition n items into k clusters, in which each item belongs to the cluster with the nearest mean [9]. To make the distribution of the items in the hash tables much more uniform, we use K-means algorithm to divide the dataset into several clusters. Then, we apply E2LSH method to each cluster to build the two-level index.

The procedure of constructing data-dependent LSH hash tables is given in Algorithm 1. We first apply the K-means algorithm to the training dataset D_T^n, and get a number of cluster centers $\mu_j (j=1,2\cdots k)$. Then we assign the points in the dataset D^n to these centers and partition the dataset into k clusters. Finally, we adopt E2LSH algorithm [4] to each cluster for building LSH tables.

Algorithm 1. Building data-dependent LSH

Input: A set of data items $\{x_i\} \in R^n$ (n is the dimension of each item.)

Output: A set of E2LSH structures $\{T_i\}$. Each E2LSH structure is composed of multiple LSH tables, and there are many hash buckets in each table. The indices of the points are preserved in corresponding buckets.

Notations: D^n, the index dataset; $x_i \in D^n$ ($i=1,2\cdots M$); D_T^n, the training set; $\mu_j (j=1,2\cdots k)$, the centers gotten by K-means algorithm on D_T^n; $A_j (j=1,2\cdots k)$, the clusters which D^n is grouped into.

Operation:

    1. $\mu_j (j=1,2\cdots k) \leftarrow$ apply K-means algorithm to D_T^n

    2. for $1 \le i \le M$ do

    3. $A_j \leftarrow \arg\min_j \| x_i - \mu_j \|^2$

    4. end for

    5. for $1 \le j \le k$ do

    6. $T_j \leftarrow$ apply E2LSH to A_j

    7. end for

Because we apply K-means algorithm to partition the dataset, the distribution of data items in our hash tables is more uniform than that in original E2LSH hash tables.

3.2   Search Algorithm for Data-Dependent LSH Index

In LSH-based methods, *short-list* search is the main bottleneck of improving search speed [8]. To reduce the time cost of *short-list* search, we present a novel pruning algorithm for our data-dependent LSH.

In the first level, we partition the dataset into a number of clusters. Intuitively, these clusters have different sizes, and the relative distances between the cluster centers are also different. Due to influence of the size of different clusters and the relative distances, we incorporate these factors into the computation of distance and propose the distance metric $D_i^{'}$ as follows:

$$D_i^{'} = \frac{d_i}{d_1} \times \frac{s_1}{s_i}. \tag{1}$$

We first compute the distances $d_i^{'}$ between the query q and all cluster centers μ_i and rank the distances in ascending order and get the sequence d_i. d_1 is the nearest center to q and s_1 is the size of corresponding cluster. In this paper, we use the amount of the items in each cluster as its size metric.

In order to perform similarity search, we first get the sequence $D_i^{'}$ as mentioned above and rank $D_i^{'}$ in ascending order to get the ordered sequence D_i. Then, we take the top m clusters $A_1, A_2, \cdots A_m$ as the candidate clusters to be probed. We apply E2LSH search algorithm to $A_1, A_2, \cdots A_m$ and perform *short-list* search to get the nearest neighbors of the query q. The procedure of the search algorithm is depicted in algorithm 2.

Algorithm 2. Search algorithm for Data-dependent LSH

Input: A query point q.

Output: the nearest neighbor of q.

Notations: $L_{D_i} (i = 1, 2, \cdots m)$, the clusters corresponding to D_i ; $m(1 \leq m < k)$, the number of probed clusters; s_i, the size of corresponding cluster.

Operation:

1.   for $1 \leq i \leq k$ do
2. $d_i^{'} = \| q - \mu_i \|_2$
3.   end for
4. $d_1 d_2 \cdots d_k \leftarrow$ rank $d_i^{'}(i = 1, 2 \cdots k)$ in ascending order
5.   for $1 \leq i \leq k$ do
6. $D_i^{'} = \frac{d_i}{d_1} \times \frac{s_1}{s_i}$
7.   end for
8. $D_1, D_2, \cdots D_k \leftarrow$ rank $D_1^{'}, D_2^{'}, \cdots D_k^{'}$ in ascending order
9. $a_1, a_2 \cdots \leftarrow$ apply E2LSH search algorithm to $L_{D_i} (i = 1, 2 \cdots m)$.
10. $n = \arg \min_i \| q - a_i \|_2$
11. return a_n

In practice, algorithm 2 is a pruning strategy. As algorithm 2 takes advantage of the information about the relative positions and sizes of all clusters, we can improve the search speed while keeping high precision, as illustrated in the next section.

4 Experimental Results

In this section, we first study the influence of different parameters. Then, we conduct multiple comparison experiments between E2LSH [4], Bi-level LSH [8] and our data-dependent LSH.

4.1   Experimental Setup

In the experiments, we use the publicly available BigANN set [10]. BigANN set is composed of four data packages including ANN_SIFT10K, ANN_SIFT1M, ANN_GIST1M and ANN_SIFT1B. We use ANN_SIFT1M, which contains 1M SIFT descriptors as index data, 100k SIFT descriptors as training set and 10k SIFT descriptors as query data, as described in Table 1. The training set is used to learn the cluster centers.

Our method is single-threaded programmed and executed on a server which has 64G main memory, Intel Xeon E5-2620*2(2.00 GHz, 7.2GT/s, 15M cache, 6cores). We will draw the conclusion in term of the comparison of accuracy and efficiency.

Table 1. Description of training set, index set and query set

Dataset	Size
training set	100,000
query set	10,000
base set	1,000,000

For experimental evaluation, we use speedup to compare our two-level index with E2LSH [4] and Bi-level LSH [8]. The speedup is defined as follows:

$$speedup = \frac{linear\_time}{ann\_time}, \tag{2}$$

where *linear_time* is time cost of linear search and *ann_time* is time cost of our method, E2LSH or Bi-level LSH. Precision is another principle to measure the performance of an index algorithm. The precision is defined as follows:

$$precision = \frac{count(AR \cap LR)}{count(LR)}, \tag{3}$$

where *AR* is the ANN search result, *LR* is the linear search result. The function *count* is used to count the number of the result set. For each query, we calculate its precision and we take the mean value over all queries.

4.2   Influences of Parameters

The performance of our data-dependent LSH is influenced by two parameters *i.e.* numbers of clusters in building index and number of probed clusters in search.

To make out how these two parameters affect the performance of our method, we plot two diagrams and two tables which show how the time cost and the number of probed clusters changes along with the number of clusters in search precision of 90% and 96%, respectively.

From Figure 2, we can see that to reach the same accuracy, the time cost declines when the number of clusters increases at the beginning, *e.g.* the number is less than 1500. But the time cost begins to increase when the number of clusters is larger than

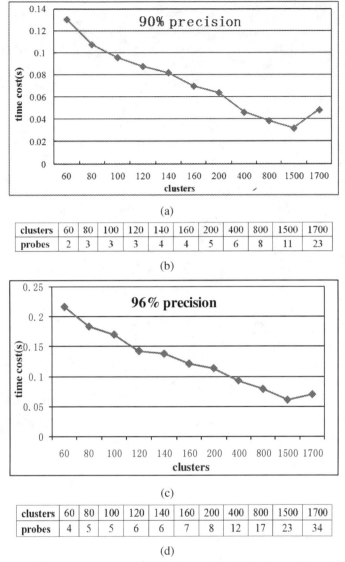

(a)

clusters	60	80	100	120	140	160	200	400	800	1500	1700
probes	2	3	3	3	4	4	5	6	8	11	23

(b)

(c)

clusters	60	80	100	120	140	160	200	400	800	1500	1700
probes	4	5	5	6	6	7	8	12	17	23	34

(d)

Fig. 2. (a), (c) The time cost changes along with the number of the clusters in precision of 90% and 96% respectively; (b), (d) the number of probed clusters with respect to different number of clusters in precision of 90% and 96% respectively

1500. The reason is that, in the beginning, with the increment of cluster number, the reduced amount of items in probed clusters is more than the increased amount of items in additional probed clusters to reach the same precision. But, the situation is reverse when cluster number is too large. Moreover, along with the increment of the number of clusters, the time cost of conducting step 1 to step 3 in algorithm 2 is also increased. So we choose 1500 clusters. As shown in Figure 2 (d), to reach high precision, we may choose 23 probed clusters.

4.3   Comparison of Overall Performance

To verify the performance of our data-dependent LSH (DP-LSH), we conduct comparative experiments among DP-LSH, E2LSH [4] and Bi-level LSH [8]. DP-LSH divides
the dataset into 1500 clusters in the first level and the number of probed clusters is set to
be 23. In the experiments, we compare the precision and speedup using different number of hash tables to testify that our algorithm can get better speedup compared to
E2LSH and Bi-level LSH in all cases. Experimental results are shown in Figure 3.

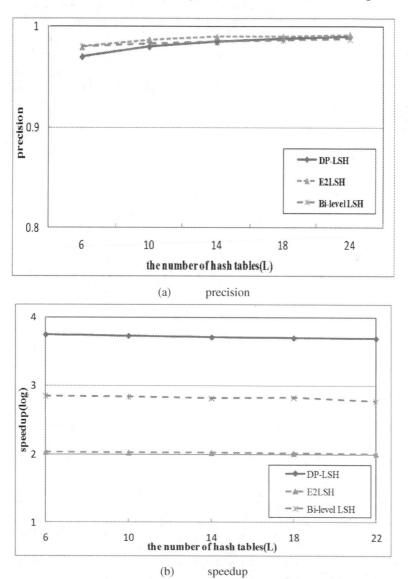

(a)   precision

(b)   speedup

Fig. 3. (a) Search precision changes along with the number of hash tables; (b) speedup ($\log^{10}$)
changes along with the number of hash tables

From Figure 3(a), we can see that the precision of DP-LSH, LSH and Bi-level LSH is varying during 0.97-1, and the precision of DP-LSH is about 1% lower than E2LSH when the number of hash tables is less than 14. Nevertheless, the precision of DP-LSH is almost equal to E2LSH and Bi-level LSH when the number of hash tables is larger than 14. So these three methods have very similar search precision.

From Figure 3(b), we can see that the speed of DP-LSH is about 48-50 times faster than E2LSH, using different number of hash tables. This efficiency superiority of DP-LSH is stable even when the number of hash tables changes. Besides, DP-LSH is much faster than Bi-level LSH. This is because that DP-LSH has a more uniform partition of dataset than Bi-level LSH, which results in better search efficiency. To our knowledge, this is the best performance of the state-of-the-art hash-based indexing methods.

Based on the above comparisons, we prove that DP-LSH can significantly improve the search speed while keeping high search precision. Moreover, the improved performance is stable.

5   Conclusions and Future Work

In this paper, we propose a new two-level hashing index structure and corresponding search algorithm. The goal is to deal with the problem that the LSH method loses efficiency when indexing non-uniform distribution datasets. Our index has two-level structure to perform ANN search in high dimensional spaces. In the first level, we divide the dataset into many clusters, to ensure the items in each cluster has near uniform distribution. In the second level, we construct LSH tables for each cluster. Besides, we also propose a novel search pruning method. In brief, the proposed method is simple and cheap, and has been proven by exhaustive experiments.

In future work, we will test our algorithm in more large scale datasets. We also need to design efficient out-of-core algorithm to handle many very large datasets (e.g. >10 billion).

Acknowledgement. This work is supported by the "Strategic Priority Research Program" of the Chinese Academy of Sciences (XDA06031000), National Nature Science Foundation of China (61303171, 61303175), Hunan province university innovation platform open fund project (14K037), National High Technology Research and Development Program (2011AA01A103).

References

1. Wan, J., Tang, S., Zhang, Y., Huang, L., Li, J.: Data Driven Multi-Index Hashing. In: IEEE International Conference on Image Processing (2013)
2. Zezula, P., Amato, G., Dohnal, V., et al.: Similarity Search: The metric space approach. Advances in Database Systems (2006)
3. Adonis, A., Indyk, P.: Near-optimal hashing algorithms for approximate nearest neighbor in high dimensions. In: Symposium on Foundations of Computer Science (2006)

4. Datar, M., Immorlica, N., Indyk, P., Mirrokni, V.S.: Locality-sensitive hashing scheme based on p-stable distributions. In: Symposium on Computational Geometry (2004)
5. Jegou, H., Amsaleg, L., Schmid, C., Gro, P.: Query adaptive locality sensitive hashing. In: IEEE International Conference on Acoustics, Speech, and Signal Processing (2008)
6. Weiss, Y., Torralba, A., Fergus, R.: Spectral Hashing. In: Advances in Neural Information Processing Systems (2008)
7. Heo, J.-P., Lee, Y.: Spherical Hashing. In: IEEE Conference on Computer Vision and Pattern Recognition (2012)
8. Pan, J., Manocha, D.: Bi-level Locality Sensitive Hashing for k-Nearest Neighbor Computation. In: Very Large Data Base (2010)
9. Bishop, C.M.: Pattern Recognition and Machine Learning. Springer (2006)
10. Jegou, H., Douze, M., et al.: Product Quantization for Nearest Neighbor Search. IEEE Transactions on Pattern Analysis and Machine Intelligence 33(1), 117–128 (2011)
11. Pauleve, L., Jegou, H., Amsaleg, L.: Locality sensitive hashing: A comparison of hash function types and querying mechanisms. Elsevier B.V. (2010)
12. Lv, Q., Josephson, W., Wang, Z., Charikar, M., Li, K.: Multi-probe LSH: efficient indexing for high-dimensional similarity search. In: Very Large Data Base (2007)
13. Bawa, M., Condie, T., Ganesan, P.: LSH forest: self-tuning indexes for similarity search. In: International Conference on World Wide Web (2005)
14. Dong, W., Wang, Z., Josephson, W., Charikar, M., Li, K.: Modeling lsh for performance tuning. In: Conference on Information and Knowledge Management (2008)
15. Babenko, A., Lempitsky, V.: The inverted multi-index. In: IEEE Conference on Computer Vision and Pattern Recognition (2012)
16. Xie, H., Zhang, Y., Tan, J., Guo, L., Li, J.: Contextual Query Expansion for Image Retrieval. IEEE Trans. on Multimedia 16(4) (June 2014)

Cosine Distance Metric Learning for Speaker Verification Using Large Margin Nearest Neighbor Method

Waquar Ahmad, Harish Karnick, and Rajesh M. Hegde

Department of Electrical Engineering,
Indian Institute of Technology Kanpur
rhegde@iitk.ac.in
http://202.3.77.107/mips/

Abstract. In this paper, a novel cosine similarity metric learning based on large margin nearest neighborhood (LMNN) is proposed for an i-vector based speaker verification system. Generally, in an i-vector based speaker verification system, the decision is based on the cosine distance between the test i-vector and target i-vector. Metric learning methods are employed to reduce the within class variation and maximize the between class variation. In this proposed method, cosine similarity large margin nearest neighborhood (CSLMNN) metric is learned from the development data. The test and target i-vectors are linearly transformed using the learned metric. The objective of learning the metric is to ensure that the k-nearest neighbors that belong to the same speaker are clustered together, while impostors are moved away by a large margin. Experiments conducted on the NIST-2008 and YOHO databases show improved performance compared to speaker verification system, where no learned metric is used.

1   Introduction

The task of speaker verification is to verify that the input utterance is by the claimed speaker. This requires checking whether the input utterance belongs to the claimant speaker or not [1]. I-vector, which stands for identity vector, represents an utterance by a fixed, low dimensional vector [2]. I-vectors are compared by calculating the dot product of the i-vectors (cosine similarity), which gives the similarity score between the i-vectors. Based on this score a verification decision is made for the input utterances. This method outperforms the one based on the Gaussian Mixture Model - Universal Background Model (GMM-UBM) [3]. Distance metric learning methods are used in machine learning to improve a system's performance by learning a metric from example data [4,5]. The learned metric transform the input feature space into a new feature space, where the separability between the data is modified such that data belong to same class are moved closer and data belong to different classes are moved away. We use the cosine similarity large margin nearest neighborhood (CSLMNN) metric that is learned from the development data. The metric calculation constrains the k

W.T. Ooi et al. (Eds.): PCM 2014, LNCS 8879, pp. 294–303, 2014.

nearest neighbors to belong to the same speaker, while other speakers are separated by a large margin [6]. The original LMNN algorithm [6] is learned by
calculating the Euclidean distance between the data. As cosine similarity gives
competitive results in state of art speaker verification system, we use cosine similarity between the data to learn the metric. The problem of metric learning is
formulated as an instance of a semidefinite programming problem to efficiently
compute the global minima.

The rest of the paper is organized as follows. Section 2 introduces the general
framework used for cosine distance metric learning for i-vector based speaker
verification. In that section, i-vector extraction and classification is discussed
which is followed by the discussion of cosine similarity large margin nearest
neighborhood metric learning. Section 3 describes the experiments conducted
on NIST 2008 and YOHO databases using the CSLMNN metric in an i-vector
framework. The performance of the system is evaluated using the detection error trade-off (DET) curves, equal error rate (EER) and minimum decision cost
function (DCF) points. Section 4 concludes with a discussion of the results obtained and some thoughts on distance metric learning in i-vector based speaker
verification.

2   Cosine Distance Metric Learning for I-Vector Based Speaker Verification

In this section, we discuss the cosine distance metric learning based on large
margin nearest neighborhood method. As cosine distance gives very competitive performance in i-vector based speaker verification system, we choose cosine
similarity as a distance measure in the proposed metric learning method. The
learned metric transform the input i-vectors into a new feature space, where
the i-vectors that belong to same speaker are moved closer and impostors are
moved away. Scoring on this new transformed space of i-vector gives better performance than the previous i-vector space. Following subsection discusses the
i-vector extraction and classification method used in current state of art speaker
verification system with a brief discussion of linear discriminant analysis as inter
session compensation and dimensional reduction technique.

2.1   I-Vector Extraction and Scoring

In this subsection, we briefly discuss the i-vector extraction and scoring for
speaker verification system. I-Vector is a compact form of speech utterance that
has been extracted using the total variability subspace [2]. It involves formation
of GMM supervector by concatenating the *means* of the MAP adapted speaker
model from the UBM model. The supervector is assumed to have the following
structure.

$$s = m + Tw \tag{1}$$

Where m is the speaker independent UBM supervector, T is the total variability
matrix and w the total variability factor that is termed as the i-vector. Matrix

T is computed from the training data in exactly the same way as the eigenvoice matrix in the JFA system with the slight difference that the speech for training belongs to different speakers. Matrix T is a low rank matrix. For the given matrix T, the i-vector w is obtained for the given utterance by:

$$w = \left(I + T^t \Sigma^{-1} N T\right)^{-1} T^t \Sigma^{-1} F \tag{2}$$

where I is an identity matrix and N is a diagonal matrix of dimension $CF \times CF$, its diagonal block are $N_c I$, $(c = 1, 2, ..C)$ is the Gaussian index and F is the supervector formed by concatenating all the centralized first order statistics. Σ is a diagonal covariance matrix of dimension $CF \times CF$. The block diagram Figure 1 illustrates the procedure for extracting the i-vector.

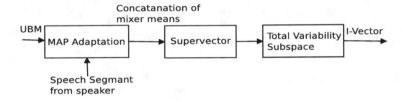

Fig. 1. Diagram illustrating the i-vector extraction from a speech dataset

Linear Discriminant Analysis. Linear Discriminant Analysis (LDA) is a popular technique to compensate for inter-session variability in the speech data. The main objective of LDA is to find new orthogonal axes such that variation between classes is maximized and within classes is minimized. The LDA transformation matrix A_{lda} consists of the eigenvectors having the largest eigenvalues for the eigenvalue problem $S_B v = \lambda S_W v$, where the between and within speaker scatter matrices, S_B and S_W respectively are calculated using:

$$S_B = \sum_{s=1}^{S} N_s \left(\mu_s - \mu\right) \left(\mu_s - \mu\right)^t \tag{3}$$

$$S_W = \sum_{s=1}^{S} \sum_{i=1}^{N_s} \left(w_i^s - \mu_s\right) \left(w_i^s - \mu_s\right)^t \tag{4}$$

nist2002 In the above formula μ_s is the mean i-vector of each speaker, S denotes the total number of speaker under consideration and N_s stands for the total number of utterances for speaker S. The matrix A_{lda} is calculated as follows:

$$A_{lda} = \arg\max_{A} \frac{|A^T S_B A|}{|A^T S_W A|} \tag{5}$$

Following this the corresponding i-vector w is transformed into the vector w' as follows:

$$w' = w * A_{lda} \tag{6}$$

I-Vector Scoring. The cosine score between the test w'_{test} and target w'_{target} i-vectors is given by the dot product between these two i-vectors,

$$Score(w'_{target}, w'_{test}) = \frac{\langle w'_{target}, w'_{test} \rangle}{\|w'_{target}\| \|w'_{test}\|} \tag{7}$$

The score obtained is then compared with a threshold value and an accept decision is taken if the score is above the threshold else it is rejected. Both the target and test i-vectors are estimated exactly in the same manner with the same UBM and the same total variability matrix T.

2.2 Cosine Similarity Large Margin Nearest Neighbor Metric Learning

Distance metric learning methods are used in classification problems to improve the performance of the system by learning a metric from example data [7,8]. This leads to better discriminability between the input data and helps to improve the performance of the classification system. Let us consider the training set of n examples of dimensionality d, $\{(w_i, w_j)\}^n_{i=1}$, where $w_i = R^d$ and $w_j \in \{1, 2, 3 .... S\}$. Here S is the total number of classes (speakers). In general, the similarity score between two inputs w_i and w_j is given by the equation.

$$Score(w'_{target}, w'_{test}) = \frac{\langle Mw'_{target}, Mw'_{test} \rangle}{\|Mw'_{target}\| \|Mw'_{test}\|} \tag{8}$$

Where M is a symmetric positive definite matrix, $M \succeq 0$. This metric is learned by the CSLMNN algorithm from the data.

The non differentiable leave-one-out and non-continuous classification error of the k nearest neighbor classifier is imitated by the CSLMNN algorithm with a convex loss function [6] by calculating the cosine similarity between the input data. The purpose of the loss function is to ensure that the local nearest neighbors around every input target i-vector having the same class label are moved closer together and inputs with different class labels are pushed further apart. One of the advantages of the CSLMNN algorithm is that the metric (global) is optimized locally. To achieve this, the CSLMNN algorithm needs prior information of nearest neighbors of the target class. We do this by measuring the Cosine distance between the i-vectors. Let $j \rightsquigarrow i$ indicate w_j is a target neighbor of w_i. CSLMNN learns the Mahalanobis metric M such that it keeps each input w_i close to its target neighbors while input vectors of different classes (impostors) are separated by a large margin. Here learning a metric is equivalent to learning a linear transformation that maps input vectors to a transformed space where the above property holds. For the input w_i, having the target w_j and impostor w_k the relation is expressed as the following equation with respect to the cosine similarity metric.

$$\frac{\langle Mw'_i, Mw'_j \rangle}{\|Mw'_i\| \|Mw'_j\|} - \frac{\langle Mw'_i, Mw'_k \rangle}{\|Mw'_i\| \|Mw'_k\|} \geq 1 \tag{9}$$

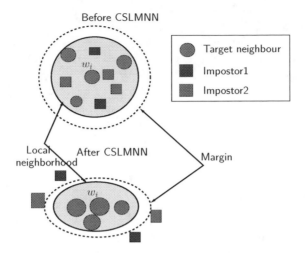

Fig. 2. Illustrates transfer of the i-vector before and after CSLMNN Metric learning. The small circles represent i-vectors with the same class label, while small rectangles represent i-vectors with different class labels.

In Figure 2 the i-vectors with the same class label are shown as small circles, while small rectangles represent i-vectors with different class labels (impostors). In the input space all points on a circle are equidistant from w_i. After learning the CSLMNN metric it can be seen that in the new transformed space the i-vectors of same speaker are moved closer, while impostors are moved far.

The semidefinite program (SDP) proposed in [6] moves the target neighbors closer by minimizing cosine distance $\sum_{j \rightsquigarrow i} \left(1 - \dfrac{\langle Mw'_i, Mw'_j \rangle}{\|Mw'_i\|\|Mw'_j\|}\right)$, while penalizing the criteria of separating the data of different class by a large margin. This problem is further solved by introducing an additive slack variable $\xi_{ijk} \geq 0$ for different class label, so that the SDP can be formulated as shown in the following Table 1. The triplet variable is $S = \{(i, j, k) : j \rightsquigarrow i, y_k \neq y_i\}$, where y_k and y_i are input data labels.

Table 1. Formulation of the convex optimization problem for the CSLMNN method

$$\min_M \sum_{j \rightsquigarrow i} \left(1 - \frac{\langle Mw'_i, Mw'_j \rangle}{\|Mw'_i\|\|Mw'_j\|}\right) + \mu \sum_{(i,j,k) \in S} \xi_{ijk}$$

Subject to : $(i.j, k) \in S$:

(1) $\dfrac{\langle Mw'_i, Mw'_j \rangle}{\|Mw'_i\|\|Mw'_j\|} - \dfrac{\langle Mw'_i, Mw'_k \rangle}{\|Mw'_i\|\|Mw'_k\|} \geq 1 - \xi_{i,j,k}$

(2) $\xi_{i,j,k} \geq 0$

(3) $M \succeq 0$

Once the CSLMNN metric M is learned from the example data, scores are obtained using the following equation for the i-vectors.

$$Score_{LMNN}(w'_{target}, w'_{test}) = \frac{\langle Mw'_{target}, Mw'_{test} \rangle}{\|Mw'_{target}\| \|Mw'_{test}\|} \qquad (10)$$

The overall block diagram of the speaker verification incorporating the CSLMNN algorithm is shown in Figure 3.

3   Performance Evaluation

In this Section we describe how the new algorithm was evaluated and compare its performance with competing approaches. Experiments were done on the NIST-2008 and YOHO databases. DET curves and EER were used to evaluate the performance of the CSLMNN metric. We compare both the raw cosine and LDA transformed i-vectors and discuss the significance of the improvements obtained in terms of the DET curves and EER using the proposed method.

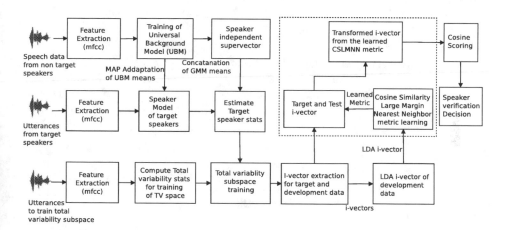

Fig. 3. Block diagram illustrate the proposed speaker verification using the learned metric. The cosine distnace metric learning is shown in dotted box.

3.1   Organization of the Various Data Sets

Two data sets NIST-2008 and YOHO were used to evaluate the proposed method. NIST 2002/2004 was used to train the background models. For development i-vectors, NIST-2005, NIST-2006 and NIST-2007 databases were used.

- **NIST 2008 Database**: The short2-short3 condition of NIST-2008 database was used here to evaluate the proposed method. The database consists of 648 male and 1140 female speakers. Short 2 condition was used for training and short 3 for testing. Short2 data contains both telephone speech as well as interview speech. The interview speech samples used in the short2 condition are of 3 minute duration extracted from excerpts of the longer interview speech. The short3 condition testing data consists of the type of speech in short2 data as well as telephone speech recorded with an auxiliary microphone [9] [10].

- **YOHO Database**: The YOHO databse consists of 108 male and 30 female speakers. The database was collected during testing of ITT'S speaker verfication system in an office environment. Speaker variation spanned a wide range over attributes like age, job description and educational background. Most of the speakers are from the New York city area with some non-native English speakers. The data was collected using a high quality telephone handset (Shure XTH-383), but did not pass through the telephone channel [11].

3.2   Experimental Conditions

A 39 dimensional MFCC (mel frequency cepstral coefficient) (13 static, 13 Δ and 13 $\Delta\Delta$) was obtained as a feature vector for i-vector extraction from the speech signal at a frame rate of 10 ms with 20 ms Hamming window. Silence was removed using VAD (Voice activity detection) and MFCC features were normalized with standard cepstral mean subtraction and variance. The UBM of 512 mixture components with diagonal covariance matrices was trained using data from non-target speakers. Maximum likelihood criteria were used for the training of the UBM model. For the i-vector system the total variability space T was trained using non-targeted speakers. I-vectors of 400 dimension were extracted from speech segments for the training and testing phases.

3.3   Experimental Results

Experiments on the NIST 2008 and YOHO database were done using the LMNN algorithm.

- Raw cosine scoring method: In this method cosine scoring is used to obtain the scores and there was no metric learning. This is the Baseline system in the i-vector framework.
- Raw cosine + Euclidean LMNN method: In this method the scoring is done on the transformed vector obtained after the LMNN matrix M is learned from the development data by using Euclidean distance between the i-vectors. .
- Raw Cosine + CSLMNN Method : In this method the matrix M is learned from the development data by measuring the cosine distance between the i-vectors.

- Raw cosine + LDA method: Here the LDA transformation matrix A_{lda} is used to reduce the dimension of the i-vector. The matrix A_{lda} is estimated using the development data. The transformed LDA i-vector is used for final cosine scoring.
- Raw cosine + LDA + Euclidean LMNN: This method involved applying LDA on the raw i-vector followed by Euclidean LMNN on the reduced LDA i-vectors.
- Raw cosine + LDA + CSLMNN: This method involved applying the CSLMNN metric scoring on LDA i-vectors. This method gives the best result compared to all above methods.

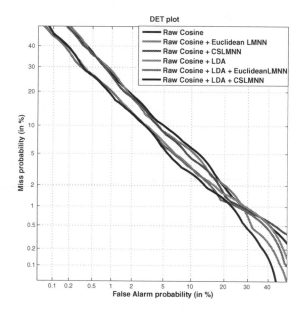

Fig. 4. DET plots of speaker verification for the NIST 2008 Database

The experimental results obtained using the proposed method have been presented in the form of DET [12] curves where a lower curve is interpreted as better performance. The corresponding EER values from both databases are also presented here. The DET plots for the NIST-2008 and YOHO databases using the LMNN metric are shown in Figure 4 and 5 respectively. The minimum DCF values are also evaluated for both the databases. Table 2 and Table 3 show the DCF value and EER value for NIST-2008 and YOHO databases. The DET curve is plotted using the Bosaris toolkit [13].

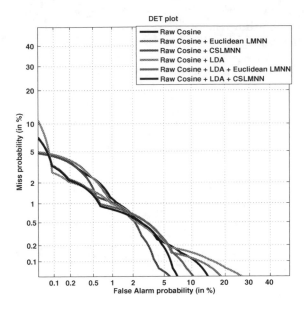

Fig. 5. DET plots of speaker verification for the YOHO Database

Table 2. DCF and EER values for NIST-2008 data

Scoring Method	EER	min DCF
Raw Cosine	7.53	0.2077
Raw Cosine + Euclidean LMNN	7.06	0.1985
Raw Cosine + CSLMNN	6.86	0.1980
Raw Cosine + LDA	5.88	0.1712
Raw Cosine + LDA + Euclidean LMNN	5.66	0.1702
Raw Cosine + LDA + CSLMNN	**5.34**	**0.1698**

Table 3. DCF and EER values for YOHO data

Scoring Method	EER	min DCF
Raw Cosine	1.09	0.0342
Raw Cosine + Euclidean LMNN	1.02	0.0324
Raw Cosine + CSLMNN	1.01	0.0258
Raw Cosine + LDA	0.99	0.0250
Raw Cosine + LDA + Euclidean LMNN	0.90	0.0251
Raw Cosine + LDA + CSLMNN	**0.85**	**0.0244**

4   Conclusions

The proposed Cosine similarity large margin nearest neighborhood metric learning method transform the input i-vectors into a new feature space. More specifically the method proposed herein learns the metric in such a way that the intra speaker variability is kept minimum by keeping the k-nearest neighbors closer to the target i-vector. Additionally, it also ensures that impostors are kept at a large margin from the target speaker. We have used cosine distance as a similarity measure for the metric learning due to competitive result obtained in i-vector based speaker verification system. The CSLMNN metric method can be fused with other compensation method to further improve the performance of speaker verification system. This is possible due to the learning methodology followed in the CSLMNN. The experimental results are encouraging and future work will focus on utilizing score normalization techniques in the cosine similarity large margin nearest neighborhood framework.

References

1. Tomi, K., Li, H.: A tutorial on text-independent speaker verification. Speech Communication 52, 12–40 (2010)
2. Dehak, N., Kenny, P.J., Dehak, R., Dumouchel, P., Ouellet, P.: Front-end factor analysis for speaker verification. IEEE Transactions on Audio, Speech, and Language Processing 19(4), 788–798 (2011)
3. R.D.A.Q.T.F.,, D.R.: Speaker verification using adapted gaussian mixture models. Digital Signal Processing 10(1), 19–41
4. S.M., J.T.: Learning a distance metric from relative comparisons. In: Advances in Neural Information Processing Systems, vol. 16, p. 41 (2004)
5. Davis, J.V., Kulis, B., Jain, P., Sra, S., Dhillon, I.S.: Information theoretic metric learning. In: Proc. Int. Conf. Mach. Learn., pp. 209–216 (2007)
6. B.J.W.K.Q., S.L.K.: Distance metric learning for large margin nearest neighbor classification. In: Advances in Neural Information Processing Systems, pp. 1473–1480 (2005)
7. Yang, L.: An overview of distance metric learning. In: Proceedings of the Computer Vision and Pattern Recognition Conference (2007)
8. Xing, E.P., Jordan, M.I., Russell, S., Ng, A.: Distance metric learning with application to clustering with side-information. In: Advances in Neural Information Processing Systems, pp. 505–512 (2002)
9. Scheffer, K.S.S.N., Graciarena, M., Shriberg, E., Stolcke, A., Ferrer, L., Bocklet, T.: The sri nist 2008 speaker recognition evaluation system. In: IEEE International Conference on Acoustics, Speech and Signal Processing, ICASSP 2009, pp. 4205–4208 (2009)
10. Li, H., Ma, B., Lee, K.-A., Sun, H., Zhu, D., Sim, K.C., You, C.: The i4u system in nist 2008 speaker recognition evaluation. In: IEEE International Conference on Acoustics, Speech and Signal Processing, ICASSP 2009, pp. 4201–4204 (2009)
11. Campbell Jr., J.P.: Testing with the yoho cd-rom voice verification corpus. In: International Conference on Acoustics, Speech, and Signal Processing, ICASSP 1995., vol. 1, pp. 341–344. IEEE (1995)
12. Martin, A., Doddington, G.: The det curve in assessment of detection task performance. In: Proc. Eurospeech, vol. 97(4), pp. 1895–1898 (1997)
13. B.N., de Villiers, E.: The bosaris toolkit: Theory, algorithms and code for surviving the new dcf. arXiv preprint arXiv, 1304.2865 (2013)

Anchor Shot Detection with Deep Neural Network

Bailan Feng, Jinfeng Bai, Zhineng Chen, Xiangsheng Huang, and Bo Xu

Interactive Digital Media Technology Research Center,
Institute of Automation, Chinese Academy of Sciences, Beijing, China
{bailan.feng,jinfeng.bai,zhineng.chen,
xiangsheng.huang,xubo}@ia.ac.cn

Abstract. Anchor Shot Detection (ASD) is a key step for segmenting news videos into stories. However, the existing ASD methods are either channel-related or channel-limited which could not satisfy the requirement for achieving effective management of large-scale broadcast news videos. Considering the variety and diversity of large-scale news videos and channels, in this paper we propose a universal scheme based on deep neural network for anchor shot detection (DNN_ASD). Firstly, DNN_ASD consists of a training procedure of deep neural network to learn the appropriate anchor shot detector. Secondly, accompanied with imbalanced sampling strategy and face-assist verification, a universal scheme of anchor shot detection for large-scale news videos and channels is available. Parallel to this, the width and depth of neural network and the transfer ability are empirically discussed respectively as well. Encouraging experimental results on news videos from 30 TV channels demonstrate the effectiveness of the proposed scheme, as well as its superiority on transfer ability over traditional ASD methods.

Keywords: Anchor Shot Detection, Deep Neural Network, Imbalanced Sampling Strategy.

1 Introduction

Due to the exploration of broadcast channels, broadcast news videos are regularly accessed by a large number of people all over the world, and an efficient medium to organize these large-scale resources is urgently required. Especially, in news videos segmenting and searching, to prepare the story units and to reinforce the accuracy of indexing images, some pre-processing procedures are necessary. Among them, anchor shot detection (ASD) can be employed as a foundational step for news videos structuring [1, 2, 3].

As a fundamental issue of news story segmentation, there have been several prior works in anchor shot detection till now, and they can be roughly grouped into two categories: model-based methods and detection-based methods.

Model-based methods focus on constructing detecting models based on machine learning techniques and detect anchor positions under some optimal criterion. For example, the authors in [4] define a set of models of anchor shot and then match

W.T. Ooi et al. (Eds.): PCM 2014, LNCS 8879, pp. 304–312, 2014.

against them all the shots of a news video in order to detect potential anchor shots. Different from above anchor model method, the authors in [5] build the key-frame model using the assumption that an anchorperson typically appears in the first five shots of a news video. The authors in [6] implement an on-line trained audio and visual model to classify the video shots into anchorperson shots and news footage shots. However, the main drawback of these methods is that these models are strongly dependent on the relational video channels, and we name this phenomenon as channel-related flaw. Since lacking powerful learning tools, it is difficult to construct a general model able to represent all the different kinds of news, and it is a severe limitation problem.

Detection-based methods aim to locate anchor positions through carefully crafted heuristic rules with a set of intuitive cues or features. For example, the authors in [7] propose to group similar key-frames into clusters on the basis of their color histograms, and then employ a two-stage pruning technique for reducing the number of falsely detected anchor shots. The authors in [3] improve above method by introducing channel configuration process. Besides that, some of researchers [8, 9] make use of face information for anchor detection as well, such as the position, the size, and the number of faces. For example, the authors in [8] make use of anchorperson's attire to reduce the computational burden of the following modules. The authors in [9] extract face regions and then utilize the color histogram within the face region and temporal coherency of the face size and position to detect anchor shots. Following these methods, however, the situation of anchorperson is very critical. In one case, the authors implicitly assume that anchorpersons in different channels should share the similar background, which is actually not satiable especially in large-scale news videos. In the other case, some methods also heavily rely on face detection results, or the performance is bounded by face detection accuracy. Both factors result in that these methods can only be suitable to small range channels, and we name this phenomenon as channel-limited flaw.

In conclusion, on one hand, the large variety of existing news channels in the world makes it impossible to construct a universal series of anchor detection rules which can be suitable for all various channels. On the other hand, the existing model-based methods fall into the dilemma that they can provide good results only for a small range of channels. Therefore, considering the variety and diversity of large-scale news videos and channels, in this paper we propose a universal anchor shot detection scheme based on deep neural network (DNN_ASD), to cope with the channel-related and channel-limited flaws mentioned above. Firstly, news videos are decomposed into shots based on the Uniform LBP feature [10], and the middle frame of each shot is extract as the key-frame for anchor shot detection. Secondly, a universal anchor model is off-line trained using the recently emergent deep neural network with an imbalanced sampling strategy. During the anchor detection process, key-frames are first given probability values by the universal anchor model above and we can obtain an anchor candidate collection. Then face-assist verification is applied to the candidate collection for further pruning false anchor shots, which jointly combines the tradeoff between channel-related flaw and channel-limited flaw. In order to test the proposed scheme, we built-up a big news database, which consists of about 51 hours with 1205 anchor shots and 47430 news report shots. More importantly, this database is constructed on 30 channels (each channel contributes to 3 videos), which is taking full account of the variety and diversity of large-scale broadcast news videos.

The remaining of this paper is organized as follows. Section 2 presents the DNN_ASD scheme in detail. In Section 3 the database used is introduced together with the experiments carried out in order to assess the performance of the proposed scheme. This is followed by conclusions in Section 4.

2   The Proposed Scheme

The proposed scheme (see Fig. 1) firstly performs a shot detection by using Uniform LBP feature to describe the basic information of images. As described in [10], this feature is more sensitive in gradual transition case than other features while it is as robustness as other features in the same shot. Therefore, a different sequence of frames is calculated based on this feature, and the shot boundary is identified accordingly.

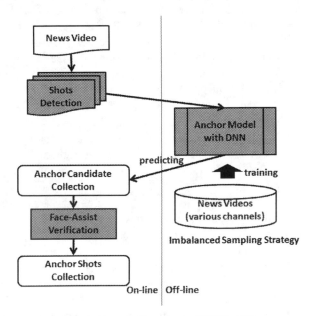

Fig. 1. Overall Flowchart of DNN_ASD

2.1   Anchor Model with DNN

Imbalanced Sampling Strategy
Inspired by the success of deep neural network models in solving challenging problems, such as speech recognition and computer vision, here we choose deep neural network as our universal anchor model. As we all know, large training data is necessary to create a robust model, especially with deep neural network. Nevertheless, on

one hand it is difficult and expensive to collect large high-quality anchor training samples. On the other hand, the negative (no anchor) training samples are abundant and easily accessible yet. According to this observation, here we propose an imbalanced sampling strategy of leveraging unlabeled (negative sample) data to expand the diversity of the precious positive ones as well as provide a robust set of negative examples for universal anchor modeling.

Specially, the training samples are generated as follows. Firstly, 50 videos from 25 channels (2 videos in each channel, 25 channels for training and testing, and the remaining 5 channels for transfer testing, more details please refer to Section 3.1) are chosen from news database as training set, which includes 728 anchor shots and 28896 news report shots. To increase the number of limited positive samples, we extract key-frames from each anchor shot with an interval of one second, and therefore totally 8966 anchor frames are achieved as our positive training samples. Secondly, to enhance the anti-interference ability of universal model, different from above positive samples, we consider the negative sampling from two aspects. On one hand, only the middle frame of each news report shot is taken out to ensure the diversity of negative samples. On the other hand, negative samples are randomly extracted from all 28896 news report shots, both containing pure scene shots and other person inside shots. This strategy can help avoid suffering some common drawbacks of the existing methods that the person shots are mistaken recognized as anchorperson shots. With this negative sampling strategy, we totally harvest 28896 no-anchor frames as our negative training samples. Finally, to leverage the imbalanced positive and negative samples, we model a clustering based pseudo-negative space for training data sampling. Similar to [11], we cluster the 28896 negative training samples into 8966 categories, and then select the nearest ones to the centroid of each category as the final negative training samples.

Feature of Anchor Frame
With the training samples, three steps are employed to extract features. Firstly, to process faster, a bi-moment normalization proposed by [12] is adopted to normalize all the anchor frames to size 32x32. Secondly, as demonstrated in [13], local gradient direction histogram features are effective in differentiating the images of Chinese characters, and therefore we transfer it to our anchor frame detection directly. Specifically, gradient directions map of each normalized anchor frame is calculated by using sobel operators, and then the map is evenly split into 64 patches of size 4x4. Based on this partition, 8 directions gradient histogram are calculated patch by patch and finally a 512-dimention vector is concatenated for deep neural network training.

Architecture of Anchor Model
Deep neural network is a multi-layer perception with multiple hidden layers. The architecture of traditional neural network is that the first layer and the last layer are input and output respectively, and in which the numbers of neurons are equal to the feature dimensions and the categories accordingly. Hidden layers located in the

middle zone are critical and the number of layers and neurons corresponds to the depth and the width in deep neural network model. In practical models, the number of hidden layers is often above 3 (3 hidden layers in our anchor model) and each layer consists of a few thousands hidden units (4096 units on each hidden layers in our anchor model). After propagation layer by layer, the output of each neuron is transformed by a non-linear function.

In our scheme, sigmod function and softmax function are chosen as the activation function of all hidden neurons and the activation function of output layer neurons respectively. The input layer has been set as 512 neurons, corresponding to the 512-dimention vector, and the output layer is set as 2 neurons corresponding to anchor category and no-anchor category. We use standard back-propagation algorithm to train the neural network, and employ new-bob learning rate schedule which do not halve the learning rate until the accuracy on cross-validation set fails to improve by more than 0.5% between each two epochs. Finally, the algorithm will terminate at the point of a predetermined minimum the CV accuracy improvement arrives. The initial learning rate and the batch size are empirically set to 0.032 and 512, and the number of hidden layer and hidden neuron is 3 and 4096 respectively. More details analysis of model architecture will be introduced in Section 3.2.

2.2   Face-Assist Verification

After model prediction, we will achieve an anchor candidate collection for each news video. By observation, we find that although above deep neural network based anchor model is powerful to most anchor situation. But some static characters shots or reporter shots which have the similar visual patch and camera status to real anchor shot may cause false alarm. Therefore, we further use Open CV [14] to detect faces in the anchor candidate collection to filter out these false shots. Especially, from the result of face detection, we utilize the number of faces, the center position of each face and the area of each face in an anchor candidate frame, and remove those do not satisfy our criterion ones from the collection. The advantages of this step are that, on one hand, the face detection methodology of Open CV is a much less computational cost than other conventional methods which is very suitable for our on-line after-handling. On the other hand, it can also help verify the output from deep neural network and achieve a better result.

3   Experimental Results

As we know, some efforts have been spent in building news video databases for benchmarking purposes in the recent past, such as the TREC Video Retrieval Evaluation in 2004. Unfortunately, these data are not publicly available outside the TRECVID contest. Therefore, we have built a new news video database.

3.1   Database and Evaluation

In our news database, as much as representative variability of the phenomenon is considered. Specially, our database is composed by 90 news videos from 30 channels, totally including 1205 anchor shots and 47430 news report shots. More in detail, the 30 channels are randomly separated into two parts: 1) 25 channels for training and testing. Concretely, 2 videos of each channel, that is 50 videos in total, are used as training set while the other 25 videos are used as testing set. 2) 5 channels for transfer testing.

In order to have a single figure of merit for comparing the performance, we calculate the F-measure value, which is a combination of Precision and Recall, as the evaluation criterion.

3.2   Influences of Width and Depth of Model

In order to confirm the architecture of anchor model, different width and depth models are tested and analyzed firstly. Table 1 and Table 2 investigate the influences of width and depth on model. From the table we can see that, on one hand, neural network with the wider and deeper architecture can obtain higher detection accuracies. On the other hand, we can also observe that the improvement is getting narrowing (or slightly descending) when the width and depth reach to 4096 and 5. So we can conclude that the architecture of <512, 4096, 4096, 4096, 2> is suitable for our anchor detection task.

Table 1. Comparison of Width of Neural Network (3 layers)

#Neurons	CV Accuracy	Test Accuracy
1024	41.03%	42.54%
2048	48.81%	49.28%
4096	54.22%	50.13%
5120	55.88%	50.04%

Table 2. Comparison of Depth of Neural Network (4096 neurons)

#Layers	CV Accuracy	Test Accuracy
3	54.22%	50.13%
4	78.12%	59.46%
5	99.61%	99.86%
6	99.70%	99.82%

3.3   Performance of Anchor Shot Detection

To give a comprehensive understanding for the effectiveness of DNN_ASD, as shown in Figure 2, the performance of our scheme are compared with the existing two ASD algorithm over 25 various channels.

From the figure, we can observe that our scheme defeats GTC_ASD [7] and Improved GTC_ASD [3] on most channels and only fails on 1 channel slightly compared with Improved GTC_ASD. The total improvements over GTC_ASD and Improved GTC_ASD are 24.3% and 5.5% on average respectively, which demonstrates the effectiveness of our DNN_ASD for large-scale anchor shot detection.

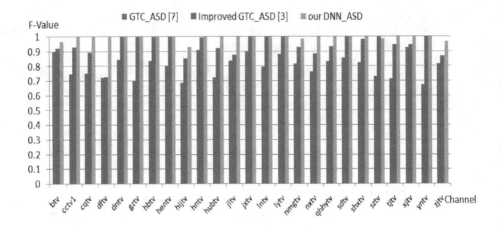

Fig. 2. Performances of Three Different Methods over 25 Channels

3.4   Transfer Analysis of the Proposed Scheme

In order to verify the transfer ability of the proposed scheme, we randomly reserve another 5 channels which are not having appeared in the training set. Then we compare the three methods introduced above, and the performances are shown in Table 3.

From the table, we can observe that on one hand, our proposed scheme can still achieve a satisfactory performance even if the test data is completely unfamiliar to the model, which confirms the good transfer ability of our proposed scheme. On the other hand, we notice that the performance of GTC_ASD fluctuates conspicuously and Improved GTC_ASD shows relatively stable. Actually, the difference of [3] compared with [7] mainly lies on the channel-configuration process, which means [3] needs to configure a series of pruning parameters for each channel so as to counteract the impact of differentiated distribution of anchor shots in different channels. This is also a restriction to the large-scale channels application, and it confirms the superiority of our scheme on the contrary.

Table 3. Transfer Comparison of Three Different Methods

Channels	GTC_ASD[7]	Improved GTC_ASD [3]	DNN_ASD
sxtv	0.7619	0.8807	1
jstv	0.9474	0.9773	1
ahtv	0.3333	0.9165	1
gdtv	0.6154	0.8563	0.8889
sctv	0.6250	0.8979	1

4   Conclusion

In this paper, we have presented a universal anchor shot detection scheme based on deep neural network (DNN_ASD). Compared with the existing methods, the contributions of DNN_ASD can be attributed to two factors. Firstly, to our best knowledge, it is the first trial to analyze the effect of DNN in anchor shot detection. Secondly, the influences of width and depth of DNN and the transfer ability of anchor model are empirically evaluated, and accompanied with imbalanced sampling strategy and face-assist verification, an appropriate anchor shot detector is achieved. Both factors accomplish a universal scheme of anchor shot detection for large-scale news videos and channels, and contribute to the significant improvement over traditional ASD methods.

Acknowledgments. This work is supported by National Nature Science Foundation of China (Grant No. 61202326, 61303175).

References

[1] Hsu, W.H., Kenney, L.S., Chang, S.F., Franz, M., Smith, J.: Columbia-IBM News Video Story Segmentation in TRECVID. In: Proc. ACM CIVR, pp. 1–11 (2005)
[2] Xu, S., Feng, B.L., Chen, Z.N., Xu, B.: A General Framework of Video Segmentation to Logical Unit based on Conditional Random Fields. In: Proc. ACM ICMR, pp. 247–254 (2013)
[3] Feng, B.L., Chen, Z.N., Zheng, R., Xu, B.: Multiple Style Exploration for Story Unit Segmentation of Broadcast News Video. Multimedia Systems 20(4), 347–361 (2014)
[4] Smoliar, S.W., Zhang, H.J., Tao, S.Y., Gong, Y.: Automatic Parsing and Indexing of News Video. Multimedia Systems 2(6), 256–265 (1995)
[5] Hanjalic, A., Lagendijk, R.L., Biemond, J.: Semi-Automatic News Analysis, Indexing and Classification System based on Topics Preselection. In: Proc. of SPIE: Electronic Imaging: Storage and Retrieval of Image and Video Databases, San Jose, CA (1999)
[6] Liu, Z., Huang, Q.: Adaptive Anchor Detection using On-line Trained Audio/Visual Model. In: Proc. of SPIE, Storage and Retrieval for Media Database (2000)

[7] Santo, M.D., Foggia, P., Percannella, G., Sansone, C., Vento, M.: An Unsupervised Algorithm for Anchor Shot Detection. In: Proc. IEEE ICPR, pp. 1238–1241 (2006)

[8] Gao, X., Tang, X.: Unsupervised Video-Shot Segmentation and Model-Free Anchorperson Detection for News Video Story Parsing. IEEE Trans. Circuits and System for Video Technology 12(9), 765–776 (2002)

[9] Hsu, W., Chang, S.F., Huang, C.W., Kennedy, L., Lin, C.Y., Iyengar, G.: Discovery and Fusion of Salient Multi-Modal Features towards News Story Segmentation. In: Proc. of SPIE: Symposium on Electronic Imaging: Storage and Retrieval of Image/Video Database, San Jose, USA (2004)

[10] Ojala, T., Pietikainen, M., Maenpaa, T.: Multiresolution Gray-scale and Rotation Invariant Texture Classification with Local Binary Patterns. IEEE Trans. Pattern Analysis and Machine Intelligence 24(7), 971–987 (2002)

[11] Tesic, J., Natsev, A., Xie, L., Smith, J.R.: Data Modeling Strategy for Imbalanced Learning in Visual Search. In: Proc. IEEE ICME, pp. 1990–1993 (2007)

[12] Liu, C.L., Sako, H., Fujisawa, H.: Handwritten Chinese Character Recognition: Alternatives to Nonlinear Normalization. In: Proc. ICDAR, pp. 524–528 (2003)

[13] Bai, J., Chen, Z., Feng, B., Xu, B.: Chinese Image Character Recognition Using DNN and Machine Simulated Training Samples. In: Wermter, S., Weber, C., Duch, W., Honkela, T., Koprinkova-Hristova, P., Magg, S., Palm, G., Villa, A.E.P. (eds.) ICANN 2014. LNCS, vol. 8681, pp. 209–216. Springer, Heidelberg (2014)

[14] Intel, Compute-Intensive, Highly Parallel Applications and Uses. Intel Technology Journal 09 (2005)

Adaptation of ANN Based Video Stream QoE Prediction Model

Jianfeng Deng[1,2], Ling Zhang[1], Jinlong Hu[1], and Dongli He[2]

[1] School of Computer Science & Engineering, South China University of Technology,
Guangzhou Higher Education Mega Centre, Guangzhou, P.R. China
[2] College of Computer Science & Information Technology, Guangxi Normal University,
No.15 Yucai Road, Guilin, P.R. China
`deng.jf@mail.scut.edu.cn`, `{ling,jlhu}@scut.edu.cn`,
`dlhe@gxnu.edu.cn`

Abstract. Pseudo-Subjective Quality Assessment (PSQA) is an effective way to prediction the Quality of experience (QoE) of video stream. The ANN-based PSQA model gives a decent QoE prediction accuracy when it is tested under the same condition as training. However, the performance of the model under mismatched conditions is little studied, and how to effectively adapt the models from one condition to another is still an open question. In this work, we first evaluated the performance of the ANN-based QoE prediction model under mismatched conditions. Our study shows that the QoE prediction accuracy degrades significantly when the model is applied to conditions different from the training condition. Further, we developed a feature transformation based model adaptation method to adapt the model from one condition to another. Experiments results show that the QoE prediction accuracy under mismatched conditions can be improved substantially using as few as five data samples under the new condition for model adaptation.

Keywords: Quality of experience (QoE), Pseudo-subjective quality assessment (PSQA), prediction model, adaptation.

1 Introduction

Video streaming based on Internet Protocol (IP) becomes increasingly popular in recent years [1]. Many IP-based video applications are real time service applications, such as video conference, remote education, etc. However, video streaming applications are sensitive to transfer bandwidth, loss, delay, etc. The quality of video transfer is seriously affected by these factors. Therefore, it is critical for network service providers to continuously monitor the quality of the transferred video.

The concept of Quality of Experience (QoE) is proposed for evaluating the human's perception of the quality of service. A set of objective and subjective methods were proposed to evaluate the QoE of video. It can be categorized into three groups[2], including subjective methods[3], Objective methods[4], and Pseudo-Subjective Quality Assessment methods[5-10]. Objective QoE evaluation methods

W.T. Ooi et al. (Eds.): PCM 2014, LNCS 8879, pp. 313–322, 2014.
© Springer International Publishing Switzerland 2014

such as Peak Signal-to-Noise Ratio (PSNR), mean squared error (MSE), Video Quality Metric (VQM)[4], etc. attempt to assess the QoE by a measured objective value. However it was shown in [6] that the objective assessment does not necessarily correlate to human's perception. Instead, the most accurate method for evaluating human's perception is subjective evaluation. However, the subjective evaluation is very labor-intensive, and makes real-time QoE evaluation impossible.

To address these issues, Pseudo-Subjective Quality Assessment (PSQA) methods were proposed. PSQA methods attempt to use a set of subjective assessment data to build a QoE prediction model, and use that model to predict the QoE automatically. The common PSQA methodology includes several steps[13], first an application condition is determined, which is defined by a group of characters, such as resolution, encoding algorithm, category of the video content, etc. Then, under that condition, a set of important factors that have significant impact on the QoE are identified. These factors usually include a set of measurable network and application parameters, such as bit rate, frame loss rate, etc., which are called Quality-Affecting Factors (QAF)[7] hereafter. In order to build the QoE prediction model, a set of subjective QoE assessment data are collected, where each data sample consists the set of measured values of QAFs as represented by a feature vector, and a QoE score assessed by human via subjective assessment; Then a regression model is trained on these data to predict the QoE score from the input feature values. Once properly trained, these PSQA methods can perform QoE assessment without human intervention, and therefore greatly reduce the cost and can provide QoE assessment in real-time.

However, most of the previous studies on PSQA assume that the QoE prediction models are tested under the same condition as the one in training. Although decent QoE prediction accuracies can be achieved when tested in the matched condition, the performance of the QoE prediction models under mismatched conditions is little studied, and how to effectively adapt the models from one condition to another is still an open question.

On the other hand, due to the complexity of the Internet, the network transfer environment varies dramatically from one application to another, and there exist an exponentially growing number of different conditions as defined by permutations of different bitrate, encoding algorithm, video genre, etc. It is prohibitively expensive to collect adequate data and train prediction model for all condition. Therefore, it is of great interest to investigate the performance of the PSQA model in mismatched conditions, and effective adaptation of a PSQA model from one condition to another is highly demanded.

We investigated these two problems in this work. We first carried out a series of experiments to evaluate the performance of an ANN-based QoE prediction model under mismatched conditions. Our study shows that the QoE prediction accuracy degrades significantly when the model is applied to conditions different from the training condition. To our best knowledge, this is the first systematic study on the problem of QoE prediction under mismatched conditions. Second, we also carried out the first investigation on the problem of adapting the QoE prediction model from one condition to a new condition. We proposed a feature transformation based model adaptation method in this paper. A series of experiments were carried out.

The experimental results show that the QoE prediction accuracy under a mismatched condition can be improved substantially using as few as five data samples for adaptation, and the accuracy improves further when more data become available for model adaptation.

The rest of the paper is organized as follows. Section 2 describes previous work that is most relevant to our work. Section 3 reviews the ANN-based QoE prediction model, and presents the empirical study on the performance of the model under different conditions. Section 4 describes the algorithm for adapting the ANN based QoE prediction model to a new condition, and presents the experimental results and analysis. The paper is summarized in section 5.

2   Related Works

Many PSQA algorithms are proposed for video quality assessment. In [5], a QoE asset model is proposed based on Artificial Neural Network. In the setting of the condition, H.263 is selected as the encoding algorithm and the resolution is CIF. QAFs in that work include the stream bit rate, frame rate, network loss rate, number of consecutively lost packets and the ratio of the encoded intra macro-blocks versus inter macro-blocks. Then, in order to build the database of distorted video clips, an environment that can simulate all kinds of quality-affecting parameters is established to produce the required distorted video. Then the subjective quality test is conducted for video clips in the database following the degradation category rating (DCR) subjective method described in ITU-T P.910[3]. Unreliable grades are removed in post-processing. Finally, an ANN is trained on that data set. Then, the model can be used to assert the QoE given the values of the QAFs. HyQoE is proposed in [7], it provides real-time video quality estimation in the wireless environment. HyQoE takes the percentage of losses of I, P and B frames, GoP sizes and several other characteristics of the video as the QAFs. It shows that I frame is important than other frames. HyQoE uses a cluster-based method to process different types of video content, and use ANN for QoE prediction. Beside ANN[5, 7, 10], Support Vector Machine (SVM) [11] and Decision Tree (DT)[11, 12] based methods are also proposed for QoE prediction. Other related methods such as Hoeffding Tree based model[14] are proposed in the literature, too.

In [15], the size of display terminal and the category of the video content are also considered. An approach called "Methods of Limits" is used to find out user's threshold determining the video quality is "acceptable" or "not acceptable". In the test, the host adjusts the Video Bitrate and the Video Frame rate to affect the quality of the video, and the observer needs to judge whether the quality is "acceptable" or "not acceptable". The "acceptable" or "not acceptable" result is used as a subjective annotation to train the QoE prediction model.

Though decent QoE prediction accuracy can be obtained, however, all the aforementioned methods assume that the model will be used under the same condition as the one where the model is trained. It is prohibitively expensive to collect adequate data and train a specific model for each possible condition. Therefore, it is of great interest to investigate the performance of the QoE prediction model under mis-matched conditions, and effectively adapting of a PSQA model from one condition to another is highly demanded.

3   Evaluation of ANN Based QoE Models under Mismatched Conditions

3.1   ANN Based QoE Prediction Models

ANN is one of the commonly used models for QoE prediction [5, 7, 10]. It is able to map from the input objective values to the output human perception of the video qual-ity. In this work, we use the ANN for QoE prediction in our study.

The ANN-based QoE prediction model is illustrated in Figure 1. The ANN used in this work has three layers, as denoted by x the input layer, h the hidden layer, and y the output layer. Correspondingly, we denote W the weight matrix between the input layer the hidden layer, and U the weight matrix between the hidden layer and the output layer. As illustrated in figure 1, we also denote f the raw objective feature vector, which consists of the measured values of the QAFs.

In a conventional ANN-based QoE model, the feature vector is fed to the ANN di-rectly, i.e.

$$x_i = f_i, \ i = 1, \dots, M \tag{1}$$

Where M is the number of nodes in the input layer, which equals to the number of QAFs. At the hidden layer, the output of a neuron is computed by:

$$h_j = \sigma\big(hnet_j\big) = \sigma\left(\sum_{i=1}^{M} w_{j,i} x_i + b_i\right), \qquad j = 1, \dots, N \tag{2}$$

Where N is the number of nodes in the hidden layer, $w_{j,i}$ is the element at the j-th row and i-th column of W, and b_i is the i-th bias. $hnet_j$ is the weighted sum of inputs for node j, $\sigma(x)$ is the activation function at the hidden layer, i.e.,

$$\sigma(x) = \frac{1}{1 + e^{-x}} \tag{3}$$

The output layer in figure 1 contains only one node, where its value is the predicted QoE, which is computed as follows:

$$y = \theta(ynet) = \sum_{j=1}^{N} u_j h_j + c_j \tag{4}$$

u_j is the element at the j-th column of U, and c_j is the j-th bias. The matrix U has only one row since the output layer has only one node. $ynet$ is the weighted sum of inputs for output node, $\theta(x)$ is the activation function at the output layer, i.e.,

$$\theta(x) = x \tag{5}$$

The data set consists of data samples in the form of (f, y^*) pairs, where f is the feature vector and y^* is the human assessment of the quality in MOS score. In training, the loss function to be minimized is:

$$E(\Lambda) = \sum_{r=1}^{R}(y_r - y_r^*)^2 \tag{6}$$

Where R is the number of data samples in the training set, $\Lambda = \{W, U, b, c\}$ is denoted by the parameter set of the ANN that needs to be learned.

Learning of the ANN is performed via the standard back-propagation (BP) algorithm [16].

3.2   Data and Experimental Settings

In data collection, we use the Evalvid tool[17] to simulate the network environment and to generate distorted video clips. In this study, the encoding bit rate, I frame loss and P frame loss are selected as the QAFs. The subjective assessment of video quality was carried out following ITU-T P.910 DCR. The MOS score is on a 5-level scale. Each distorted video clip is rated by five observers. A total of 27 human observers participated in this video quality subjective assessment. In evaluation, we predicted the QoE rating using the ANN-based model for each video clip in the test set.

In evaluation, following [11], k-fold cross-validation is used to test the effectiveness of the model. The methods are implemented based on Matlab[18].

3.3   Evaluation under Resolution-Mismatched Conditions

In this section we evaluate the performance of the ANN-based model under resolution-mismatched conditions. In our experiment, we use we use the same video clip with three resolution settings: 720p (1280x720), 4CIF (704x576) and CIF (352x288). We built the model for each of the three conditions. First we use k-fold method to test the accuracy of each model under the same condition, and then tested the model with the data of the other two conditions.

Given the original video clips, distorted clips are generated by adjusting the value of the three QAFs. The value and the range of each QAF are carefully determined based on two considerations. First, the range is determined such that the quality of the video covers the full range of the MOS score in the subjective assessment. Second, in order to minimize the human labor cost when sampling the value of the QAF over the desired range, different values of a QAF are set apart from each other such that there is an observable difference in terms of the video quality. The values of each QAF are sampled as follows, for the 720p condition, the value of the bit rate is sampled at: 4000kbps, 1000kbps and 512kbps, the values of the I, P frame loss rates are sampled at: 0.1%, 0.5%, 1%, 2%, 3%, 4%, 5%, 7%, 10%. After permutation, there are a total of 243 distorted video clips generated. For the conditions of 4CIF and CIF resolution, the values of the I, P frame loss rates are sampled at: 0.1%, 1%, 3%, 5%, 8%, 11%, while the values of bit rate for the 4CIF condition are sampled at: 1024kbps, 512kpbs and 256kbps, and the values of the bit rate of the CIF condition are sampled at: 512bps, 256bps and 128 bps. In total there are 108 distorted video clips for the condition of 4CIF and CIF, respectively.

Table 1. Diverse resolution model correct rate in mismatch-condition

Model\test data	High(720p)	Mid(4CIF)	Low(CIF)
High Model(720p)	76.8%	61.6%	46.3%
Mid Model(4CIF)	64.1%	68.7%	54.5%
Low Model(CIF)	46.9%	48.5%	64.3%

The experiment results are tabulated in table 1. From the results presented in table 1, we observed that the performance of the ANN-based QoE prediction model degrades substantially when tested under a mismatched condition. For example, the model of the 720p condition achieves an accuracy of 76.8% when tested under the same 720p condition. However, the prediction accuracy drops to 61.6% when tested on the 4CIF condition, which is also significantly lower than the accuracy of the model that is trained on data in the 4CIF condition and tested in the same 4CIF condition (68.7%). We also observed that, the performance degradation becomes even worse when the degree of the condition mismatch gets larger. For example, The accuracy of the 720p model drops to an even lower 46.3% in the CIF condition, far lower than that under the 4CIF condition. This is because that, compared to the mismatch between 720p (1280x720) and 4CIF (704x576), the mismatch between 720p and CIF (352x288) is much more significant.

Results in indicates that the accuracy of conventional ANN-based QoE prediction model degrades substantially under mismatched conditions.

4   Adaptation of ANN-Based QoE Prediction Models

Inspired by recent model adaptation methods in speech recognition and information retrieval [19, 20], we strive to address this issue through a model adaptation approach, i.e., instead of re-training the whole ANN on the data collected under the new condition, we will adapt the ANN-based QoE prediction model that is well trained under an source condition to the new condition using only a small amount of adaptation data.

4.1   Adaptation of the ANN-Based QoE Prediction Model

First we assume adequate data samples are collected and the ANN is well-trained under a source condition. Then, when given a new condition, instead of re-training the whole parameter set of the ANN under the new condition, we propose a feature transformation based method to adapt the ANN-based QoE prediction model. The method is illustrated in Figure 1. Compared to the conventional model there are two key differences. First, the parameters of the ANN are fixed. Second, instead of feeding the raw feature values of the QAFs directly to the input layer of the ANN, we apply a linear transformation to adapt the feature vector under the new condition to an adapted feature vector fitting the source condition. Moreover, we further assume that the QAFs are independent with each other, and therefore we only need to scale and shift each feature value independently. This greatly reduces the number of free parameters, and requires only a small amount of adaptation data to train.

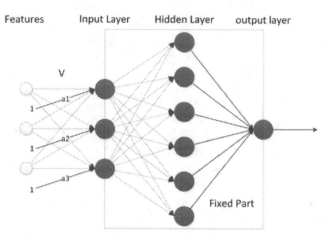

Fig. 1. Illustration of the feature transformation based ANN adaptation

The feature adaptation is formally defined as follows,

$$x = Vf + a \qquad (7)$$

$$V = \begin{pmatrix} v_{11} & \cdots & 0 \\ \vdots & v_{22} & \vdots \\ 0 & \cdots & v_{33} \end{pmatrix} \qquad (8)$$

The f is the feature vector. V is the transformation matrix. Since we only need to scale and shift each feature value independently, V is a diagonal matrix, and a is the bias vector accounting for the shift of the feature values. x is the adapted feature vector. Once x is computed, the hidden layer and the output are computed in the forward propagation as in the regular ANN following equation (2) – (5). The free parameters in equation (8) include $\Lambda = \{V, a\}$, which can be trained via the back-propagation algorithm[16] with certain modification, which we describe in following sections.

In the back propagation, the transformation matrices of input layer and hidden layer are fixed. Only the transformation matrix V should be trained. In the back propagation process, we still calculate the delta values at each layer, which is used to calculate the gradient w.r.t. V.

At the Output Layer:

$$\delta_Y = \frac{-\partial E}{\partial ynet} = \frac{-\partial E}{\partial y} \frac{\partial y}{\partial ynet} = (y^* - y)\theta'(ynet) = (y^* - y) \qquad (9)$$

At the Hidden Layer:

$$\delta_{H,i} = \frac{-\partial E}{\partial hnet_i} = \frac{-\partial E}{\partial hnet}\frac{\partial hnet}{\partial h_i}\frac{\partial h_i}{\partial hnet_i} = \delta_Y u_i \sigma'(hnet_i) = \delta_Y u_i h_j(1 - h_j) \quad (10)$$

And at the Input Layer:

$$\delta_{X,i} = \frac{-\partial E}{\partial xnet_i} = \sum_{j=0}^{N_2} \frac{-\partial E}{\partial hnet_j}\frac{\partial hnet_j}{\partial x_i}\frac{\partial x_i}{\partial xnet_i} = \sum_{j=0}^{N_2} \delta_{H,j} w_{j,i} \quad (11)$$

Then we can compute the gradient of the V and a, respectively:

$$\Delta v_{i,j} = \delta_{X,i} f_j \qquad \Delta a_i = \delta_{X,i} \quad (12)$$

It should be noted that, when a large amount of data are available and the whole ANN can be trained reliably, this linear transformation will be merged into the non-linear transformation between the input layer and the hidden layer of the ANN. However, when data is limited, fixing the pre-trained ANN and just training this linear feature transformation is more effective, as the experimental results show in following sections.

4.2   Experiment

We carried out a series of experiments to evaluate the effectiveness of the proposed model adaptation method. First, we build the ANN-based QoE prediction model under the source condition using a large amount of training data. This model is called the base model. Under a new condition, we first collect a small amount of data, a.k.a. adaptation data, and then adapt the base model to the new condition by training the feature transformation parameters based on that adaptation data following the approach described in section 4.1.

We also carried out a "direct-train" experiment for comparison, in which an ANN-based model is trained from scratch using the small amount of adaptation data.

We evaluated the effectiveness of adapting the ANN-based QoE prediction model from one condition to another with a different resolution. The results are presented in Figure 2.

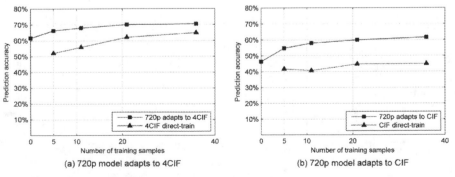

Fig. 2. Results of adapting the 720p Base Model to 4CIF and the CIF conditions

In Figure 2, the base model is trained under the 720p condition, all the 720p experiment data, e.g. 243 samples are used to train the baseline model. The model is then adapted to the 4CIF condition and the CIF condition, respectively. From Figure 2, first we observed that the performance of the base QoE prediction model under a new condition can be substantially improved using only a small amount of adaptation data. For example, Figure 2 (a) shows that, without any adaptation, the 720p base model give an accuracy of 61.6% under the 4CIF condition. After using only 5 adaptation data samples, the QoE prediction accuracy under the 4CIF condition is improved substantially from 61.6% to 66.2% (4.6% absolute improvement). When more data are available for adaptation, the accuracy keeps improving. Second, we observed that the improvement of the QoE prediction accuracy is even larger when the degree of condition mismatch is more severe. For example, Figure 2 (b) shows that adapting the 720p (high resolution) model using five samples in the CIF (low resolution) condition improves the QoE prediction accuracy by 8.4 % absolutely (from 46.3%, without adaptation, to 54.7%, using five data samples for adaptation), which is larger than the 4.6% absolute improvement in the 4CIF condition (middle resolution). Third, as shown in figure 2, the adaptation approach consistently outperforms the "direct-train" approach. Further analysis shown that, compared to "direct-train", the adaptation approach gives much better performance when only a small amount of data under the new condition are available for training. This strongly demonstrates the effectiveness of the proposed QoE model adaptation method.

5   Conclusion

In this work, we carried the first systematic study on the problem of applying the PSQA based QoE prediction model under mismatched conditions. Based on the commonly used ANN-based PSQA model, our study shows that the prediction accuracy degrades significantly when the model is applied to conditions different from the training condition. Further, in order to adapt the ANN-based QoE prediction model from one condition to another using a small amount of data, we developed a feature transformation based QoE model adaptation method. We have carried out a series of experiments and the evaluation results shown that the QoE prediction accuracy under mismatched conditions can be improved substantially using as few as five data samples for adaptation, and the prediction accuracy improves further when more adaptation data become available.

Acknowledgements. This work is supported by large-scare high definition video conference system application demonstration of China Next Generation Internet (Grant No.CNGI2008-118), and The Open Fund of Communication & Computer Network Lab of GuangDong (Grant No.201108).

References

1. cisco: Cisco Visual Networking Index: Forecast and Methodology, 2012-2017 (2013)
2. Lin, C., Hu, J., Kong, X.Z.: Survey on Models and Evaluation of Quality of Experience. Chinese Journal of Computers 35, 1–15 (2012) (in Chinese)

3. R. ITU-T, P910: Subjective video quality assessment methods for multimedia applications (2008)

4. Pinson, M.H., Wolf, S.: A new standardized method for objectively measuring video quality. IEEE Transactions on Broadcasting 50, 312–322 (2004)

5. Mohamed, S.G.: A study of real-time packet video quality using random neural networks. IEEE Transactions on Circuits and Systems for Video Technology 12, 1071–1083 (2002)

6. Venkataraman, M., Chatterjee, M.: Inferring video QoE in real time. IEEE Network 25, 4–13 (2011)

7. Aguiar, E., Riker, A., Abelém, A., Cerqueira, E., Mu, M.: Video quality estimator for wireless mesh networks. In: IEEE 20th International Workshop on Quality of Service (IWQoS), pp. 1–9. IEEE (2012)

8. Menkovski, V., Exarchakos, G., Liotta, A.: Online QoE prediction. In: 2010 IEEE Second International Workshop on Quality of Multimedia Experience (QoMEX), pp. 118–123. IEEE (2010)

9. Chen, K.T., Tu, C.C., Xiao, W.C.: OneClick: A framework for measuring network quality of experience. In: INFOCOM 2009, pp. 702–710. IEEE (2009)

10. Piamrat, K., Viho, C., Bonnin, J.M., Ksentini, A.: Quality of experience measurements for video streaming over wireless networks. In: Sixth International Conference on Information Technology: New Generations, pp. 1184–1189. IEEE (2009)

11. Menkovski, V., Oredope, A., Liotta, A., Sánchez, A.C.: Predicting quality of experience in multimedia streaming. In: Proceedings of the 7th International Conference on Advances in Mobile Computing and Multimedia, pp. 52–59. ACM (2009)

12. Balachandran, A., Sekar, V., Akella, A., Seshan, S., Stoica, I., Zhang, H.: Developing a predictive model of quality of experience for internet video. In: Proceedings of the ACM SIGCOMM 2013, pp. 339–350. ACM (2013)

13. Agboma, F., Liotta, A.: QoE-aware QoS management. In: Proceedings of the 6th International Conference on Advances in Mobile Computing and Multimedia, pp. 111–116. ACM (2008)

14. Menkovski, V., Exarchakos, G., Liotta, A.: Machine learning approach for quality of experience aware networks. In: 2010 2nd International Conference on Intelligent Networking and Collaborative Systems (INCOS), pp. 461–466. IEEE (2010)

15. Agboma, F., Liotta, A.: Addressing user expectations in mobile content delivery. Mobile Information Systems 3, 153–164 (2007)

16. Mitchell, T.M.: Machine learning. McGraw-Hill Science/Engineering/Math. (1997)

17. Klaue, J., Rathke, B., Wolisz, A.: Evalvid – A framework for video transmission and quality evaluation. In: Kemper, P., Sanders, W.H. (eds.) TOOLS 2003. LNCS, vol. 2794, pp. 255–272. Springer, Heidelberg (2003)

18. Matlab, http://www.mathworks.com/products/matlab

19. Lei, X., Hamaker, J., He, X.: Robust feature space adaptation for telephony speech recognition. In: INTERSPEECH (2006)

20. Wang, H., He, X., Chang, M.W., Song, Y., White, R.W., Chu, W.: Personalized Ranking Model Adaptation for Web Search. In: Proceedings of the 36th International ACM SIGIR Conference on Research and Development in Information Retrieval, pp. 323–332. ACM, Dublin (2013)

Leveraging Color Harmony and Spatial Context for Aesthetic Assessment of Photographs

Hong Lu[1,*], Jin Lin[1], Bohong Yang[1], Yiyi Chang[2], Yuefei Guo[1],
and Xiangyang Xue[1]

[1] Shanghai Key Laboratory of Intelligent Information Processing,
School of Computer Science, Fudan University
honglu@fudan.edu.cn
[2] School of Computer and Communication Engineering,
Tianjin University of Technology

Abstract. Computer aesthetic assessment of pictures is aimed at automatically computed aesthetic values of pictures. It has potential wide areas of application in real world. We apply color harmony, one of the most important aesthetic standards, and explore the spatial context of features. Based on the framework of [9], we provide a color harmony descriptor which includes the circular region sampling method, and follow the principle of Ordered-Bag-of-Features to explore the spatial context. And we conduct experiments on a public and large-scale aesthetic assessment dataset. Experimental results demonstrate the effectiveness of the proposed method.

Keywords: Computer aesthetic assessment, Color harmony, Circular region sampling, Spatial context.

1 Introduction

Facing more and more complex image data, it is a big challenge to classify the useful and beautiful photos from large-scale image data. The purpose of computer aesthetic assessment is to handle the challenge. It aims at automatically computing aesthetic values out of a picture based on some aesthetic rules. Besides, common aesthetic standards (i.e. spatial composition and objects appearance [4]), color harmony plays an important role in aesthetic assessment. However, [9] points out that color harmony is largely ignored in current researches, and has only a few simple applications. Based on [9], we focus on color harmony and propose a modified method for aesthetic assessment of photos. Our main work includes:

- Circular Region Sampling, a method for local region sampling, which is more effective.
- Based on the proposed method above, demonstrating how the spatial context matters through experiments, and following the guidance of Ordered Bag-of-Features to explore the spatial context of the photo.

---

* Corresponding author.

W.T. Ooi et al. (Eds.): PCM 2014, LNCS 8879, pp. 323–332, 2014.

— Experiments that demonstrate different performances of proposed method on different categories of pictures, and score distributions of training sets.

2   Related Works

According to [5], instead of directly using original low level aesthetic features (i.e. color histogram and gray), some high level features tend to be used more widely in current work of aesthetic assessment. Color harmony is a kind of high level feature.

As for the work based on color harmony, before the work in [9], some low level color features, such as color intensity and global color histogram, are often used as aesthetic features. Specifically, [3] uses Matsuda Model to improve the degree of color harmony of a given picture, but [3] is not a work belongs to aesthetic assessment. Also, [5] uses a term 'Opposing Colors', which is one situation of color harmony. The work in [9] uses a color harmony model provided by study of color science in aesthetic assessment.

There are many color harmony models based on different criteria [10]. [9] points out that there are only two color harmony models which are suitable for implementation on computer. These two color harmony models are Moon-Spencer Model introduced in [7] and Matsuda Model introduced in [6]. Moon-Spencer Model is an old model, which measures the harmony on the basis of the difference between two colors in Munsell Color Space. As stated in [3], Matsuda Model computes a histogram of the picture based on hue and saturation values in HSV Color Space, and uses the specific template of the histogram to measure the color harmony of the whole picture.

The design of local descriptor largely relies on the color harmony model. According to Moon-Spencer Model, people can judge harmony by difference of colors. Moon-Spencer Model provides a simple but effective way to quantize this kind of difference, thus it is used in [9].

3   Method

Basically, it is common to follow these steps to get a representation of a photo in such kind of work: sample local regions from the photo, extract local descriptors from each region, then represent the whole photo using these descriptors. [9] also followed those steps. Thus the way of extracting local descriptors from a local region, and of representing the whole photo will affect the final performance. In order to explore a more effective color harmony feature, we focus on these aspects in our work.

3.1   Extracting Local Descriptors: Circular Region Sampling

Based on Moon-Spencer Model, [9] selects the dominant color to be the chosen color. And the Munsell hue, value and chroma of each pixel, which are relative to the chosen color, are computed to form the local descriptor. In [9], the chosen color is the dominant color determined by hue values; the Munsell value (or lightness uesd in [9]) and chroma used to make up the local descriptor are respectively the average of Munsell value and chroma of the pixels with the dominant color; besides, linear interpolation technique is used to transfer a RGB color to Munsell color, which can bring some errors. Then the average of lightness and chroma makes the errors to be spread. Thus we use the color of the central pixel as the chosen color to avoid further errors in computation.

As for the sampling technique, we use Circular Region Sampling to extract local regions, while [9] uses the grid-sampling technique (as shown in Fig. 1). Since in a circular region, the distance between each pixel and the center would be limited in a certain value, so the chosen color of the central pixel can have a balanced impact on all pixels in the region. Thus we use Circular Region Sampling.

The overall comparison between our work and [9] is depicted in Fig. 2.

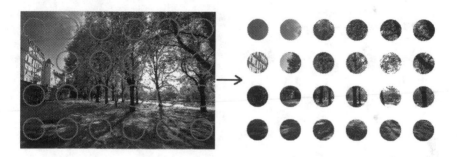

Fig. 1. Circular Region Sampling. Extract circular local regions according to a certain region size and sampling dignity. In the figure above, we give a simple example which extracts only 24 (4x6) circular regions. The regions can be overlapped, and we do not distinguish local regions with/without color boundaries as [9] does.

chosen color (dominant color) ← [ ] → [ ] → chosen color (color of central pixel)

Fig. 2. Comparison of sampling technique and chosen color. (a) grid-sampling used in [9], and the dominant color determined by hue is selected as the chosen color. **(b)** circular region sampling used in our method, and the color of central pixel is selected as the chosen color.

3.2   Representing the Whole Picture: Ordered-Bag-of-Features

It is common to explore the spatial context in order to avoid the shortage of traditional Bag-of-Features which ignores the spatial information of local features. For example, in [13], the image is divided into 1,2,3,5, or 10 rows to explore the spatial context. [9] has also explored the spatial context by dividing the image into several rectangular segments and applying BoF in each segment, but there are few details.

Based on the proposed method in Sec. 3.1, we also explore the spatial context using Ordered-Bag-of-Features (Ordered-BoF) as mentioned in [1][1]. As depicted in Fig. 3, there are two types of projections, i.e. linear projection and circular projection. The linear projection is used in our method based on the experiment comparison results. The experiments and results are introduced in detail below.

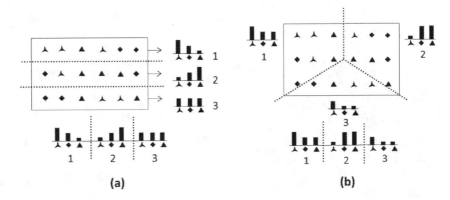

Fig. 3. Ordered-Bag-of-Features. Different shapes in the figure represent different kinds of features. **(a)** Linear projection: in our work we use horizontal lines to divide the photo, and here is an example of three bins. **(b)** Circular projection: we choose the center of the picture as the center of circular projection, and here is an example of three bins.

3.3   Summary

In summary, we explore the local region sampling, the choice of chosen color when applying Moon-Spencer Model, and spatial context of the whole photo. According to the discussion above, we follow these steps to implement our method based on the framework of that in [9]:

1. Sample local regions from the photo using Circular Region Sampling technique. We do not divide local regions into two sets of regions with/without color boundaries as [9] does.

---

[1] [1] also proposed a further improved Spatial-BoF which takes translation invariance, rotation invariance and scaling invariance into consideration. We think the Spatial-BoF mainly focuses on the object detection, and its new features are not suitable in such context of color harmony.

2. Extract local descriptors. For each local region, choose the color of the center as the chosen color. Then use hue, lightness and chroma values in Munsell Color System, which is relative to the chosen color, of each pixel to represent the color harmony of the local region.
3. Use Ordered-BoF model with linear projection to get a vector representing color harmony of the whole picture.

4   Experiments

4.1   Experimental Setup

We choose AVA dataset provided by [8] as experimental data. We also collect images from DPChallenge[2] through the guidance of AVA dataset. [9] also collects its dataset from DPChallenge. AVA dataset is a public dataset for aesthetic assessment. It includes a total of 250,000 pictures, and each picture in the dataset has an aesthetic score in the range from 1 to 10. A higher score indicates a better aesthetic value of the picture. In our work, we label the picture with the aesthetic score larger/smaller than 5.5 as high/low quality.

In our experiments, we choose several subsets of AVA dataset. Inspired by [8], in one subset, we select 20,000 pictures with the aesthetic score falls in the top and bottom 12% for training, and *randomly* select another 20,000 pictures for testing. The aesthetic values of the top and bottom 12% pictures are less ambiguous, which are therefore more consistent by different human judgers [8], and they are also not too unambiguous. The experimental analysis of such selection is stated in Sec.4.4.

Support Vector Machine (SVM) [2] with radial basis function (RBF) kernel is used for classification. Cross-validation is used for parameter determination. For BoF, the k-means and kd-tree implemented in VLFeat [12] are used. The maximum size of image is 800×800, and the minimum size is 160×160. Besides, the local regions can be overlapped.

Besides, for circular region sampling, we set the radius of the circular region to 4, 8, 16, 24, 32, 48 pixels, respectively. And accordingly we set the step between the centers of the neighboring circular regions to 4, 8, 12, 16, 20, 24, 28, 32, 48 pixels. Finally, we empirically set the radius and the step both to be 4 pixels.

4.2   Experiments on the Method

Experiments on Circular Region Sampling. We randomly select ten subsets of photos which exclude ambiguous photos whose aesthetic score lies between 5.00 and 5.99 for training. The ten subsets contain 100, 200, 300, $\cdots$, 1k photos, respectively. Other 1,000 photos with randomly distributed aesthetic scores are used for testing. We use grid-sampling technique with distinction of regions with/without color boundaries in [9] and the proposed Circular Region Sampling

---

[2] DPChallenge. http://www.dpchallenge.com/

introduced in Sec. 3.1 for comparison. And the chosen color is determined using dominant color or central pixel color respectively. Both implementations are without any concerns about spatial context. The experimental result is shown in Fig. 4. With the increase of training data size, the proposed method can obtain a higher performance with much less number of training data.

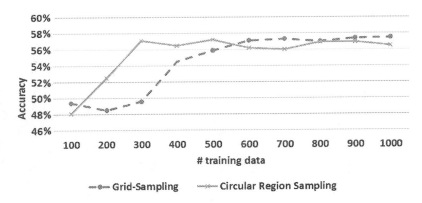

Fig. 4. Comparison of grid-sampling and circular region sampling

Experiments on Ordered-BoF. We also conduct an experiment of Ordered-BoF introduced in Sec. 4.2. In our work, we use horizontal lines for linear projection, and the center of the picture as the center of circular projection. The result is depicted in Fig. 5. The performance of linear ordered-BoF is always higher than the performance of circular ordered-BoF and that of traditional Bag-of-Features (the horizontal dot line in Fig. 5), and it reaches the maximum when the number of bins is 4. So the linear ordered-BoF is effective, and we use linear ordered-BoF with 4 bins in our method.

4.3 Experimental Results on Different Categories

As stated in [8], aesthetic features can have different performances over different categories of pictures. In [9], the experimental results are based on the average of different categories, and the performances in each category are not shown. Thus we do experiments on different categories of pictures.

AVA dataset provides some images which have already been classified into several categories, such as animal, architecture, cityscape, floral, food drink, landscape, portrait and still life. Each category contains about 2,500 pictures for training, and another 2,500 for testing. And the aesthetic scores of pictures in each category are randomly distributed. The experimental result on different categories is shown in Fig. 6.

From Fig. 6, we can observe that the proposed method shows different performances over different categories. Specifically, the method performs better on images in food drink, cityscape, landscape, portrait, etc. than those in architecture, animal.

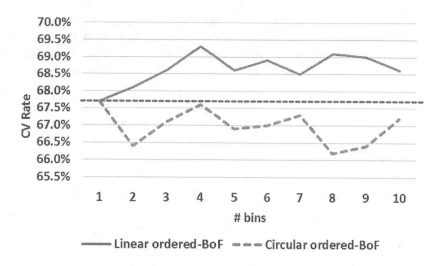

Fig. 5. Experimental result on ordered-BoF. The red line above represents the performance with original Bag-of-Features. And we use Cross-Validation Rate (CV Rate) to measure the performance of Ordered-BoF with different number of bins.

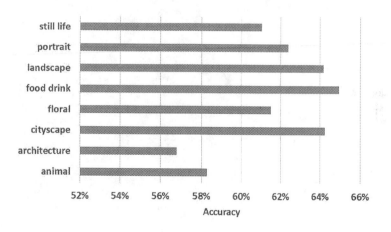

Fig. 6. Experimental results on different categories

4.4   Influence of Aesthetic Score Distribution

An **ambiguous** image is defined as an image whose aesthetic score is very close to the mean score. According to [8], an ambiguous image often implies the disputed judgements on the aesthetic value, and empirically plays a bad role in aesthetic classification.

We select about 4,500 pictures from subset of AVA used in Sec. 4.3, and divide them into five training sets (numbered from 1 to 5) based on their score distributions (see Table 1). Each training set contains 1,000 pictures, except set

5 which has 319 pictures subject to their extreme scores. Besides, we *randomly* select another 1,000 pictures as the testing images. In this experiment, we respectively use the five training sets to train five SVM classifiers, and test them on the same testing set described before respectively.

Table 1. Score distributions of training sets used in Section 4.4. The aesthetic score in this table is defined and used in AVA dataset as described in Section 4.1. With the increase of training set ID, the scores of images in the training set are more far away from the average score (it is about 5.5 in AVA dataset), therefore less ambiguous.

Training Set ID	Score Distribution
1	[5.0,6.0]
2	[4.5,5.0) ∪ (6.0,6.5]
3	[4.0,4.5) ∪ (6.5,7.0]
4	[3.5,4.0) ∪ (7.0,7.5]
5	[1.0,3.5) ∪ (7.5,10]

Fig. 7. Experimental result on influence of score distribution

According to Fig. 7, aesthetic score distribution does influence performances of color harmony features, but a training set containing the most unambiguous images may result in a bad performance (like training set 5 in Fig. 7). On these grounds, we choose pictures with aesthetic scores falling in the top and bottom 12% (as described in Sec. 4.1) in the training set used in comparison experiment below.

4.5   Comparison Experiment

We conduct a comparison experiment of our proposed method with the method in [9]. In this experiment, we use the 40,000 pictures of AVA dataset as the experimental dataset. Three features are extracted, including RGB histogram on the whole picture, color harmony features extracted using method in [9], and features extracted using proposed method in this paper.

As mentioned in Sec. 3.2, [9] contains few details about how it explores spatial context. Thus we apply horizontal linear ordered-BoF in the implementations of both the proposed method and the method in [9]. Besides, because of the large size of the training data, we perform cross validation for SVM parameter selection on a subset of training data extracted both for the proposed method and the method in [9] respectively.

The experimental result is shown in Fig. 8. From Fig. 8, we can observe that the performance of our method is better than the method in [9], and much better than RGB histogram in our experiment. The experimental result shows that our proposed method is effective.

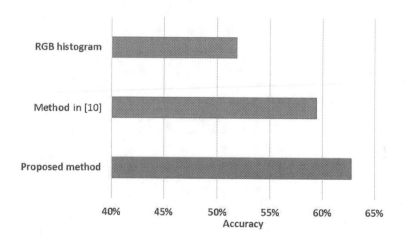

Fig. 8. Result of comparison experiment

5   Future Work

[10] gives an empirical framework for aesthetic assessment of color combinations proposed by [11]. According to this framework, current models for aesthetic assessment of color combinations only focus on the objective harmony of color combinations, but neglect the human preference, which is much more subjective. It could be a break-through point for the future work of aesthetic assessment based on color combinations.

Acknowledgements. This work was supported in part by the Natural Science Foundation of China (No. 61170094), Shanghai Committee of Science and Technology (14JC1402202 and 14441904403), and 863 Program 2014AA015101.

References

1. Cao, Y., Wang, C., Li, Z., Zhang, L., Zhang, L.: Spatial-bag-of-features. In: 2010 IEEE Conference on Computer Vision and Pattern Recognition (CVPR), pp. 3352–3359. IEEE (2010)
2. Chang, C.-C., Lin, C.-J.: Libsvm: a library for support vector machines. ACM Transactions on Intelligent Systems and Technology (TIST) 2(3), 27 (2011)
3. Cohen-Or, D., Sorkine, O., Gal, R., Leyvand, T., Xu, Y.-Q.: Color harmonization. ACM Transactions on Graphics (TOG) 25, 624–630 (2006)
4. Datta, R., Joshi, D., Li, J., Wang, J.Z.: Studying aesthetics in photographic images using a computational approach. In: Leonardis, A., Bischof, H., Pinz, A. (eds.) ECCV 2006, Part III. LNCS, vol. 3953, pp. 288–301. Springer, Heidelberg (2006)
5. Dhar, S., Ordonez, V., Berg, T.L.: High level describable attributes for predicting aesthetics and interestingness. In: 2011 IEEE Conference on Computer Vision and Pattern Recognition (CVPR), pp. 1657–1664. IEEE (2011)
6. Matsuda, Y.: Color design. Asakura Shoten 2(4) (1995)
7. Moon, P., Spencer, D.E.: Geometric formulation of classical color harmony. JOSA 34(1), 46–50 (1944)
8. Murray, N., Marchesotti, L., Perronnin, F.: Ava: A large-scale database for aesthetic visual analysis. In: 2012 IEEE Conference on Computer Vision and Pattern Recognition (CVPR), pp. 2408–2415. IEEE (2012)
9. Nishiyama, M., Okabe, T., Sato, I., Sato, Y.: Aesthetic quality classification of photographs based on color harmony. In: 2011 IEEE Conference on Computer Vision and Pattern Recognition (CVPR), pp. 33–40. IEEE (2011)
10. Palmer, S.E., Schloss, K.B., Sammartino, J.: Visual aesthetics and human preference. Annual Review of Psychology 64, 77–107 (2013)
11. Schloss, K.B., Palmer, S.E.: Aesthetic response to color combinations: preference, harmony, and similarity. Attention, Perception, & Psychophysics 73(2), 551–571 (2011)
12. Vedaldi, A., Fulkerson, B.: Vlfeat: An open and portable library of computer vision algorithms. In: Proceedings of the International Conference on Multimedia, pp. 1469–1472. ACM (2010)
13. Vogel, J., Schiele, B.: Semantic modeling of natural scenes for content-based image retrieval. International Journal of Computer Vision 72(2), 133–157 (2007)

A Multi-exposure Fusion Method
Based on Locality Properties

Yuanchao Bai[1], Huizhu Jia[1,*], Hengjin Liu[1], Guoqing Xiang[1], Xiaodong Xie[1],
Ming Jiang[2], and Wen Gao[1]

[1] National Engineering Laboratory for Video Technology,
Peking University, Beijing 100871, China
{yuanchao.bai,hzjia,hjliu,donxie,wgao}@pku.edu.cn
[2] Department of Information Sciences, School of Mathematical Sciences,
Peking University, Beijing 100871, China
ming-jiang@ieee.org

Abstract. A new method is proposed for fusing a multi-exposure sequence of images into a high quality image based on the locality properties of the sequence. We divide the images into uniform blocks and use variance to represent the information of blocks. The richest information (RI) image is computed by piecing together blocks with largest variances. We assume that images in the sequence are high dimensional data points lying on the same neighbourhood and borrow the idea from locally linear embedding (LLE) to fuse a result image which is closest to the RI image. The result is comparable to the state-of-art tone mapping operators and other exposure fusion methods.

Keywords: Exposure fusion, High dynamic range, Locality properties, LLE.

1 Introduction

Digital cameras are developing quickly in producing higher-resolution images, but the limited dynamic range is still a challenge. Currently, the vast majority of images are represented in 8 bits per pixel for each colour channel, which is called low dynamic range (LDR) image, in contrast with the high dynamic range (HDR) in real world scenes. As a result, images captured by cameras in HDR scenes are either under- or over-exposed, missing parts of the information in dark or high-light regions. In order to achieve the full dynamic range, a bracketed exposure sequence is often made and is turned into a single image for convenient display.

There are several techniques to create such multi-exposure images. One of these techniques, reviewed in details by Erik Reinhard et al. [8], is to create a kind of HDR format images that uses floating point number to represent pixel channels and then to tone map the HDR format images into displayable gamut.

---

* Corresponding author.

W.T. Ooi et al. (Eds.): PCM 2014, LNCS 8879, pp. 333–342, 2014.

However, the obvious drawback of this approach is that it needs to create the HDR format image first, which is complex and costs lots of memory for the users who only want displayable multi-exposure results. Besides, the camera parameters that are required to calculate the radiometric response function [1] are often unknown by common users.

Another kind of techniques is called exposure fusion, which allocates a normalized weight to every image and then fuses the bracketed sequence into a multi-exposure image.

Tom Mertens et al. [7] proposed a technique for fusing images. The technique blends multiple exposures, guided by simple quality measures like saturation and contrast. Images are fused in a multiresolution fashion to deal with the seams.

Goshtasby [4] presented an block-wise fusion techniques. The technique first calculates the entropy of each block in each image and selects the blocks that contains the most information. Then the blocks are fused to a high quality image by a monotonically decreasing blending function. An algorithm to determine the optimal block size is also proposed. However, the optimization algorithm is slow and may be not necessary because the entropy is changed by uniform block size and is not the larger, the better.

Kotwal et al. [6] defined exposure fusion as a multi-objective optimization problem based on desired characteristics of the output and provided the solution using an Euler-Lagrange technique. The proposed technique yields visually appealing fused images with a high value of contrast. The slow iteration is the inevitable drawback of the algorithm.

Ashish et al. [11] proposed a low complexity detail preserving multi-exposure image fusion method. The fusion is performed pixel-by-pixel and does not involve any filtering or transformation. However, the saturation and the contrast of the multi-exposure method is not that good.

In this paper, we take advantage of the locality properties, the locality of pixels among their neighbourhoods and the locality of images with different exposures. Based on these properties, we make a model to optimize the information in the final image and also take the visual quality into consideration. The proposed algorithm is block-wised and each block can run parallel. The experiments show that quality of the results is comparable with tone mapping operators and other exposure fusion methods.

The rest of the paper is organized as follows. In section 2, we will introduce our algorithm in details. In section 3, many experiments and comparisons are done. In section 4, we make a conclusion and the future expectation.

2   Exposure Fusion Based on Locality Properties

Our exposure fusion method aims at allocating a normalized weight to every pixel of every image in the multi-exposure image sequence and then fuses these images to a final image. To achieve this goal, we choose the parts with rich information in the multi-exposure image sequence and use the locality properties to obtain the weights. In our method, we assume that the images are well aligned, possibly using a registration algorithm [13],[5].

2.1 Linear Fusion Model

Most of the exposure fusion method use the Linear Fusion Model (LFM), which means that the result is computed by a weighted average at each pixel in N images. To make sure that the fused pixels is in the 0 to 255 display range, the weights of N images at each pixel (i, j) are normalized:

$$F_{ij} = \sum_{k=1}^{N} W_{ij}^k I_{ij}^k, \quad \text{s.t.} \quad \sum_{k=1}^{N} W_{ij}^k = 1 \tag{1}$$

where i, j, k mean the pixel (i, j) of k-th input image in the sequence. I is the pixel value, W is the weight and F is the result. An example of a multi-exposure sequence is shown in Fig. 1. According to this model, we only need to compute the weight for each pixel, but it is not straightforward when rich information and visual quality are taken into consideration. Our method is based on LFM, which takes advantage of locality properties to guide the fusion.

(a)   (b)

(c)

Fig. 1. Multi-exposure image sequence. The photos are taken in different exposure time.

2.2   Information Measure with Locality Property

Each area of a scene has a corresponding best exposure. In the other word, one of the images in the exposure sequence contains the richest information of the area [4]. Inspired by this idea, we find the bridge between the visual quality and countable quantity. Although the fusion process is pixel-wised, one pixel cannot represent the quantity of information. We consider one pixel and its neighbourhood. When the neighbourhood is not very large, the relation between the visual quality and information quantity is monotonic, which means that the better the visual quality is, the more information the neighbourhood has. However, the monotonicity does not hold when the neighbourhood is large, because the neighbourhood may contain areas that have different best exposures. We conclude this result as the first locality property that guide our algorithm.

We use the variance to measure the information, which is a widely used concept in data analysis. If an area is well-exposed, the variance will reach its largest. When an area is under- or over-exposed, the area will have less details and the variance will be small. To simplify the calculation, we compute the variance on the $m \times m$ uniform blocks:

$$\mathrm{Var}(\mathbf{b}) = \frac{1}{m^2} \sum_{i=1}^{m^2} (b_i - \bar{b})^2 \tag{2}$$

where the $\mathrm{Var}(\mathbf{b})$ is the variance of a block $\mathbf{b}$. b_i is the i-th pixel value and $\bar{b}$ is the mean of pixels in the block. We divide the images into $m \times m$ sized blocks and compare the variances of blocks at the same place in different images in the sequence. The blocks with largest variances are extracted and pieced together to form an initial starting-point image. This image contains the richest information but is not comfortable for viewing, as shown in Fig. 2. The richest information image is the important variable of our next fusion step.

(a)   (b)

Fig. 2. The richest information image. The grayscale one is shown in (a) with the block size 16×16. The colored one is shown in (b).

2.3   Fusion with Locality Property

As the richest information (RI) image has been obtained, we fuse the multi-exposure sequence according to this image. We could directly smooth the RI image, but undesirable halos around the edge or shades are present, as demonstrated in Fig. 3(a).

(a)   (b)

Fig. 3. (a) is the result from directly smoothing RI image with halos. (b) is our algorithm with one weight per image.

To address the halos-shades problem and keep the information as much as possible, we borrow the idea from locally linear embedding (LLE) [10], which is an important algorithm in manifold learning. The LLE algorithm is based on simple geometric intuitions. Suppose there are N points in a high dimensional data space of dimensionality D. The main concept of LLE is that each data point and its neighbours are expected to lie on or close to locally linear patch of the manifold, so the data point can be reconstructed from its neighbours. In exposure fusion, the images in exposure sequence and the RI image are the high dimensional data points. As these images represent the same scene, these points are assumed to lie on a locally linear patch and can be reconstructed by neighbours. We propose a fusion model according to this assumption:

$$\hat{\mathbf{W}} = \underset{\mathbf{W}}{\arg\min} \left\| \sum_{k=1}^{N} W^k \mathbf{I}^k - \mathbf{R} \right\|, \quad \text{s.t.} \quad \sum_{k=1}^{N} W^k = 1 \qquad (3)$$

where $\mathbf{R}$ is the RI image, $\mathbf{I}^k$ is the k-th image in the exposure sequence, which is the neighbour of RI image. Our goal is to optimize the weights $\hat{\mathbf{W}}$ to fuse the sequence and make the fused image closest to the RI image. The solution to (3) can be derived below:

$$\hat{\mathbf{W}} = \left(\mathbf{I}^T \mathbf{I} \right)^{-1} \cdot \left[\mathbf{I}^T \mathbf{r} - \mathbb{1} \cdot \frac{\mathbb{1}^T (\mathbf{I}^T \mathbf{I})^{-1} \mathbf{I}^T \mathbf{r} - 1}{\mathbb{1}^T (\mathbf{I}^T \mathbf{I})^{-1} \mathbb{1}} \right] \qquad (4)$$

where $\mathbf{I} = (\mathbf{i}^1, \mathbf{i}^1, \cdots, \mathbf{i}^N)$, $\mathbf{i}^k = \mathrm{vec}(\mathbf{I}^k)$. The variable $\mathbf{r} = \mathrm{vec}(\mathbf{R})$ and $\mathbb{1}$ is the vector whose all entries are 1. However, the answer can not be found when the inverse matrix of $\mathbf{I}^T\mathbf{I}$ does not exist. To solve (3) in any cases and more efficiently, gram matrix has been involved, let $\mathbf{G}_{jk}$ be the (j, k) entry of gram matrix:

$$\mathbf{G}_{jk} = \left(\langle \mathbf{i}^j - \mathbf{r}, \mathbf{i}^k - \mathbf{r}\rangle\right)_{j,k\in[1,N]} \tag{5}$$

Then we compute the equation:

$$\mathbf{W} = (\mathbf{G} + \delta\mathbf{E})^{-1} \cdot \mathbb{1} \tag{6}$$

δ is a small number and $\mathbf{E}$ is the identity matrix. $\mathbf{G} + \delta\mathbf{E}$ is to make the matrix a positive-definite matrix and invertible. After that, the answer $\hat{\mathbf{W}}$ is solved by normalizing $\mathbf{W}$.

$$\hat{\mathbf{W}} = \frac{\mathbf{W}}{\sum_{k=1}^{N} W^k} \tag{7}$$

The result is shown in Fig. 3(b). The halo does not exist and the visual quality is improved. Fig. 3(b) contains more information than any of the images in Fig. 1 but the information is less than that in Fig. 3(a).

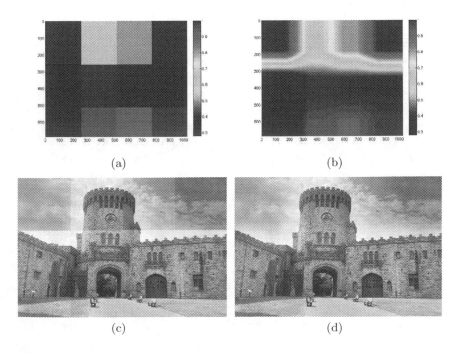

(a)   (b)

(c)   (d)

Fig. 4. (a) shows the un-smoothed weight map. (b) shows the weight map that has been smoothed. (c) is the result fused by un-smoothed weight map. (d) is the result fused by smoothed weight map.

To achieve more information, one weight per image is not satisfying. We do the fusion on $M \times M$ sized windows, which is much larger than $m \times m$ sized blocks. By this way, we can keep more local information. Then we use the mean filter to smooth the edges of weight map between windows. The 2-dimensional mean filter can be decomposed into two 1-dimensional mean filter and is able to run parallel. Besides, only pixels around the window edges need to be smoothed, the filter process can be quite efficient. According to the analysis of [12], fusion in groups of two results in lower computation complexity. In our fusion, we also take advantage of this concept to compute faster and get better result than fusing all images together at a time. For example, we first fuse the images in Fig. 1(a) and Fig. 1(b) to get the "image12". Secondly, we fuse "image12" and Fig. 1(c) to achieve the result. Fig. 4 shows the second step.

3 Experiments

Our experiments are done on the JPG-encoded photographs. We use 16×16 sized blocks and 256×256 sized windows for all the examples. In our exposure fusion, we do not need any information about the camera, except in Fig. 6. In Fig. 6, we need the exposure time of every image to create the HDR format image to test the tone mapping algorithm.

In Fig. 5, we compare our result to Tom Mertens et al.'s approach [7], our result is a little bit brighter than theirs but we both contains all the details and information in the exposure sequence. The halo does not exist in the experiment.

(a) Mertens et al. [7]   (b) Our algorithm

Fig. 5. Comparison with other exposure fusion. [7] is shown in (a). Our algorithm is shown in (b).

Fig. 6 is the multi-exposure sequence. We generate a HDR format image with [1] from this sequence. Then we use the HDR format image to run Durand et al. [3], Reinhard et al. [9] and Drago et al. [2] tone mapping methods. Three tone mapping results and our exposure fusion result are shown in Fig. 7. Our algorithm result looks better than these tone mapping results but in the area of neon, the word is not that clear.

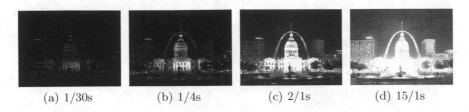

(a) 1/30s   (b) 1/4s   (c) 2/1s   (d) 15/1s

Fig. 6. Images with different exposure time for HDR recovering

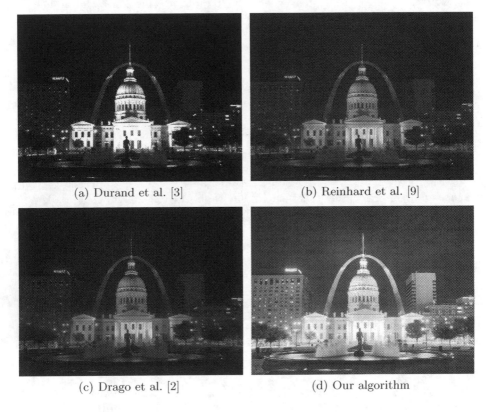

(a) Durand et al. [3]   (b) Reinhard et al. [9]

(c) Drago et al. [2]   (d) Our algorithm

Fig. 7. Comparison with tone mapping operators. [3] is shown in (a). [9] is shown in (b). [2] is shown in (c). Our algorithm is shown in (d).

In Fig. 8, we also compare the histogram of the images in the sequence to our result. Histogram of Fig. 6 are shown from Fig. 8(a) to Fig. 8(d) and the histogram of our result is shown in Fig. 8(e). We can see that our algorithm can center the pixel values and under- or over-exposure pixels decrease a lot.

From these comparisons, the proposed algorithm successfully extracts well-exposure parts from each image and performs well in visual quality.

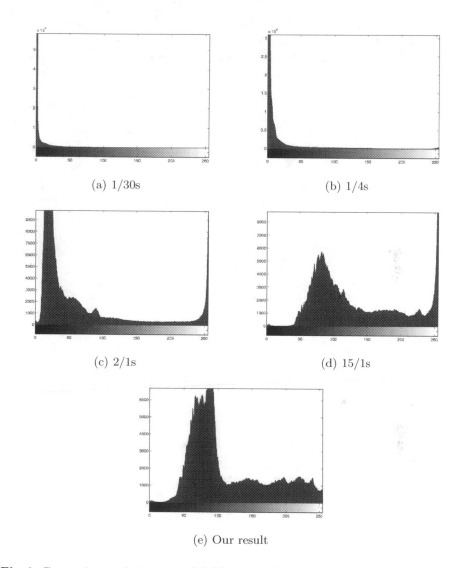

(a) 1/30s

(b) 1/4s

(c) 2/1s

(d) 15/1s

(e) Our result

Fig. 8. Comparison in histograms. (a)-(d) are the histograms of images in Fig. 6. (e) is the histogram of our result.

4   Conclusion

In this paper, we exploit the locality properties in the multi-exposure sequence to extract information from different exposure images. Also, we borrow the concept from LLE to model our fusion and achieve good results. In our exposure fusion, no camera information has to be known and no HDR format images will be generated.

As our algorithm can run parallel in block-level, it is more suitable to take advantage of this parallelism and implement our algorithm on GPU. Finally, as HDR video cameras are developing quickly, e.g. RED-EPIC camera, we are trying to apply this technique on multi-exposure video sequences for the next step.

Acknowledgements. This work is partially supported by grants from the Chinese National Natural Science Foundation under contract No.61171139 and National High Technology Research and Development Program of China (863 Program) under contract No.2012AA011703.

References

1. Debevec, P.E., Malik, J.: Recovering high dynamic range radiance maps from photographs. In: ACM SIGGRAPH 2008 Classes, p. 31. ACM (2008)
2. Drago, F., Myszkowski, K., Annen, T., Chiba, N.: Adaptive logarithmic mapping for displaying high contrast scenes. Computer Graphics Forum 22, 419–426 (2003)
3. Durand, F., Dorsey, J.: Fast bilateral filtering for the display of high-dynamic-range images. ACM Transactions on Graphics (TOG) 21, 257–266 (2002)
4. Goshtasby, A.A.: Fusion of multi-exposure images. Image and Vision Computing 23, 611–618 (2005)
5. Grosch, T.: Fast and robust high dynamic range image generation with camera and object movement. In: Vision, Modeling and Visualization, pp. 277–284. RWTH Aachen (2006)
6. Kotwal, K., Chaudhuri, S.: An optimization-based approach to fusion of multi-exposure, low dynamic range images. In: 2011 IEEE International Conference on Information Fusion (FUSION), Chicago, IL, USA, July 5-8, p. 7 (2011)
7. Mertens, T., Kautz, J., Van Reeth, F.: Exposure fusion. In: 15th Pacific Conference on Computer Graphics and Applications, PG 2007, pp. 382–390. IEEE (2007)
8. Reinhard, E., Heidrich, W., Debevec, P., Pattanaik, S., Ward, G., Myszkowski, K.: High dynamic range imaging: acquisition, display, and image-based lighting. Morgan Kaufmann (2010)
9. Reinhard, E., Stark, M.M., Shirley, P., Ferwerda, J.A.: Photographic tone reproduction for digital images. In: SIGGRAPH, pp. 267–276 (2002)
10. Roweis, S.T., Saul, L.K.: Nonlinear dimensionality reduction by locally linear embedding. Science 290(5500), 2323–2326 (2000)
11. Vanmali, A.V., Deshmukh, S.S., Gadre, V.M.: Low complexity detail preserving multi-exposure image fusion for images with balanced exposure. In: 2013 National Conference on Communications (NCC), pp. 1–5. IEEE (2013)
12. Wang, J., Xu, D., Li, B.: Exposure fusion based on steerable pyramid for displaying high dynamic range scenes. Optical Engineering 48(11), 117003 (2009)
13. Ward, G.: Fast, robust image registration for compositing high dynamic range photographs from handheld exposures. Journal of Graphics Tools 8, 17–30 (2003)

Image Abstraction Using Anisotropic Diffusion Symmetric Nearest Neighbor Filter

Zoya Shahcheraghi[1,2], John See[1], and Alfian Abdul Halin[2]

[1] Faculty of Computing and Informatics, Multimedia University,
Cyberjaya, Malaysia
[2] Faculty of Computer Science & Information Technology, Universiti Putra Malaysia,
Serdang, Malaysia

Abstract. Image abstraction is an increasingly important task in various multimedia applications. It involves the artificial transformation of photorealistic images into cartoon-like images. To simplify image content, the bilateral and Kuwahara filters remain popular choices to date. However, these methods often produce undesirable over-blurring effects and are highly susceptible to the presence of noise. In this paper, we propose an image abstraction technique that balances region smoothing and edge preservation. The coupling of a classic Symmetric Nearest Neighbor (SNN) filter with anisotropic diffusion within our abstraction framework enables effective suppression of local patch artifacts. Our qualitative and quantitative evaluation demonstrate the significant appeal and advantages of our technique in comparison to standard filters in literature.

Keywords: Image abstraction, artistic stylization, edge-preserving filters, anisotropic diffusion.

1 Introduction

Image and video abstraction, through Image-based Artistic Rendering (IB-AR) [1] have become increasingly popular in contemporary films, computer generated animation, computer games and visual communication. IB-AR involves artificially incorporating cartoon-like effects to photorealistic images. Although various image processing filters have been used, few have produced artistically appealing and interesting results. This is probably because the filters are mainly applied for photorealistic enhancement, restoration and recovery. In contrast, IB-AR generally aims towards simplification of image content. Basically, image abstraction enables details in low-contrast regions to be removed without filtering across discontinuities, leaving the overall image structure unaffected. Two prominent techniques are the bilateral and Kuwahara filters. The bilateral filter excels in smoothing low-contrast regions while preserving high-contrast edges. It however fails in high-contrast images where either no abstraction is produced or salient visual features are totally removed. Furthermore, iterative filtering using bilateral filter over-blurs edges causing an undesirable washed-out appearance. On the other hand, the Kuwahara filter is able to cope better with high-contrast

W.T. Ooi et al. (Eds.): PCM 2014, LNCS 8879, pp. 343–352, 2014.

images, but it is rather unstable in the presence of noise and suffers from blocky artifacts. Another classical edge-preserving filter called the Symmetric Nearest Neighbor (SNN) filter has relations to the mean and median filters with desirable edge-preserving properties. Unfortunately, the SNN is also susceptible to local patch-like artifacts. Surprisingly, the SNN filter has yet to be explored in literature for image abstraction. In this paper we propose an Anisotropic Diffusion Symmetric Nearest Neighbour Filter (AD-SNN) which balances region smoothing and edge preservation, while enforcing artifact suppression.

2   Related Work

In this section, we provide an overview of relevant image/video abstraction works in literature that utilize edge-preserving filters.

The bilateral filter [2] is by far the most popular edge-preserving filter and it has been widely used for image smoothing, noise reduction and segmentation. The landmark work by Winnemller et al. [3] established an effective framework for abstraction, with the use of an isotropic bilateral filter with Difference-of-Gaussians (DoG) edge extraction to produce highly stylized video frames. Kyprianidis and Döllner [4] came up with a two-pass separated implementations of the bilateral filter along the gradient and tangent directions, which estimates the local orientation based on structure tensors. Elsewhere, Kang et al. [5] also improved the original technique by introducing flow-based image abstraction, where the shapes of the bilateral and DoG filters were deformed to follow a vector field derived from salient image features. An anisotropic bilateral filter was used as opposed to an isotropic bilateral filter.

In another application, Cong et al. [6] introduced a mixed-reality selective image abstraction tool, where a new diffusion model was proposed by combining nonlinear diffusion and bilateral filter. More recently, Zang et al. [7] presented a novel image smoothing approach by using a 1-D space-filling curve to perform EMD-based filtering. Gaussian and bilateral filtering were used to consolidate the edge structure in their space filling curve based image smoothing. Generally, the bilateral filter can sometimes produce inconsistent abstraction in high-contrast images whereby any of the two extremities could occur: no abstraction performed, or over-removal of fine details.

The Kuwahara filter [8] on the other hand is a classical edge-preserving image smoothing filter that performs well in high contrast regions. The generalized Kuwahara filter was used for image and video abstraction by Papari et al. [9] whereby the original Kuwahara filter was improved by replacing the rectangular sub-regions with smooth weighting functions constructed over sectors of a circular area. Kyprianidis et al. [10] introduced an anisotropic Kuwahara filter where the weighting functions are alternatively defined over sectors of ellipses. This modification improved its resistance towards noise and artifacts. More recently, the same authors further introduced another variant that performs at multiple scales [11]. Generally, the Kuwahara filter causes unusual blocky artifacts, and is highly susceptible towards image noise. Many recent methods attempt to resolve these problems but its method of filtering with overlapping windows could be the main cause of these problems.

3   Proposed Method

The proposed method resides within an image abstraction framework that is inspired by the works of Winnemöller et al. [3] and Kang et al. [5]. The architecture employed by these well-known works can be succinctly summarized into two distinct components or processes: (i) *region smoothing*, and (ii) *edge contouring*, in which the contributions of this work involves the former. State-of-the-art methods for edge contouring [5,4] are incorporated into our framework. An overview of the processes in our framework is shown in Figure 1.

In the first step of abstraction, we perform edge-preserving region smoothing on the perceptually uniform CIELab colorspace. The edge contours are then extracted after a certain number of smoothing iterations, while color quantization is applied on the luminance channel as described in [3]. Finally, the edge contours are combined with the color-quantized output in the original RGB colorspace to produce the abstracted output. In this work, we proposed to utilize a classic Symmetric Nearest Neighbor (SNN) filter that is both edge-preserving and simplistic in computation, while adding further artifact suppression from anisotropic diffusion.

3.1   Symmetric Nearest Neighbor (SNN) Filter

Symmetric Nearest Neighbor (SNN) filter [12,13] is an edge-preserving smoothing filter that is arguably one of the most all-rounded filters in terms of both noise immunity (smoothing effect) and edge preservation. It balances both aspects equally as compared to filters such as the median filter (more smoothing), and Kuwahara filter (more edge-preserving).

The SNN filter uses a neighborhood selection technique whereby it compares the 8-connected neighboring pixels in a symmetric fashion. By example of a 3×3 filter, the symmetric pairs of neighbor pixels around the center pixel, *i.e.* N-S, W-E, NW-SE, NE-SW, are inspected by selecting the intensity values of those closest (intensity-wise, not distance-wise) to the center pixel (see Figure 2) from each pair. In other words, the pixel in each symmetric pair that is closest to the value of the center pixel is selected. Finally, the value of the center pixel is computed by the mean value of the four selected neighborhood pixels. Concisely,

$$
\begin{aligned}
\hat{P}_{N,S} &= min\{p_{x,y}, \{p_{x,y+h}, p_{x,y-h}\}\}, \quad \text{for N-S pair} \\
\hat{P}_{W,E} &= min\{p_{x,y}, \{p_{x+h,y}, p_{x-h,y}\}\}, \quad \text{for W-E pair} \\
\hat{P}_{NW,SE} &= min\{p_{x,y}, \{p_{x+h,y+h}, p_{x-h,y-h}\}\}, \quad \text{for NW-SE pair} \\
\hat{P}_{SW,NE} &= min\{p_{x,y}, \{p_{x+h,y-h}, p_{x-h,y+h}\}\}, \quad \text{for SW-NE pair}
\end{aligned}
\tag{1}
$$

where $h = 1$ for the case of a 3×3 SNN filter, while $min\{\cdot\}$ selects a neighborhood pixel (from each symmetric pairing) that has an intensity value that is closest

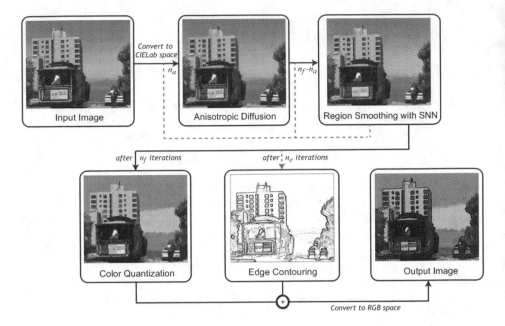

Fig. 1. Schematic overview of the proposed image stylization framework

to that of the center pixel $p_{x,y}$. In this case, the output pixel value is simply the mean of the four values obtained from Eq. 1, *i.e.*

$$f_{x,y} = \{\hat{P}_{N,S} + \hat{P}_{W,E} + \hat{P}_{NW,SE} + \hat{P}_{SW,NE}\}/4 \qquad (2)$$

To further generalize this concept for a $w \times w$ SNN filter (where w should be an odd number and $w > 1$), we consider $\eta_w = \{\sum_{i=1}^{(w-1)/2} 4i\}$ number of symmetric pairs within the square filter neighborhood of size w. The example in Figure 2 illustrates a few possible symmetric pairs within a 5×5 neighborhood. Concisely, the SNN-filtered pixel value at location (x, y) can be determined as,

$$f_{x,y} = \frac{1}{\eta_w} \sum_{j=1}^{\eta_w} min\{p_{x,y}, \hat{S}_j\} \qquad (3)$$

where $\hat{S}_j$ is the j-th symmetrically opposing pair of pixels. To retain the original image size, zero-padding is applied before performing filtering.

Geometrically, this concept is edge-preserving as the filter "votes" for the neighboring pixels that most closely resemble the center pixel before taking the average value of the selected neighbors. This compensates the smoothing effect (averaging) with retention of intensity gradients (closest value voting).

3.2 Anisotropic Diffusion

Anisotropic diffusion is another classical technique proposed by Perona and Malik [14] to selectively enhance contrast in parts of an image by using a modified

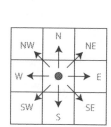

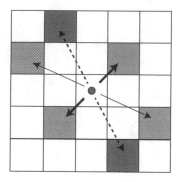

Fig. 2. Symmetric pairs of pixels in SNN filter. The 3×3 *(left)* filter has four pairs; The color-coded pixels in the 5×5 *(right)* window indicate some possible pairs.

heat diffusion equation. Anisotropic diffusion filters have the desirable property of blurring small discontinuities and sharpening edges, which is most useful for image abstraction (the iterative bilateral filter is an example of a nonlinear diffusion filter [3] where the Gaussian blur is weighted nonlinearly). The function of the SNN filter closely resembles a typical rank order mean filter, which will result in the appearance of visual artifacts at corner edges. Hence, we propose to apply the nonlinear diffusion effect on the original image features before performing filtering with SNN, as this maintains the simplicity of the SNN computation without incurring computationally expensive nonlinear filtering.

The 2-D discrete implementation of anisotropic diffusion is summarized as,

$$\frac{\partial}{\partial t} I(\bar{x}, t) = \text{div} \left[c(\bar{x}, t) * \nabla I(\bar{x}, t) \right] \tag{4}$$

where $\bar{x}$ represents the spatial coordinates (x, y), div$[\cdot]$ denotes the divergence operator and ∇ is the gradient operator *w.r.t.* the spatial coordinates.

The intensity values are then updated locally by the sum of the flow contributions along all four directions (north, south, east, west)

$$I(\bar{x}, t + \Delta t) \approx I(\bar{x}, t) + \Delta t * \frac{\partial}{\partial t} I(\bar{x}, t) \tag{5}$$

For better isotropy, signal flow can also be computed between diagonally neighboring pixels (8-connectedness), with consideration for longer distances ($\sqrt{2}$) between diagonal neighbors. The conduction (diffusion) coefficient function,

$$c(\bar{x}, t) = e^{(-(\|\nabla I\|/K)^2)} \tag{6}$$

which prioritizes high-contrast edges over low-contrast edges, is used in our work. Good parameter values were determined empirically; we fix the threshold constant that controls the level of conduction to $K = 30$, while $\Delta t = 0.143$ (or 1/7) for numerical stability. A subsequent work by Gerig et al. [15] suggested for the diffusion process to be repeated for better convergence towards stable smoothened regions. Further details on its mathematical principles and formulation can be found in the original paper [14] and [15]. The coupling of the anisotropic diffusion and the SNN filter in our framework gives rise to the proposed Anisotropic Diffusion Symmetric Nearest Neighbor (AD-SNN) technique.

3.3   Edge Contouring

The second component of the framework involves edge contouring. Generally, this is an essential abstraction element that creates cartoon-like effects and increases the visual distinctiveness of edges at important locations such as object boundaries and background textures. In our work, we do not propose any new edge contouring methods since the scope of our contribution is in the first component. As such, two prominent methods from literature are applied to our framework, both of which are used in our experiments:

– **Multi-image gradient (MiG).** Di Zenzo [16] proposed an approach that computes the gradient image of a multi-image (or multi-spectral image) by calculating the maximum rate of change in the intensity values along the spectral dimensions. An approximation to the directional derivatives are first determined using Sobel filter. Then, the structure tensor for each point (x, y) is set in quadratic form (with g being the tensor coefficients), measuring the maximum squared rate of change of a vector $n = (n_x, n_y)$:

$$E_n = g_{xx}n_x^2 + 2g_{xy}n_xn_y + g_{yy}n_y^2 \tag{7}$$

– **Flow-based Difference-of-Gaussians (FDoG).** Kang et al. [17] introduced a method of extracting spatially coherent lines, guided by Edge Tangent Flow (ETF). The Difference-of-Gaussians (DoG) is given by,

$$E_{\sigma_e} = G_{\sigma_e} - \tau \cdot G_{\sigma_r} \tag{8}$$

where G indicate the two Gaussian blurring functions with their respective values; σ_r is set to $1.6 \cdot \sigma_e$ to approximate the Laplacian-of-Gaussian, σ_e controls the spatial scale of edges. The threshold τ determines the sensitivity of the edge detector. We use $\sigma_e = 1.0$ and $\tau = 0.99$ in our experiments.

The final thresholding step that uses a smoothed step function similar to that described in [3],

$$T(E_{\sigma_e}) = \begin{cases} 1 & \text{if } E_{\sigma_e} > 0, \\ 1 + tanh(\varphi_e \cdot E_{\sigma_e}) & \text{otherwise} \end{cases} \tag{9}$$

where φ_e tunes the sharpness of the edges E_{σ_e}

3.4   Iterative Processing

In our framework, a total of n_f iterations are performed for region smoothing. This is a common practice in most works [3,4,5] as it provides sufficient region contrasting and a more desirable level of abstraction. Anisotropic diffusion is applied for a lesser number of iterations to soften the effect of artifact suppression— the first n_a iterations where $n_a < n_f$. Similarly, the edges are extracted after n_e iterations (where $n_e < n_f$) to reduce the presence of noise. In our experiments, we tested and used the following values: $n_f \in \{3, 4\}$, $n_a = 1$, $n_e = 2$. More precisely, both the Kuwahara and AD-SNN filters use $n_f = 3$, while the bilateral filter uses $n_f = 4$ (similar to the setting from [4]).

4 Results and Analysis

Experimental studies remain one of the most difficult problems in the area of non-photorealistic rendering (NPR) when evaluating the artistic stylization of images or videos, as acknowledged by Hertzmann [18]. It is common to perform visual assessment of the aesthetic qualities of stylized images through subjective evaluation by human observers, as employed by many previous works [3,19,20].

The proposed approach was evaluated using a mix of different quantitative and qualitative evaluation techniques. Firstly, a qualitative inspection compares our stylized images with that of other competing approaches: bilateral filter (BF) and Kuwahara filter (KF). Second, a systematic visual assessment was conducted on human observers to produce a quantitative score on the aesthetic appeal of the approaches.

4.1 Qualitative Visual Inspection

By visually examining the effects of region smoothing on the evaluated approaches, the proposed AD-SNN method appears to provide a middle ground between the BF and the KF. This is most noticeable from the sample image reviewed in Figure 3, where the strong preservation of gradients on the BF image renders it too closely to the original photograph. Meanwhile, the blurring sensation in the KF image causes obvious distortions to the shape of objects, which is a rather undesirable characteristic. Thus, the AD-SNN filter strikes a balanced compromise between the two extremes, while offering elegant and natural patch shadings. On a closer look (Figure 4), we can observe irregularities at a finer level. The color-quantized regions extracted after performing BF tends to produce incomplete edge contouring results after re-combination. Also, the blocky effect of the KF immensely deteriorates the quality of abstraction.

4.2 Quantitative Assessment

We conduct a systematic experiment using 20 popular images used in various works in literature [4,10,21]; 19 of which are taken from Philip Greenspun's photograph database[1] while one large size photograph of the Grand Canal of Venice was downloaded from the Internet. We handpicked the images based on their usage as examples in previous works, and also to provide a balanced

---

[1] http://philip.greenspun.com/photography/

Fig. 3. Edge-preserving region smoothing using bilateral filter *(top)*, Kuwahara filter *(middle)* and the proposed AD-SNN filter *(bottom)* on a photograph of the Grand Canal in Venice.

Fig. 4. Observations: Less than complete edge contouring on the color-quantized regions found by bilateral filter *(left pair)*, and obvious loss of details using Kuwahara filter *(right pair)*. The right-side image of each pair uses the proposed method.

Table 1. Mean and standard deviation of the relative amount of votes obtained each evaluated method through visual assessment

Method	Mean	Standard deviation
Bilateral	0.720	0.230
Kuwahara	0.408	0.162
AD-SNN	**0.876**	**0.098**

coverage of different types of photographs. There are 4 animal images, 6 single-person or group photographs, 4 landscape photos and 6 building/object photos. A total of 25 observers (10 of which have reasonable expertise in media arts and graphics, while the rest are novices) were given 10 sets of images; each set contains three stylized images (corresponding to the bilateral, Kuwahara and AD-SNN approaches) of a randomly selected photograph in random order. To alleviate possible bias from the edge style, we also randomly select between the MiG and FDoG (see Section 3.3) for the generated sets.

In contrast to our earlier work [20] which is primarily based on scoring or point-based feedback, we utilize a different methodology here, similar to that recently proposed in [19]. For each set of images, the observers were asked to cast two votes for the two most aesthetically pleasing stylizations. The relative amount of votes received by an approach for each participant is computed by $V_{rel} = V_{cast}/V_{total}$, where V_{cast} is the number of votes cast for the approach and V_{total} is the total number of image sets shown to the observer. The average voting score across all observers, $\overline{V_{rel}}$ is then computed to give the overall score.

Table 1 shows the overall result of our visual assessment experiment for the three evaluated filters. The AD-SNN filter provides the best artistic stylization, while its low standard deviation score highlights the consistency of opinion by the observers in favor of this method. It is unsurprising that the Kuwahara filter is visually unappealing due to the presence of artifacts discussed earlier. The bilateral filter remains competitive, but its high standard deviation offers some doubt over its attractiveness when applied to a highly variable set of 20 images and 2 different edge contours.

5   Conclusion

Concretely, we have presented a new technique called Anisotropic Diffusion Symmetric Nearest Neighbor (AD-SNN) filter for region smoothing within an image abstraction framework. With appropriate color quantization and a good choice of edge contouring, our method is able to produce aesthetically appealing images through a balanced measure of region smoothing and edge preservation. More importantly, blocky artifacts can be effectively mitigated by the effect of anisotropic diffusion. An extensive evaluation conducted shows greater appeal over other evaluated methods. In future work, the SNN filter can be further extended to an isotropic (circular) or anisotropic (elliptical) filter shape which will encourage smoother boundaries and corners in object regions.

References

1. Kyprianidis, J.E., Collomosse, J., Wang, T., Isenberg, T.: State of the 'Art': A taxonomy of artistic stylization techniques for images and video. IEEE Trans. on Visualization and Computer Graphics 19(5), 866–885 (2013)
2. Tomasi, C., Manduchi, R.: Bilateral filtering for gray and color images. In: Sixth International Conference on Computer Vision 1998, pp. 839–846. IEEE (1998)
3. Winnemöller, H., Olsen, S.C., Gooch, B.: Real-time video abstraction. ACM Transactions on Graphics (TOG) 25(3), 1221–1226 (2006)
4. Kyprianidis, J.E., Döllner, J.: Image abstraction by structure adaptive filtering. In: Theory and Practice of Computer Graphics, pp. 51–58 (2008)
5. Kang, H., Lee, S., Chui, C.K.: Flow-based image abstraction. IEEE Transactions on Visualization and Computer Graphics 15(1), 62–76 (2009)
6. Cong, L., Tong, R., Dong, J.: Selective image abstraction. Vis. Comput. 27(3), 187–198 (2011)
7. Zang, Y., Huang, H., Zhang, L.: Efficient structure-aware image smoothing by local extrema on space-filling curve. IEEE Transactions on Visualization and Computer Graphics (in press, 2014)
8. Kuwahara, M., Hachimura, K., Eiho, S., Kinoshita, M.: Processing of ri-angiocardiographic images. Digital Processing of Biomedical Images, 187–202 (1976)
9. Papari, G., Petkov, N., Campisi, P.: Artistic edge and corner enhancing smoothing. IEEE Transactions on Image Processing 16(10), 2449–2462 (2007)
10. Kyprianidis, J.E., Kang, H., Döllner, J.: Image and video abstraction by anisotropic kuwahara filtering. Computer Graphics Forum 28, 1955–1963 (2009)
11. Kyprianidis, J.E.: Image and video abstraction by multi-scale anisotropic kuwahara filtering. In: Proceedings of the ACM SIGGRAPH/Eurographics Symposium on Non-Photorealistic Animation and Rendering, pp. 55–64. ACM (2011)
12. Harwood, D., Subbarao, M., Hakalahti, H., Davis, L.S.: A new class of edge-preserving smoothing filters. Pattern Recognition Letters 6(3), 155–162 (1987)
13. Pietikäinen, M., Harwood, D.: Segmentation of color images using edge-preserving filters. In: Advances in Image Processing and Pattern Recognition, pp. 94–99 (1986)
14. Perona, P., Malik, J.: Scale-space and edge detection using anisotropic diffusion. IEEE Transactions on Patt. Anal. and Mach. Intell. 12(7), 629–639 (1990)
15. Gerig, G., Kubler, O., Kikinis, R., Jolesz, F.A.: Nonlinear anisotropic filtering of mri data. IEEE Transactions on Medical Imaging 11(2), 221–232 (1992)
16. Di Zenzo, S.: A note on the gradient of a multi-image. Computer Vision, Graphics, and Image Processing 33(1), 116–125 (1986)
17. Kang, H., Lee, S., Chui, C.K.: Coherent line drawing. In: Proc. of 5th Int. Symposium on Non-photorealistic Animation and Rendering, pp. 43–50 (2007)
18. Hertzmann, A.: Non-photorealistic rendering and the science of art. In: Proceedings of the 8th International Symposium on Non-Photorealistic Animation and Rendering, pp. 147–157. ACM (2010)
19. Alencar Júnior, J.D.O., de Queiroz, J., Gomes, H.: An approach for non-photorealistic rendering that is appealing to human viewers. In: 26th Conference on Graphics, Patterns and Images (SIBGRAPI), pp. 242–249. IEEE (2013)
20. Shahcheraghi, Z., See, J.: On the effects of pre-and post-processing in video cartoonization with bilateral filters. In: 2013 IEEE International Conference on Signal and Image Processing Applications (ICSIPA), pp. 37–42. IEEE (2013)
21. DeCarlo, D., Santella, A.: Stylization and abstraction of photographs. ACM Transactions on Graphics (TOG) 21, 769–776 (2002)

Reduction of Multichannel Sound System Based on Spherical Harmonics

Shanshan Yang[1], Xiaochen Wang[1,2,*], Dengshi Li[1,3],
Ruimin Hu[1,2], and Weiping Tu[1,2]

[1] National Engineering Research Center for Multimedia Software,
School of Computer, Wuhan University, Wuhan, China
{yangssgood,clowang}@163.com, reallds@126.com,
{hrm1964,echo_tuwp}@163.com
[2] Research Institute of Wuhan University in Shenzhen, China
[3] School of Computer and Mathematics, Jianghan University, Wuhan, China

Abstract. In order to meet people's demand for 3D audio in family, it's a critical problem to recreate a 3D spatial sound field with few loudspeakers. In this paper, we introduce a L- to (L-1)-channel reduction method based on spherical harmonic decomposition and sound field of head can be perfectly reproduced. When loudspeakers are too few to perfectly reproduce the sound field of head, we ensure low distortions of sound field at ears. On this basis, multichannel reduction algorithm from L- to M-channel system is proposed. As an example, reduction of NHK 22.2 system has been implemented and eleven loudspeaker arrangements from 22 to 6 channels are derived. Results show the sound field of head can be reproduced perfectly until 10 channels and 8-, 6-channel systems can keep low distortions at ears. Compared with Ando's multichannel conversion method by subjective evaluation, our proposed method is better in terms of sound localization.

Keywords: 3D audio, multichannel system, reduction, sound field reproduction, spherical harmonics.

1 Introduction

Three-dimensional (3D) spatial sound with loudspeakers brings us more and more joviality and experiences. On one hand, methods targeted at the accurate physical reconstruction of sound fields over an extended listening area have been developed and implemented in recent decades. The typical methods are higher order ambisonics (HOA) [1–4] and wave field synthesis (WFS) [5, 6]. Due to the employment of large number of loudspeakers, which can reach several hundred or even more, these reconstruction methods are referred to as *massive multichannel* sound reproduction methods. On the other hand, practical techniques for spatial

* This work is supported by National Nature Science Foundation of China (No. 61231015, 61201340, 61201169, 61102127, 61272278), Nature Science Foundation of Hubei Province (No. 2012FFB04205) and Guangdong-Hongkong Key Domain Breakthrough Project of China (No. 2012A090200007).

W.T. Ooi et al. (Eds.): PCM 2014, LNCS 8879, pp. 353–362, 2014.

reproduction have been developed further in parallel to the theoretical methods. It is worth mentioning that NHK 22.2 multichannel sound system [7] (proposed by NHK laboratory of Japan) considers the perceptive characteristic of human, resulting in good sense of spatial sound impressions. However, loudspeakers of 22.2 multichannel sound system are still too large to be applied in home cinemas. In order to satisfy people's demand for 3D spatial sound in family environment, it is necessary to reduce the number of loudspeakers.

1.1 Related Work

In 2011, Akio Ando proposed a M- to M'-channel conversion method [8] which maintained the physical properties of sound at the listening point (i.e., the center of head) in the reproduced sound field. Signals of 22.2 multichannel sound system without two low-frequency channels could be converted into those of 10-, 8- or 6-channel sound systems. Although subjective evaluations showed that this method could reproduce the spatial impression of the original 22-channel sound with eight loudspeakers, people's two ears can not be on the listening point at the same time. Thus, an important practical question naturally appears about how to reproduce the sound field over a given region.

Higher order Ambisonics (HOA) based on cylindrical two-dimensional (2D or horizontal plane) harmonic [2] or spherical three-dimensional (3D or full sphere) harmonic [1] decomposition of a sound field was proposed to perfectly reproduce the sound field within a given region. Although high reproduction frequencies and large reproduction regions are considered, a few 3D reproduction systems were implemented based on HOA since it required loudspeakers to be placed on a sphere which surrounds the target reproduction region uniformly [1]. However, The uniform deployment of a spherical loudspeaker array is impractical in reality.

Therefore, this paper introduces a L- to $(L-1)$-channel reduction based on spherical harmonics and the sound field of head region can be perfectly reproduced by $(L-1)$-channel sound system. When loudspeakers are too few to reproduce the sound field over head region, we ensure low distortions of sound field at ears. By iteration on this basis, we propose an automatic reduction algorithm from L- to M-channel sound system. Since NHK 22.2 system considers the human auditory perception properties and can create spatial sense greatly, we used it as an example to implement its reduction and obtained eleven loudspeaker placements from 22- to 6-channel systems. Simulations show sound field of head can be reproduced until 10 channels and low distortions at ears can be kept in 8-, 6- channel systems. Subjective results indicate that the proposed method is better than Ando's 22-channel conversion method for the sound localization. This method can get reduced sound systems quickly and automatically.

Notation. Throughout this paper, we use the following notational conventions: 1) vectors are represented by lower case bold face, e.g., $\mathbf{x}$; 2) A unit vector in the direction $\mathbf{x}$ is denoted by $\hat{\mathbf{x}}$, i.e., $\hat{\mathbf{x}} = \mathbf{x}/|\mathbf{x}|$; 3) polar coordinate originates at the origin $\mathbf{o}$ and represents a position of $\mathbf{y}$ as (σ, ϕ, θ), where σ is the distance from the origin, ϕ is the azimuthal angle, and θ is the elevation angle; and 4) The symbol $i = \sqrt{-1}$ is used to denote the imaginary part of a complex number.

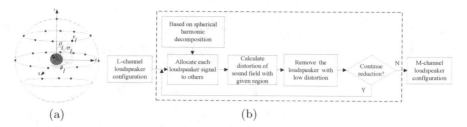

(a)   (b)

Fig. 1. Our proposed method.(a)Formulation of reduction from *L-* to *(L-1)*-channel;
(b)Block diagram of proposed method.

2   Formulation of Reduction

Let the *l*th loudspeaker $\mathbf{y}_l = (\sigma_l, \phi_l, \theta_l)$ be a point source. It produces a sound
field within a source-free region at an arbitrary receiving point $\mathbf{x} = (\gamma, \varphi, \vartheta)$
(shown in Fig. 1(a)), which can be expressed as a linear combination of these
spherical harmonics [3],

$$T_l(\mathbf{x}; k) = \frac{e^{-ik|\mathbf{y}_l - \mathbf{x}|}}{|\mathbf{y}_l - \mathbf{x}|} = -ik4\pi \sum_{n=0}^{\infty} \sum_{m=-n}^{n} h_n^{(2)}(k\sigma_l) j_n(k\gamma) \overline{Y_n^m(\hat{\mathbf{y}}_l)} Y_n^m(\hat{\mathbf{x}}) \quad (1)$$

where 1) $k = 2\pi\lambda^{-1} = 2\pi f c^{-1}$ is the wave number (with c the speed of wave
propagation, f the frequency and λ the wavelength); 2) $j_n(\cdot)$ and $h_n^{(2)}(\cdot)$ are the
first kind spherical Bessel function and second kind spherical Hankel function of
the *n*th degree respectively.

The spherical harmonics are defined as

$$Y_n^m(\hat{\mathbf{x}}) = \sqrt{\frac{(2n+1)}{4\pi} \frac{(n-|m|)!}{(n+|m|)!}} P_n^{|m|}(\cos\vartheta) e^{im\varphi} \quad (2)$$

where $P_n^m(\cdot)$ is the associated Legendre function (which reduces to the Legendre
function for $m = 0$). The subscript n is referred to as the *order* of the spherical
harmonic, and m is referred to as the *mode*. For each order n, there are $2n+1$
modes (corresponding to $m = -n, ..., n$).

The following orthogonality property is exhibited on spherical harmonics.

$$\int Y_n^m(\hat{\mathbf{x}}) \overline{Y_{n'}^{m'}(\hat{\mathbf{x}})} d\hat{\mathbf{x}} = \delta_{nn'} \delta_{mm'} \quad (3)$$

where $\delta_{nn'}$ denotes the Kronecker delta function, and integration is over the unit
sphere.

Supposed that there are L loudspeakers in the original sound field, whose lo-
cations are $\mathbf{y}_l = (\sigma_l, \phi_l, \theta_l), l = 1, 2, ..., L$ and *L-1* loudspeakers in the reproduce
sound field (shown in Fig. 1(a)), the aim is to use *L-1* loudspeakers (the *l*th
loudspeaker is $\mathbf{y}_l, l = 1, 2, ..., L-1$) to reproduced the original sound field at the
receiving point $\mathbf{x}$.

The original sound field produced by L loudspeakers (the lth loudspeaker is $\mathbf{y}_l, l = 1, 2, ..., L$) at the receiving point $\mathbf{x}$ is given by

$$S(\mathbf{x}; k) = \sum_{l=1}^{L} T_l(\mathbf{x}; k) = \sum_{l=1}^{L-1} T_l(\mathbf{x}; k) + T_L(\mathbf{x}; k) \qquad (4)$$

Applying a complex-valued frequency-dependent weighting function $w_l(k) = 1 + a_l(k)$ to the lth loudspeaker, the reproduced sound field produced by L-1 loudspeakers at the receiving point $\mathbf{x}$ is

$$\tilde{S}(\mathbf{x}; k) = \sum_{l=1}^{L-1} w_l(k) T_l(\mathbf{x}; k) = \sum_{l=1}^{L-1} T_l(\mathbf{x}; k) + \sum_{l=1}^{L-1} a_l(k) T_l(\mathbf{x}; k) \qquad (5)$$

Thus, to exactly reproduce the original field at the receiving point $\mathbf{x}$ by L-1 loudspeakers (the lth loudspeaker is $\mathbf{y}_l, l = 1, 2, ..., L - 1$), it requires equating Eq. 4 with Eq. 5, i.e.,

$$T_L(\mathbf{x}; k) = \sum_{l=1}^{L-1} a_l(k) T_l(\mathbf{x}; k) \qquad (6)$$

From Eq. 6, when the sound field of Lth loudspeaker can be accurately reproduced by left L-1 loudspeakers whose weights are $a_l(k), l = 1, 2, ...L - 1$, the original field of L loudspeakers at the receiving point $\mathbf{x}$ can be exactly reproduced. If we want to reproduce the sound field within the region χ represented as a dark grey shaded area (shown in Fig. 1(a)), the receiving point $\mathbf{x}$ could be integrated over the sphere with respect to $\hat{\mathbf{x}}$. In this paper, the given region is limited within man's head.

This paper mainly introduces an iterative process of multichannel reduction from L channels to M channels with low distortion of sound field within head. Fig. 1(b) shows the flow diagram of proposed method which includes three parts: 1) extracting L-channel loudspeaker configuration in original sound field; 2) the iteration of multichannel reduction (shown in the dotted box); and 3) deriving a set of reduced loudspeaker configurations in reproduced sound field. The iteration includes two main points: allocating the signal of each loudspeaker in L-channel sound system to others through calculating weight coefficients based on spherical harmonics; finding which one of L loudspeakers could maintain low distortion of sound field within head and reducing it from L loudspeakers.

3   Reduction from L to L-1 Channels

3.1   Weight Coefficients of L-1 Channels

We assume that 1) the receiving point $\mathbf{x}$ satisfies $|\mathbf{x}| < |\mathbf{y}_l|$, where $\mathbf{y}_l$ is the location of the lth loudspeaker; and 2) all loudspeakers are located on a sphere of radius σ, i.e., $\sigma_l = \sigma, l = 1, 2, ...L$. After substituting Eq. 1, multiplying each side by $\overline{Y_n^m(\hat{\mathbf{x}})}$, and integrating over the unit sphere with respect to $\hat{\mathbf{x}}$, Eq. 6 could be derived

$$j_n(k\gamma)\overline{Y_n^m(\hat{\mathbf{y}}_L)} = j_n(k\gamma) \times \sum_{l=1}^{L-1} a_l(k)\overline{Y_n^m(\hat{\mathbf{y}}_l)} \tag{7}$$

where we have used the orthogonality property in Eq. 3. When substituting Eq. 2, Eq. 7 is further derived

$$P_n^{|m|}(\cos\theta_L)e^{-im\phi_L} = \sum_{l=1}^{L-1} a_l(k)P_n^{|m|}(\cos\theta_l)e^{-im\phi_l}$$
$$n = 0, ..., \infty, m = -n, ..., n \tag{8}$$

However, it would require an infinite number of loudspeakers (one for each n and m term) to satisfy Eq. 8 exactly for every term in the spherical harmonics expansion. Literature [1] indicates that the sound field of the Lth loudspeaker could be accurately reproduced only by equating the terms for $n = 0, 1, ..., N$ if most of the power of the sound field within the chosen reproduction region is contained in the first N orders, which is referred to as an Nth order reproduction system. According to the method introduced in [1], the normalized truncation error[1] ε_N is given by

$$\varepsilon_N = 1 - \sum_{n=0}^{N} (2n+1)(j_n(k\gamma))^2 \tag{10}$$

It is shown as a function of $k\gamma$ for various order expansions in Fig. 2(a). For any given N, the error decreases monotonically when $k\gamma$ is below a certain value, i.e., $\varepsilon_N(k\gamma) < \varepsilon_N(k\gamma_0), \forall \gamma < \gamma_0$. Only when error is very high (above about 50%) the error is not monotonically decreasing. Thus, if an Nth order expansion can accurately represents the sound field on a sphere of radius γ_0, the sound field within the sphere can also be accurately reproduced. So Eq. 8 can be truncated at order N, i.e.,

$$P_n^{|m|}(\cos\theta_L)e^{-im\phi_L} = \sum_{l=1}^{L-1} a_l(k)P_n^{|m|}(\cos\theta_l)e^{-im\phi_l}$$
$$n = 0, ..., N, m = -n, ..., n \tag{11}$$

where

$$(L-1) \geq (N+1)^2 \tag{12}$$

Weight coefficients $a_l(k)$ can be calculated in Eq. 11 and the weight of each loudspeaker $w_l(k) = 1 + a_l(k), l = 1, ..., L - 1$ is obtained.

---

[1] Assume a sound field of $T_L(\mathbf{x}; k)$ is approximated to that of $T_L^N(\mathbf{x}; k)$, obtained by truncating the infinite series Eq. 1 at order N, i.e., the outer summation in Eq. 1 is only taken over $n = 0, 1, ..., N$. Define the normalized truncation error ε_N as

$$\varepsilon_N \triangleq \int \frac{\left|T_L(\mathbf{x}; k) - T_L^N(\mathbf{x}; k)\right|^2}{|T_L(\mathbf{x}; k)|^2} d\hat{\mathbf{x}} \tag{9}$$

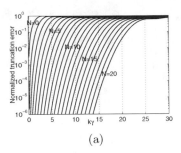
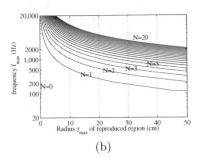

(a)   (b)

Fig. 2. Relationship of the factors in the sound field $T_L^N(\mathbf{x}; k)$.(a)Truncation error of the sound field $T_L^N(\mathbf{x}; k)$; (b)the relationship of the order N, upper frequency f_{max} and upper spheral radius γ_{max} of accurate reproduced sound field.

3.2   Low Distortions of Reproduced Sound Field

When we adopt a rule of thumb given in [4], the relationship of the upper radius γ_{max} of the accurate reproduced sound field, the upper frequency f_{max} and the order N can be approximated as

$$N = \frac{e\pi}{c} f_{\max}\gamma_{\max} \tag{13}$$

Fig. 2(b) shows the relationship of the order N, upper frequency f_{max} and upper spheral radius γ_{max} of accurate reproduced sound field. For any given γ_{max} of accurate reproduced sound field, the higher the order N is, the higher the upper frequency f_{max} is. Similarly, for any given f_{max} of accurate reproduced sound field, the higher the order N is, the greater the upper radius γ_{max} is. From Eq. 12, the order N could be given when number of loudspeakers in reproduced sound field is confirmed. If the frequency is regarded as a certain constant, the upper radius γ_{max} could be obtained.

Since we focus on the sound field region χ of man's head, we assume that head is a standard sphere with spheral radius R. Center of head is located on origin and both ears are symmetric with YOZ plane whose locations are **Left** $= (R, 180, 90)$ and **Right** $= (R, 0, 90)$ respectively. If $\gamma_{max} \geq R$, the distortion of reproduced sound field is defined as

$$\varepsilon(\chi; k) \triangleq \int_\chi \frac{\left| S(\mathbf{x}; k) - \tilde{S}(\mathbf{x}; k) \right|^2}{|S(\mathbf{x}; k)|^2} d\chi \tag{14}$$

If $\gamma_{max} < R$, the distortion of reproduced sound field is represented by errors of sound field at ears.

$$\varepsilon(\mathbf{Left}; k) \triangleq \frac{\left| S(\mathbf{Left}; k) - \tilde{S}(\mathbf{Left}; k) \right|^2}{|S(\mathbf{Left}; k)|^2} \tag{15}$$

Similarly, $\varepsilon(\mathbf{Right}; k)$ can be calculated. Reduction from L- to $(L\text{-}1)$-channel should ensure low distortions mentioned above.

Algorithm 1. Pseudo-code of the iterative multichannel reduction

---

1: **Input:** L-loudspeaker positions in the original field, the frequency of operation f, and radius R of head

2: **Output:** M-channel loudspeaker arrangements found in the reproduced field

3: **repeat**

4:     determine the expansion order N by Eq. 12 and γ_{max} of accurate reproduced region by Eq. 13

5:     **for** l=1:L **do**

6:         allocate the signal of lth loudspeaker to those of other (L-1) loudspeakers *Section 3.1*

7:         **if** $\gamma_{max} \geq R$ **then**

8:             calculate the distortion of reproduced field $\varepsilon(\chi; k)$ by Eq. 14

9:         **else**

10:             calculate the distortions of reproduced field at ears $\varepsilon(\textbf{Left}; k)$ and $\varepsilon(\textbf{Right}; k)$ by Eq. 15

11:         **end if**

12:     **end for**

13:     delete the loudspeaker with the minimum distortion of reproduced field and $L = L - 1$

14: **until** stop criterions are met

---

4 Reduction Algorithm from L to M

The procedure of multichannel reduction is summarized in Algorithm 1, which includes two stop criterions: 1) all three minimum distortions $\varepsilon(\chi; k)$, $\varepsilon(\textbf{Left}; k)$ and $\varepsilon(\textbf{Right}; k)$ exceed a given threshold; and 2) none of loudspeakers can be reduced.

5 Experiments

5.1 Simulation

To evaluate the performance of the proposed multichannel reduction method, we used the 22.2 multichannel system without LFE (shown in Fig. 3(a)) as an example. Free-field source conditions are assumed and sound field resulting from the loudspeaker is a spherical wave. Let the center of head be located on the origin. Distance from each loudspeaker to the origin is 200cm, i.e., $\sigma = 200$. Distance from each ear to the center of head is 8.5cm, i.e. $R = 8.5$. Original sound sources are 22-channel white noise [9] and the original sound field is created by 22-channel audio system. Considering locations of 22 loudspeakers are symmetrical based on the median plane, we delete the selected loudspeaker and its symmetric one at the same time in the process of reduction.

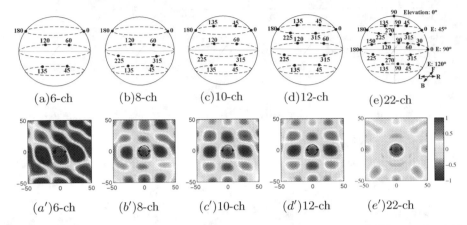

Fig. 3. Simulation results. $(a) \sim (e)$ are loudspeaker arrangements of different channel systems and their corresponding sound fields are shown in $(a') \sim (e')$.

According to Eq. 12 and Eq. 13, sound field within the region of head could be perfectly reproduced by more than nine loudspeakers assuming that the frequency f is 1000Hz. Thus, we use 22-channel white noise with 1000Hz to calculate errors of sound field at ears, i.e. $\varepsilon(\mathbf{Left}; k)$ and $\varepsilon(\mathbf{Right}; k)$. We select 0.3 as a given distortion threshold of sound field at ears according to distortions at ears of 8-channel system in 22-channel conversion method which could reproduce the spatial impression. The given threshold of $\varepsilon(\chi; k)$ is 0.04 which is selected according to literature [1].

Eleven layouts are derived from 22 channels to 6 channels. Fig. 3(a-e) show the 6-, 8-, 10-, 12-, 22-channel loudspeaker arrangements and their corresponding sound fields are shown in Fig. 3(a'-e'). We can get that the sound field of 12-channel systems is the same as 22-channel original sound field, but a few distortions come into being in 10-channel reproduced sound field and they become more and more evident in 8-, 6-channel reproduced sound fields. In other words, the sound fields over the region of head can be reproduced perfectly from 21- to 12-channel systems. Distortions begin generating from 10-channel system and increasing gradually in 8-, 6-channel system.

5.2 Subjective Experiment

Because sound field inside region of head can not be reproduced accurately by 6-, 8-, 10-channel systems (Fig. 3(a', b', c')), subjective experiments about them have been done by the RAB paradigm [10] to evaluate whether they are acceptable in subjective evaluation. Two stimuli (A and B) and a third reference (R) are presented and the difference threshold is measured. The reverberation time at 500Hz in the soundproof room is 0.18s and background noise is 30dB(A). In this experiment, R is the original 22-channel white noise with 10s, one stimuli (A or B) is the sound of 6- (8-, or 10-) channel of Andos conversion method in [8] and the other one is the sound of 6- (8-,or 10-) channel in Fig. 3(a, b, c). Testers are

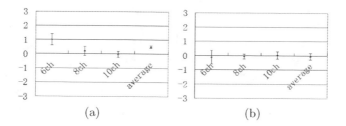

Fig. 4. Results of the Comparative Mean opinion score (CMOS) in subjective experiment of 10-, 8- and 6-channel between the method of [8] and our proposed method(95% confidence limits. (a)Sound Localization; (b)Sound Intensity.

asked to compare the difference between A and B relative to R and give scores according to the continuous seven-grade impairment scale. Two aspects of the sound are evaluated: sound localization and sound intensity.

In this experiment, subjects are 20 students all major in audio signal processing. Fig. 4 shows the Comparative Mean Opinion Score (CMOS) of sound localization and sound intensity given by the subjects with 95% confidence limits. The average CMOS of sound localization is 0.5 in Fig. 4(a). The CMOS of sound localization reproduced by 6- and 8-channel system are 1.005 and 0.2 respectively, higher than zero and the average scales of 10-channel is about zero. It shows that sound localization effects of 6- and 8-channel systems reproduced by our method are better than those by Ando's conversion method. The average CMOS of sound intensity shown in Fig. 4(b) is about zero, which indicates that sound intensity effects of 6-, 8- and 10-channel systems reproduced by our method are almost comparable with those by Ando's conversion method.

Since 22.2 multichannel sound system considers the perceptive characteristic of people and creates a good sense of spatial sound impression, we use it as a reduction example of our proposed method. According to results mentioned above, eleven loudspeaker layouts can be derived by reducing from 22 to 6 channels. The sound field over head can be reproduced until 10-channel system, but distortions begin to increase gradually in 8- and 6-channel systems. Although 8- and 6-channel systems can not reproduce the sound field over head perfectly, the subjective evaluation shows that their sound localization effects are better than those of Ando's conversion method and their sound intensity are comparable.

6   Conclusion

A method of multichannel system reduction is proposed for reproducing 3D sound fields with few loudspeakers in family. The reduction is implemented by the iteration of L- to $(L\text{-}1)$-channel reduction based on spherical harmonic decomposition. The sound field over head can be reproduced perfectly. And low distortion at ears are ensured when loudspeakers are too few to reproduce field. As an example, the reduction of NHK 22.2 multichannel sound system has been implemented and eleven loudspeaker arragements from 22 to 6 channels can be got. The sound field over head can be reproduced until 10 channels. And 8-, 6-channel systems can ensure low distortions at ears. Compared with Ando's multichannel conversion method by subjective evaluation, our proposed method is better for sound localization and comparable for sound intensity. To some extent, the proposed method can save the cost and time spent by subjective experience effectively and get the loudspeaker arrangements with different channels quickly and automatically.

References

1. Ward, D.B., Abhayapala, T.D.: Reproduction of a plane-wave sound field using an array of loudspeakers. IEEE Transactions on Speech and Audio Processing 9(6), 697–707 (2001)
2. Wu, Y.J., Abhayapala, T.D.: Theory and design of soundfield reproduction using continuous loudspeaker concept. IEEE Transactions on Audio, Speech, and Language Processing 17(1), 107–116 (2009)
3. Zhang, W., Abhayapala, T.: Three Dimensional Sound Field Reproduction using Multiple Circular Loudspeaker Arrays: Functional Analysis Guided Approach. IEEE Transactions on Audio, Speech, and Language Processing 22(7), 1184–1194 (2014)
4. Kennedy, R.A., Sadeghi, P., Abhayapala, T.D., Jones, H.M.: Intrinsic Limits of Dimensionality and Richness in Random Multipath Fields. IEEE Transactions on Signal Processing 55(6), 2542–2556 (2007)
5. Ahrens, J., Spors, S.: Applying the ambisonics approach to planar and linear distributions of secondary sources and combinations thereof. Acta Acustica United with Acustica 98(1), 28–36 (2012)
6. Spors, S., Ahrens, J.: Analysis and improvement of pre-equalization in 2.5-dimensional wave field synthesis. In: Audio Engineering Society Convention 128. Audio Engineering Society (2010)
7. Hamasaki, K., Matsui, K., Sawaya, I., Okubo, H.: The 22.2 Multichannel Sounds and its Reproduction at Home and Personal Enviroment. In: Audio Engineering Society Conference: 43rd International Conference: Audio for Wirelessly Networked Personal Devices. Audio Engineering Society (2011)
8. Ando, A.: Conversion of multichannel sound signal maintaining physical properties of sound in reproduced sound field. IEEE Transactions on Audio, Speech, and Language Processing 19(6), 1467–1475 (2011)
9. Sawaya, I., Oode, S., Ando, A., Hamasaki, K.: Size and Shape of Listening Area Reproduced by Three-dimensional Multichannel Sound System with Various Numbers of Loudspeakers. In: Audio Engineering Society Convention 131. Audio Engineering Society (2011)
10. Recommendation, ITU-R BS. 1284-2: General Methods for the Subjective Assessment of Sound Quality. International Telecommunications Union (2002)

Automatic Multichannel Simplification with Low Impacts on Sound Pressure at Ears

Dengshi Li[1], Ruimin Hu[1,2,*], Xiaochen Wang[1,2], Weiping Tu[1,2], and Shanshan Yang[1]

[1] National Engineering Research Center for Multimedia Software,
School of Computer, Wuhan University, Wuhan, China
reallds@126.com, {hrm1964,clowang,echo_tuwp,yangssgood}@163.com
[2] Research Institute of Wuhan University in Shenzhen, China

Abstract. People hope to use minimum arrangement of loudspeakers to reproduce the experience of the film 3D sound at home. Although the Ando's conversion can convert n- to m-channel sound system by maintaining the sound pressure at the origin, it is time-consuming and expensive to use lots of subjective evaluations to distinguish the experience of spatial sound reproduced by m-loudspeaker arrangements. To solve this problem, we not only ensure the sound pressure at the origin invariably, but also limit the absolute error of sound pressure at ears to a given threshold or less while simplifying from n- to $(n\text{-}1)$-channel, and an automatic simplification algorithm from n- to m-channel sound system is proposed. The 22.2 multichannel sound system without two low-frequency effect channels as an example can be simplified to eight channels automatically and the total loudspeaker arrangements is 23. The subjective evaluation is comparable to that of the Ando's conversion and the cost of subjective evaluation is saved.

Keywords: sound field reproduction, three-dimensional sound, multichannel audio.

1 Introduction

The 3D audio technology brings people a convenient and wonderful sense of spatial sound impression. From a practical standpoint, one aims to provide the most convincing experience with the minimum equipment and channels to generate the experience of spatial sound. By comparing different methods of spatial sound reconstruction, on one hand there are binaural techniques [1], which deliver a convincing experience over two channels by presenting necessary binaural cues. On the other hand there are wave-field synthesis (WFS) [2,3] and Higher-order

* This work is supported by National Nature Science Foundation of China (No. 61231015, 61201340, 61201169, 61102127, 61272278), Nature Science Foundation of Hubei Province (No. 2012FFB04205) and Guangdong-Hongkong Key Domain Breakthrough Project of China (No. 2012A090200007).

W.T. Ooi et al. (Eds.): PCM 2014, LNCS 8879, pp. 363–372, 2014.

Ambisonics(HOA) [4,5], which typically use a large number of channels to accurately reproduce the wave front generated by a virtual source. In addition to these, there are systems with five or more channels. One of the most famous reconstruction is a 22.2 multichannel sound system [6], which uses 24 loudspeakers located on three layers to produce three-dimensional spatial impressions of sound. Although loudspeakers can be arranged optimally in the theater, the number of loudspeakers in 22.2 multichannel system is still large and inconvenient for household experience of spatial sound ambience. Moreover, MPEG also mentioned to reproduce the film 3D sound at home when it proposed a possible call (N12610) for 3D audio technology in the home cinema and personal TV environment in 2012.

1.1 Related Work

Once each location of loudspeakers in the reproduced sound field is given, one method is to be formulated as a standard numerical optimization problem by minimizing the difference of sound pressure between the original sound field and the reproduced field in a given region. For instance, literature [7] measured a set of sampling points within the given region and then minimized the error of the spatial covariances at these points. Although this approach is a useful design technique in specific cases, it is not practical to apply this kind of professional measuring method in the household environment. Another method is to select a special point within the given region and the sound pressure at this point in the reproduced field will be equivalent with that in the original field. For instance, [8] proposed a multichannel conversion method while maintaining the sound pressure at the origin in the reproduced field and succeeded in converting 22-channel signals of the 22.2 multichannel sound system without two low-frequency effect channels into 10- and 8-channel signals.

However, in most cases we don't know how to place these loudspeakers in the reproduced field. The most common approach is to use the subjective evaluation to obtain the layouts of these loudspeakers although it is a time-consuming and laborious work. For example, the converted layout of 10-, 8- or 6-channel in [8] were obtained by lots of subjective experiments based on the 22-channel sound system. The natural question is how to find a easier way to obtain loudspeaker arrangements in the reproduced sound field while conforming to subjective experiences as much as possible.

Therefore, an approach of multichannel simplification is presented in this paper to obtain a set of loudspeaker arrangements automatically. During the process of automatic simplification, the basic concept is to limit the absolute error of sound pressure at ears to a given threshold or less on the premise of maintaining the sound pressure at the origin. The subjective evaluation showes the sound localization and intensity reproduced by 10- and 8-channel of the proposed method are comparable with those of the 22-channel conversion method. Therefore, the proposed method is an effective method to save the cost of subjective evaluations.

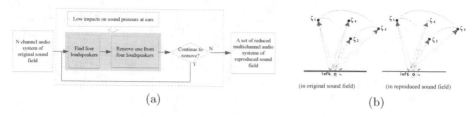

(a)

(b)

Fig. 1. Our proposed method. (a) Block diagram of proposed method; (b) Simplification of one from four loudspeakers (located at ζ_1, ζ_2, ζ_3 and ζ_4 in original field) by generating the phantom source at ζ_4 with the rest of loudspeakers in reproduced field.

Notation. Throughout this paper, we use the following notational conventions: 1) vectors are represented by lower case bold face, e.g., $\mathbf{x}$; 2) polar coordinate originates at the origin $\mathbf{o}$ and represents a position of ζ as $(\sigma, \theta, \varphi)$, where σ is the distance from the origin, θ is the azimuthal angle, and φ is the elevation angle; and 3) The symbol $i = \sqrt{-1}$ is used to denote the imaginary part of a complex number.

2   Formulation of Simplification

If the input signal of a loudspeaker $\zeta = (\sigma, \theta, \varphi)$ is $s(t)$ and the location of receiving point $\mathbf{x}$ is on XOY plane(think of this plane as a plane between human ears), the Fourier transform of sound pressure at the receiving point $\mathbf{x} = (\gamma, \vartheta, 0)$ can be written as

$$p(\mathbf{x}, \omega) = A \frac{e^{-ik|\zeta - \mathbf{x}|}}{|\zeta - \mathbf{x}|} s(\omega) \tag{1}$$

where assumptions 1) each loudspeaker can be modeled as a point source; 2) the reflected sound can be neglected in the original and reproduced fields; 3) the sound pressure at a unit distance from a loudspeaker is proportional to the input to the loudspeaker (the proportional coefficient is denoted as A); and 4) $k = 2\pi\lambda^{-1} = 2\pi f c^{-1}$ is the wave number (with c the speed of wave propagation, f the frequency and λ the wavelength).

The method we proposed is to automatically simplify loudspeaker arrangements from n-channel audio system in original sound field with low impact of sound pressure at ears in the reproduced field. Fig. 1(a) shows the diagram of proposed method including three parts: 1) extraction of the loudspeaker arrangement of n-channel audio system in original sound field; 2) the iteration of multichannel simplification (shown in the dotted box); and 3) setting up a set of simplified loudspeaker arrangements in reproduced sound field. The iteration of multichannel simplification needs to find four loudspeakers firstly and then delete one of them while maintaining low impacts on sound pressure at ears. The iteration ends when the absolute error of sound pressure at ears exceeds a given threshold or none of loudspeakers could be deleted. The key point of the iteration is how to use three loudspeakers to represent four loudspeakers meanwhile the absolute error of sound pressure at ears should be below a given threshold.

Suppose that there are four loudspeakers ζ_1, ζ_2, ζ_3 and ζ_4 in the original sound field, whose locations are $(\sigma_j, \theta_j, \varphi_j)$, $j=1,2,3,4$ and three loudspeakers ζ_1, ζ_2 and ζ_3 in the reproduced sound field (shown in Fig. 1(b)), the aim is to use three loudspeakers ζ_1, ζ_2 and ζ_3 to reproduce the original sound pressure at the origin **o** while maintaining low impact on sound pressure at ears.

Let the signal of each loudspeaker ζ_j in the original field be $s_j(t)$, $j=1,2,3,4$, and that of each loudspeaker ζ_j in the reproduced field is $q_j(t) = s_j(t) + w_j s_4(t)$, $j=1,2,3$, the Fourier transform of the sound pressure at the origin **o** $= (0,0,0)$ in the original field is

$$p(\mathbf{o}, \omega) = \sum_{j=1}^{4} A \frac{e^{-ik\sigma_j}}{\sigma_j} s_j(\omega) = A \frac{e^{-ik\sigma_4}}{\sigma_4} s_4(\omega) + \sum_{j=1}^{3} A \frac{e^{-ik\sigma_j}}{\sigma_j} s_j(\omega) \qquad (2)$$

and that in the reproduced field is

$$\tilde{p}(\mathbf{o}, \omega) = \sum_{j=1}^{3} A \frac{e^{-ik\sigma_j}}{\sigma_j} q_j(\omega) = \sum_{j=1}^{3} A \frac{e^{-ik\sigma_j}}{\sigma_j} w_j s_4(\omega) + \sum_{j=1}^{3} A \frac{e^{-ik\sigma_j}}{\sigma_j} s_j(\omega) \qquad (3)$$

where $w_j, j = 1, 2, 3$ is the jth loudspeaker signal weighting coefficient of the reproduced field.

In order to generate the phantom source of ζ_4 with three loudspeakers in the reproduced sound field, it requires the directions of three loudspeakers ζ_1, ζ_2 and ζ_3 to generate a spherical triangle including that of the loudspeaker ζ_4 in the original field. Moreover, the sound pressure at the origin **o** in the reproduced field could be maintained when

$$p(\mathbf{o}, \omega) = \tilde{p}(\mathbf{o}, \omega) \qquad (4)$$

Such an approach was considered in [8]. However, human ears can not be on the origin **o** at the same time actually. Thus, it is important to lower the impact on sound pressure at ears on the premise of maintaining the sound pressure at the origin in the reproduced field.

Assuming that the influence of head is ignored and the symmetry of the human ears is considered. Let the distances from loudspeaker ζ_j to the right and left ear be represented as ρ_j and ρ_j' respectively, $j = 1, 2, 3, 4$. The Fourier transform of the sound pressure at the right ear **right** $= (\alpha, 0, 0)$ and the left ear **left** $= (\alpha, 180, 0)$ in the original filed is

$$p(\mathbf{right}, \omega) = \sum_{j=1}^{4} A \frac{e^{-ik\rho_j}}{\rho_j} s_j(\omega) \,, p(\mathbf{left}, \omega) = \sum_{j=1}^{4} A \frac{e^{-ik\rho_j'}}{\rho_j'} s_j(\omega) \qquad (5)$$

and that in the reproduced field is

$$\tilde{p}(\mathbf{right}, \omega) = \sum_{j=1}^{3} A \frac{e^{-ik\rho_j}}{\rho_j} q_j(\omega) \,, \tilde{p}(\mathbf{left}, \omega) = \sum_{j=1}^{3} A \frac{e^{-ik\rho_j'}}{\rho_j'} q_j(\omega) \qquad (6)$$

Considering to evaluate the impact of sound pressure at both ears and at the origin $\mathbf{o}$, the relative errors of sound pressure at right ear and left ear in the original field are defined

$$\eta(\mathbf{right} - \mathbf{o}, \omega) \triangleq \frac{p(\mathbf{right}, \omega) - p(\mathbf{o}, \omega)}{p(\mathbf{o}, \omega)} \ , \eta(\mathbf{left} - \mathbf{o}, \omega) \triangleq \frac{p(\mathbf{left}, \omega) - p(\mathbf{o}, \omega)}{p(\mathbf{o}, \omega)} \qquad (7)$$

and those in the reproduced field are

$$\tilde{\eta}(\mathbf{right} - \mathbf{o}, \omega) \triangleq \frac{\tilde{p}(\mathbf{right}, \omega) - \tilde{p}(\mathbf{o}, \omega)}{p(\mathbf{o}, \omega)} \ , \tilde{\eta}(\mathbf{left} - \mathbf{o}, \omega) \triangleq \frac{\tilde{p}(\mathbf{left}, \omega) - \tilde{p}(\mathbf{o}, \omega)}{p(\mathbf{o}, \omega)} \qquad (8)$$

The impacts on sound pressure at ears are analysed by the absolute errors which are represented as

$$\begin{aligned} \varepsilon(\mathbf{right} - \mathbf{o}, \omega) &= |\tilde{\eta}(\mathbf{right} - \mathbf{o}, \omega) - \eta(\mathbf{right} - \mathbf{o}, \omega)|^2 \ , \\ \varepsilon(\mathbf{left} - \mathbf{o}, \omega) &= |\tilde{\eta}(\mathbf{left} - \mathbf{o}, \omega) - \eta(\mathbf{left} - \mathbf{o}, \omega)|^2 \end{aligned} \qquad (9)$$

Hence, the absolute error of sound pressure at ears should be below a given threshold when removing one from four loudspeakers. A set of simplified loudspeaker arrangements in reproduced sound field could be obtained automatically when the principles for multichannel simplification satisfy:

- it maintains the sound pressure at the origin in the reproduced field;
- the absolute error of sound pressure at each ear is less than a given threshold.

3 Multichannel Simplification

3.1 Multichannel Simplification from Four to Three

In order to lower the impacts on sound pressure at ears after removing one from four loudspeakers (shown in Fig. 1(b)), two important practical questions naturally arise: 1) the relationship of loudspeaker locations between the removed one and the rest; 2) the weighting coefficients of the rest three loudspeakers. When only considering the directions of four loudspeakers, we assume that all loudspeakers are located on a sphere of radius R.

Considering the symmetry of head, we focus on the absolute error of sound pressure at the right ear $\mathbf{right} = (\alpha, 0, 0)$ only. From Eq. 7 and Eq. 8, Eq. 9 could be derived

$$\begin{aligned} \varepsilon(\mathbf{right} - \mathbf{o}, \omega) &= |\tilde{\eta}(\mathbf{right} - \mathbf{o}, \omega) - \eta(\mathbf{right} - \mathbf{o}, \omega)|^2 \\ &= \left| \frac{\tilde{p}(\mathbf{right}, \omega) - p(\mathbf{right}, \omega)}{p(\mathbf{o}, \omega)} + \frac{p(\mathbf{o}, \omega) - \tilde{p}(\mathbf{o}, \omega)}{p(\mathbf{o}, \omega)} \right|^2 \end{aligned} \qquad (10)$$

Once the sound pressure at the origin in the reproduced field is equal to that in the original field(i.e., Eq. 4 is satisfied), the absolute error of sound pressure at the right ear $\varepsilon(\mathbf{right} - \mathbf{o}, \omega)$ in Eq. 10 could be represented as

$$\varepsilon(\textbf{right} - \textbf{o}, \omega) = \left| \frac{\tilde{p}(\textbf{right}, \omega) - p(\textbf{right}, \omega)}{p(\textbf{o}, \omega)} \right|^2 = \left| \frac{(\sum\limits_{j=1}^{3} \frac{e^{ik\rho_j}}{\rho_j} w_j - \frac{e^{ik\rho_4}}{\rho_4}) s_4(\omega)}{\sum\limits_{j=1}^{4} \frac{e^{ik\sigma_j}}{\sigma_j} s_j(\omega)} \right|^2$$

(11)

Where $\varepsilon(\textbf{right} - \textbf{o}, \omega)$ in Eq. 11 should be dependent on the sound pressures at the right ear **right** both in the original and reproduced field as well as the sound pressure at the origin in the original field.

Assuming that 1) the signal of each loudspeaker ζ_j in the original field satisfies $s_1(t) = s_2(t) = s_3(t) = s_4(t)$; 2) $w_1 + w_2 + w_3 = 1$; and 3) only a single frequency of sound waves at the observation point, the impact of $\varepsilon(\textbf{right} - \textbf{o}, \omega)$ is mainly determined by δ which is defined

$$\delta \triangleq \left| \sum_{j=1}^{3} \frac{e^{ik\rho_j}}{\rho_j} w_j - \frac{e^{ik\rho_4}}{\rho_4} \right|^2 = \left| e^{ik\rho_4} (\sum_{j=1}^{3} \frac{e^{ik(\rho_j - \rho_4)}}{\rho_j} w_j - \frac{1}{\rho_4}) \right|^2$$

(12)

If the spherical triangle area formed by three loudspeakers(i.e., ζ_1, ζ_2 and ζ_3) is as small as possible, the smaller the difference between $\rho_j, j = 1, 2, 3$ and ρ_4 in Eq. 12 is, the smaller the impact of $\varepsilon(\textbf{right} - \textbf{o}, \omega)$ in Eq. 11 will be. The same result has been returned in the impact of $\varepsilon(\textbf{left} - \textbf{o}, \omega)$.

Assuming that the selected loudspeaker is located at ζ_4, it is important that the three loudspeaker positions should satisfy the following conditions: 1) a spherical triangle formed by the three loudspeaker directions includes the direction of the removed loudspeaker whose location is ζ_4; and 2) the spherical triangle is the minimum among those subscribe to 1).

Using the notation in Fig. 1(b), we calculate four spherical areas:

S_{124}: formed by three loudspeakers ζ_1, ζ_2 and ζ_4;
S_{134}: formed by three loudspeakers ζ_1, ζ_3 and ζ_4;
S_{234}: formed by three loudspeakers ζ_2, ζ_3 and ζ_4;
S_{123}: formed by three loudspeakers ζ_1, ζ_2 and ζ_3.

We can determine the position ζ_4 be inside the spherical triangle formed by ζ_1, ζ_2 and ζ_3 if

$$S_{123} = S_{124} + S_{134} + S_{234}$$

Which is used to verify condition 1).

In order to use three loudspeakers to represent four loudspeakers while maintaining the sound pressure at the origin **o**, Eq. 4 can be derived

$$\frac{e^{-ik\sigma_4}}{\sigma_4} = \frac{e^{-ik\sigma_1}}{\sigma_1} w_1 + \frac{e^{-ik\sigma_2}}{\sigma_2} w_2 + \frac{e^{-ik\sigma_3}}{\sigma_3} w_3$$

(13)

In the sequel, the method of literature [8] which can derive the weighting coefficients of three loudspeakers are obtained.

Algorithm 1. Pseudo-code of the iteration of multichannel simplification

---

1: **Input**: N-loudspeaker positions, N-loudspeaker signals
2: **Output**: M-channel loudspeaker arrangements found in the reproduced field
3:    $S \leftarrow$ all of the 3-combinations of loudspeakers from n loudspeakers
4:    $S1 \leftarrow$ subset of S where positions of 3 loudspeakers form a spherical triangle
5:    $S2 \leftarrow$ subset of $S1$ where each triangle region includes at least one of n loudspeakers

6:

7: **repeat**
8:       $S3 \leftarrow$ subset of $S2$ where only one loudspeaker is included within the triangle region
9:       $S3 \leftarrow$ sorted by the spherical triangle area
10:      obtain signals of three loudspeakers whose positions could form the smallest spherical triangle
11:      allocate the signal of the loudspeaker within triangle to signals of three loudspeakers
12:      remove the loudspeaker within the triangle region
13:      $S2 \leftarrow$ delete the associated elements of the removed loudspeaker in $S2$
14: **until** a stop criterion is met

3.2 Automatic Multichannel Simplification from N to M

This section mainly introduces the process of iteration for automatic simplification from N to M channels. The main body of the iteration includes two parts: 1) finding all spherical triangles formed by three of N-loudspeaker positions, where each triangle region includes only one loudspeaker; and 2) uninterruptedly removing the loudspeaker within the smallest triangle region until the stop criterion is met. Algorithm. 1 shows the pseudo-code of the iteration of automatic multichannel simplification.

4   Experiment

In order to test the performance of the proposed method, we used the 22.2 multichannel sound system without two low-frequency effect channels(shown in Fig. 2(a)) as an example where: 1) the distance from each loudspeaker to the origin **o** is 200cm ($\sigma = 200$); 2) the locations of both ears are **right** $= (8.5, 0, 0)$ and **left** $= (8.5, 180, 0)$ respectively; 3) the sound source is 22-channel white noise (described in [9]); and 4) the original sound field at the receiving point is produced by 22-channel audio system.

As for 10-, 8- and 6-channel sound system converted by the 22-channel system, the absolute errors of sound pressure at both ear can be calculated by Eq. 9 where the frequency of sound wave is 1000Hz at the observing point. Since the positions of 10,8 and 6 loudspeakers are symmetrical and the signals of all loudspeakers are the same, the absolute error of sound pressure at the right ear is equal to that

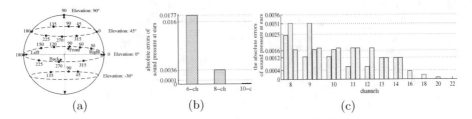

(a)   (b)   (c)

Fig. 2. The absolute errors of sound pressures at ears. (a) 22-channel loudspeaker arrangement without two **LFE** channels in original field in [8]; (b) the absolute errors of sound pressures at ears where three loudspeaker arrangement are 10-, 8- and 6-channel converted by 22-channel conversion method in [8]; (c)the absolute errors of sound pressures at ears reproduced by our proposed method.

at the left ear. From Fig. 2(b), the absolute errors of sound pressure at ears may increase with the decrease of the amount of loudspeakers. Considering that the converted 10- and 8-channel sound keeps the spatial impression of the original 22-channel sound described in the subjective experiments of [8], the absolute error 0.0036 of 8-channel is used as the threshold. Therefore, the termination criteria of Algorithm. 1 will be: 1) the absolute error of sound pressure at ears exceeds the threshold; or 2) none of loudspeakers could be removed.

Considering the locations of 22 loudspeakers are symmetrical based on the median plane, the automatic simplification will remove the selected loudspeaker and its symmetric one. Using the proposed algorithm of multichannel simplification (shown in Algorithm. 1), there are 23 layouts simplified from 22- to 8-channel whose absolute errors of sound pressure at ears Fig. 2(c) shows the absolute errors of sound pressure at ears are under the given threshold(0.0036). Table 1 shows the layouts of 10- and 8-channel loudspeaker arrangements in the reproduced field by proposed method.

In order to further evaluate the proposed method, the subjective experiments are carried out using the RAB paradigm [10], a method of determining a different threshold by presenting two stimuli (A and B) and a third (R) that is the reference, in a soundproof room where the reverberation time at 500Hz was

Table 1. The loudspeaker arrangements of layouts (10- and 8-channel) in the reproduced field by the proposed method

layout	loudspeaker arrangement in reproduced space (azimuthal angle, elevation angle)									
10-1	(0,0)	(30,0)	(90,0)	(150,0)	(180,0)	(270,0)	(90,45)	(270,45)	(0,90)	(90,-30)
10-2	(0,0)	(90,0)	(180,0)	(270,0)	(0,45)	(90,45)	(180,45)	(270,45)	(0,90)	(90,-30)
10-3	(0,0)	(30,0)	(150,0)	(180,0)	(270,0)	(45,45)	(90,45)	(135,45)	(270,45)	(90,-30)
8-1	(0,0)	(90,0)	(180,0)	(270,0)	(90,45)	(270,45)	(0,90)	(90,-30)	-	-
8-2	(0,0)	(90,0)	(180,0)	(270,0)	(90,45)	(225,45)	(315,45)	(90,-30)	-	-
8-3	(0,0)	(180,0)	(270,0)	(0,45)	(90,45)	(180,45)	(270,45)	(90,-30)	-	-

0.18s and 30dB(A) background noise. The R is the original 22-channel white noise with 10s, one stimuli (A or B) is the sound of layouts in [8] and the other is the sound of layouts in Table 2. The subject is asked to compare the difference betwween A and B relative to R, according to a continuous seven-grade impairment scale shown in Table 2. The impairment is assessed from the following two viewpoints: sound localization and sound intensity.

Table 2. Scales used for subjective evaluation

Comparison of the stimuli	Much better	Slightly better	Better	The same	Slightly worse	Worse	Much worse
Scores	3	2	1	0	-1	-2	-3

In the experiment, subjects are 20 students whose majors are all audio signal processing. They should be able to compare the sound localization and intensity reproduced by each 10- and 8-channel sound system of the proposed method with those by the same channels system of 22-channel conversion method. Fig. 3 shows the Comparative Mean opinion score (CMOS) of sound localization and sound intensity given by the subjects with 95% confidence limits. Fig. 3(a) represents the result of sound localization and Fig. 3(b) is the result of sound intensity. From Fig. 3, the CMOS of sound localization and intensity reproduced by each 10- or 8-channel system of the proposed method are comparable to those reproduced by the same channels system of 22-channel conversion method. Moreover, the sound localization reproduced by layout 8-3 is slight better than that by 8-channel system of 22-channel conversion method (shown in Fig. 3(a)).

The simulation result shows that all of 23 layouts, produced by the proposed method of automatic multichannel simplification, can successfully lower the impact on sound pressure at ears if the absolute error of sound pressure at ears are under 0.0036. To some extent, the proposed method of automatic simplification is an effective method to save the cost of subjective evaluations since the subjective evaluation is comparable to that of 22-channel conversion.

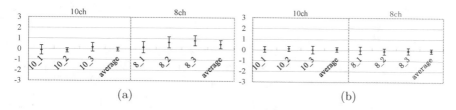

Fig. 3. Results of the Comparative Mean opinion score (CMOS) in the experiment (95% confidence limits).(a)sound localization; (b)sound intensity.

5   Conclusion

In this paper, an automatic simplification of n multichannel sound system is proposed on the basic concept of the 22-channel conversion. Moreover, the proposed method considers the impact on the sound pressures at both ears, rather than only at the receiving point. Meanwhile, every one of loudspeaker arrangements simplified by our proposed method should satisfy two conditions: 1)it maintains the sound pressure at the receiving point in the reproduced field; and 2) the absolute errors of sound pressure at both ears are less than a given threshold. The 22.2 multichannel sound system without two low-frequency effect channels as an example can be simplified from 22 channels to eight channels by this method and the total number of loudspeaker arrangements is 23. Moreover, the subjective evaluation results show that this method is comparable to 22-channel conversion method, but this method can effectively save the cost of subjective evaluations.

References

1. Linkwitz, S.: A Model for Rendering Stereo Signals in the ITD-Range of Hearing. In: Audio Engineering Society Convention 133. Audio Engineering Society (2012)
2. Ahrens, J.: Analytic Methods of Sound Field Synthesis. Springer (2012)
3. Spors, S., Wierstorf, H., Geier, M.: Comparison of modal versus delay-and-sum beamforming in the context of data-based binaural synthesis. In: Audio Engineering Society Convention 132 Audio Engineering Society (2012)
4. De Sena, E., Hacihabiboglu, H., Cvetkovic, Z.: On the Design and Implementation of Higher Order Differential Microphones. IEEE Transactions on Audio, Speech, and Language Processing 20(1), 162–174 (2012)
5. De Sena, E., Hacihabiboglu, H., Cvetkovic, Z.: Analysis and design of multichannel systems for perceptual sound field reconstruction. IEEE Transactions on Audio, Speech, and Language Processing 21(8), 1653–1665 (2013)
6. Hamasaki, K., Matsui, K., Sawaya, I., Okubo, H.: The 22.2 Multichannel Sounds and its Reproduction at Home and Personal Enviroment. In: Audio Engineering Society Conference: 43rd International Conference: Audio for Wirelessly Networked Personal Devices. Audio Engineering Society (2011)
7. Takahashi, Y., Ando, A.: Down-mixing of multi-channel audio for sound field reproduction based on spatial covariance. Applied Acoustics 71(12), 1177–1184 (2010)
8. Ando, A.: Conversion of multichannel sound signal maintaining physical properties of sound in reproduced sound field. IEEE Transactions on Audio, Speech, and Language Processing 19(6), 1467–1475 (2011)
9. Sawaya, I., Oode, S., Ando, A., Hamasaki, K.: Size and Shape of Listening Area Reproduced by Three-dimensional Multichannel Sound System with Various Numbers of Loudspeakers. In: Audio Engineering Society Convention 131. Audio Engineering Society (2011)
10. Recommendation, ITU-R BS. 1284-2: General Methods for the Subjective Assessment of Sound Quality. International Telecommunications Union (2002)
11. Report ITU-R BS.2159-4.: Multichannel sound technology in home and broadcasting applications (2012)

Acoustic Beamforming with Maximum SNR Criterion and Efficient Generalized Eigenvector Tracking

Toshihisa Tanaka* and Mitsuaki Shiono

Tokyo University of Agriculture and Technology
Nakacho, Koganei-shi, Tokyo, 184–8588, Japan
tananat@cc.tuat.ac.jp
http://www.sip.tuat.ac.jp/

Abstract. A recently proposed adaptive acoustic beamformer based on the maximization of the output SNR (Max-SNR beamformer) has an advantage of requiring no information of transfer functions. A key technology to implement Max-SNR beamformers is to estimate generalized eigenvector (GEV) of covariance matrices of target signal and noise, which are basically unknown. We develop a novel GEV tracking algorithm with decaying time windows that enable Max-SNR beamformer to adapt rapidly moving sources. Simulation results support the analysis.

1 Introduction

Acoustic information is one of the crucial elements in multimedia. Extracting and enhancing a target speech signal by multiple sensors (microphones) have a wide range of multimedia applications including videoconferencing, acoustic multimedia systems, hands-free communication, mobile speech recognition, hearing aid, and so forth [19], [3], [9], [7]. For removing interference and noise from the microphone array signals, beamforming techniques exploit spatial and spectral diversities to discriminate between the target and the other components.

A number of efficient beamforming techniques have been proposed; maximum likelihood beamformer [9], [18], minimum variance distortionless response [5], [12], and generalized sidelobe canceler [4], [2]. To extract a target signal from microphone signals, these beamformers need to know transfer functions from the source to an array of the microphones. However, the transfer functions are often not only unknown but also varying in dynamic environment (microphones on mobile handsets, for example).

A beamformer that do not need the knowledge of transfer functions is the maximum signal-to-noise ratio (Max-SNR) beamformer [14], [6], which finds the filter coefficients of beamformer by maximizing the SNR of the output signal. In theory, the filter coefficients are given as the principal generalized eigenvector

---

* This work is supported in part by JSPS Graint-in-Aid for Scientific Research (B), 23300069 and 26280054.

W.T. Ooi et al. (Eds.): PCM 2014, LNCS 8879, pp. 373–382, 2014.

(GEV) of a pair of spatial correlation matrices of microphone signals and noise signals. However, these correlation matrices are usually unknown and even dynamic. This makes it difficult to impliment Max-SNR beamformers; and thus, tracking the principal GEV is a key to establish this technique. Warsitz et al. proposed an adaptive algorithm based on gradient ascent for maximization of the output SNR [14]. However, a gradient ascent leads to a trade-off between convergence speed and stability. Inadequate selection of the step size parameter may cause inaccurate estimation and/or slow convergence of the filter coefficients.

In this paper, we focus on finding the principal GEV that strongly governs the performance in Max-SNR beamforming. Recently, several efficient adaptive algorithms for GEVs have been proposed [17], [13], [16]. Based on these recent results, we establish a tracking algorithm of the principal GEV suitable for Max-SNR beamforming. The underlying idea behind the proposed method is applying decaying windows to a series of past samples to improve the tracking performance for changes of signals [17]. The proposed algorithm achieves fast and stable convergence of filter coefficients, and enables us to adapt sudden changes of signals, such as changes of the target position or source.

Notation. The following notation and mathematical operations are used in the paper. A vector is denoted by a bold-faced letter, e.g., $\mathbf{X}$. The correlation matrix of $\mathbf{X}$ is denoted by $\mathbf{R}_X$. The expected value is denoted by $E[\cdot]$. The transpose and the Hermitian transpose are denoted by $\cdot^T$ and $\cdot^H$, respectively. The trace of a matrix is denoted by $tr(\cdot)$.

2   Signal Observation Model

We consider an array of M microphones, where the m-th microphone receives signal $x_m(t)$, $m = 1, \ldots, M$. Let $X_m(n, k)$ be the short-time Fourier transform of $x_m(t)$, where n denotes the frame index, $k = 0, \ldots, L-1$ denotes the frequency bin and L denotes the size of the discrete Fourier transform. We assume that the microphone signal $X_m(n, k)$ consists of two components: target signal $S_m(n, k)$ and noise signal $N_m(n, k)$. The microphone signal $X_m(n, k)$ is modeled as

$$X_m(n, k) = S_m(n, k) + N_m(n, k) = H_m(k)U(n, k) + N_m(n, k), \qquad (1)$$

where $H_m(k)$ is the transfer function from the target source to the m-th microphone and $U(n, k)$ is the source signal. Using the vector notation, we can rewrite (1) as $\mathbf{X}(n, k) = \mathbf{S}(n, k) + \mathbf{N}(n, k)$, where $\mathbf{X}(n, k) = (X_1(n, k), \ldots, X_M(n, k))^T$, $\mathbf{S}(n, k) = (S_1(n, k), \ldots, S_M(n, k))^T$, and $\mathbf{N}(n, k) = (N_1(n, k), \ldots, N_M(n, k))^T$. Output $Y(n, k)$ is a linear combination of the microphone signals given as

$$Y(n, k) = \mathbf{F}^H(k)\mathbf{X}(n, k), \qquad (2)$$

where $\mathbf{F}(k) = (F_1(k), \ldots, F_M(k))^T$ is a called beamformer. Beamformer $\mathbf{F}(k)$ is obtained by the following methods.

3   Max-SNR Beamformer

A beamformer that maximize the SNR of beamformer output is called a Max-SNR beamformer [14]. The advantage of the Max-SNR beamformer is that information of transfer function $\mathbf{H}(k)$ is not needed. Only the knowledge of spatial correlation matrix $\mathbf{R}_N(k)$ is required.

Assuming that $S_m(n, k)$ and $N_m(n, k)$ are uncorrelated, we obtain the mean power of the output signal $P_Y(k)$ given as

$$P_Y(k) = E[|Y(n, k)|^2] = \mathbf{F}^H(k)\mathbf{R}_S(k)\mathbf{F}(k) + \mathbf{F}^H(k)\mathbf{R}_N(k)\mathbf{F}(k). \qquad (3)$$

SNR of $Y(n, k)$ is given as

$$\mathrm{SNR}(k) = \frac{\mathbf{F}^H(k)\mathbf{R}_S(k)\mathbf{F}(k)}{\mathbf{F}^H(k)\mathbf{R}_N(k)\mathbf{F}(k)} = \frac{\mathbf{F}^H(k)\mathbf{R}_X(k)\mathbf{F}(k)}{\mathbf{F}^H(k)\mathbf{R}_N(k)\mathbf{F}(k)} - 1. \qquad (4)$$

The Max-SNR beamformer is given as a solution of the optimization problem: $\max_{\mathbf{F}(k)} \mathrm{SNR}(k)$. The maximizer satisfies $\nabla_{\mathbf{F}}\mathrm{SNR}(k) = 0$, that is,

$$\mathbf{R}_X(k)\mathbf{F}_{\mathrm{opt}}(k) = \frac{\mathbf{F}_{\mathrm{opt}}^H(k)\mathbf{R}_X(k)\mathbf{F}_{\mathrm{opt}}(k)}{\mathbf{F}_{\mathrm{opt}}^H(k)\mathbf{R}_N(k)\mathbf{F}_{\mathrm{opt}}(k)}\mathbf{R}_N(k)\mathbf{F}_{\mathrm{opt}}(k) = \lambda_{\max}(k)\mathbf{R}_N(k)\mathbf{F}_{\mathrm{opt}}(k),$$
$$(5)$$

which is a generalized eigenvalue problem of a matrix pair of $\mathbf{R}_X(k)$ and $\mathbf{R}_N(k)$, where $\lambda_{\max}(k)$ is the largest generalized eigenvalue and $\mathbf{F}_{\mathrm{opt}}(k)$ is the corresponding eigenvector (principal GEV).

Warsitz et al. proposed an adaptive algorithm based on gradient ascent that adaptively update estimate $\hat{\mathbf{F}}(k)$ [14]. The underlying idea is to reduce $\max_{\mathbf{F}(k)} \mathrm{SNR}(k)$ to the following constrained maximization problem:

$$\max_{\mathbf{F}(k)} \mathbf{F}^H(k)\mathbf{R}_X(k)\mathbf{F}(k), \quad \text{subject to } \mathbf{F}^H(k)\mathbf{R}_N(k)\mathbf{F}(k) = 1. \qquad (6)$$

The resulting update rule for $\hat{\mathbf{F}}(k)$ is given as

$$\hat{\mathbf{F}}(n, k) = \hat{\mathbf{F}}(n - 1, k) + \frac{c(k) - \hat{\mathbf{F}}^H(n - 1, k)\mathbf{G}(n, k)}{2\mathbf{G}^H(n, k)\mathbf{G}(n, k)}\mathbf{G}(n, k)$$
$$+ \mu Y^*(n, k)\left[\mathbf{X}(n, k) - \frac{A(n, k)\mathbf{G}(n, k)}{2\mathbf{G}^H(n, k)\mathbf{G}(n, k)}\right], \qquad (7)$$

where μ is the step size for the gradient ascent, and

$$\mathbf{G}(n, k) = \hat{\mathbf{R}}_N(n, k)\hat{\mathbf{F}}(n - 1, k) \qquad (8)$$

$$A(n, k) = \frac{Y(n, k)}{Y^*(n, k)}\mathbf{X}^H(n, k)\mathbf{G}(n, k) + \mathbf{G}^H(n, k)\mathbf{X}(n, k). \qquad (9)$$

$$\hat{\mathbf{R}}_N(n, k) = \epsilon\hat{\mathbf{R}}_N(n - 1, k) + (1 - \epsilon)\mathbf{N}(n, k)\mathbf{N}^H(n, k) \qquad (10)$$

with forgetting factor $0 < \epsilon < 1$. See [14] for details of the derivation.

It should be remarked that practically, it is hard to employ (10), since the microphone signal generally includes noise signal $\mathbf{N}(n, k)$ as well as the target signal. A possible way to estimate the noise is to use voice activity detection (VAD) that obtains $\mathbf{N}(n, k)$ while the target signal is absent. VAD is a classifier that discriminates whether or not the microphone signal include the target signal [14]. In [14], $\hat{\mathbf{R}}_N(n, k)$ is updated if the VAD recognizes that the microphone signal consists of only noise. Otherwise, $\hat{\mathbf{R}}_X(n, k)$ is updated.

4   Adaptive Max-SNR Beamformer with Efficient Principal Generalized Eigenvector Tracking

Although the adaptive beamformer described in the previous session is achieved by a gradient ascent algorithm, the step size parameter greatly affects both the convergence speed and stability. Moreover, in the case of a sudden change of location of a speech source, this beamformer exhibits poor performance in tracking the source as we will see in Section 5.

We develop a GEV tracking algorithm as an extension of the recursive least squares (RLS), that is well-known adaptive filter algorithm based on the Newton method, which generally provides faster convergence than the gradient ascent. Indeed, the work presented in this paper is obtained by alleviating an adaptive algorithm for GEV derived from the modified weighted criterion [17]. The underlying idea behind the proposed tracking algorithm is introducing decaying windows that enables us to track a dynamically changing signal source.

4.1   Tracking the Principal Generalized Eigenvector with Decaying Windows

Since MWC considers all the past samples like infinite impulse response (IIR) filters to estimate the correlation matrices, it is not suitable for beamforming under the situation that the target and noise are varying, such as moving speaker. By applying the decaying windows for the past samples, the proposed algorithm enable us to track the principal GEV fast and robust against changes of input signals.

First of all, it should be noted that the principal GEV amounts to the minimizer of the following cost function [17];

$$J[\hat{\mathbf{F}}(n, k)] = tr[\mathbf{C}^H \hat{\mathbf{R}}_X(n, k)\mathbf{C}] - 2tr[\hat{\mathbf{F}}^H(n, k)\hat{\mathbf{R}}_X(n, k)\hat{\mathbf{F}}(n, k)]$$
$$+ tr[(\hat{\mathbf{F}}^H(n, k)\hat{\mathbf{R}}_X(n, k)\hat{\mathbf{F}}(n, k))(\hat{\mathbf{F}}^H(n, k)\hat{\mathbf{R}}_N(n, k)\hat{\mathbf{F}}(n, k))], \quad (11)$$

where $\mathbf{C}$ is a matrix represented by $\hat{\mathbf{R}}_N = (\mathbf{C}^{-1})^T \mathbf{C}^T$. A key behind the proposed method is to estimate the spatial correlation matrices $\hat{\mathbf{R}}_X(n, k)$ and $\hat{\mathbf{R}}_N(n, k)$ with past L_X and L_N samples. Specifically,

$$\hat{\mathbf{R}}_X(n,k) = \beta\hat{\mathbf{R}}_X(n-1,k) + \mathbf{X}(n,k)\mathbf{X}^H(n,k) - \beta^{L_X}\mathbf{X}(n-L_X,k)\mathbf{X}^H(n-L_X,k),$$
(12)

$$\hat{\mathbf{R}}_N(n,k) = \alpha\hat{\mathbf{R}}_N(n-1,k) + \mathbf{N}(n,k)\mathbf{N}^H(n) - \alpha^{L_N}\mathbf{N}(n-L_N,k)\mathbf{N}^H(n-L_N,k),$$
(13)

where β and α are the forgetting factors in the range between 0 to 1, L_X and L_N are the numbers of samples of the decaying windows.

Instead of directly minimizing J, which is of fourth-order, the projection approximation [15] is applied to the cost function $J[\hat{\mathbf{F}}(n,k)]$ to convert it into a quadratic function of $\hat{\mathbf{F}}(n,k)$. The projection approximation is achieved by $\hat{\mathbf{R}}_X(n,k)\hat{\mathbf{F}}(n,k) \approx \hat{\mathbf{R}}_X(n,k)\hat{\mathbf{F}}(n-1,k)$, and thus $\mathbf{r}_{XZ}(n,k)$ becomes

$$\mathbf{r}_{XZ}(n,k) = \beta\mathbf{r}_{XZ}(n,k) + \mathbf{X}(n,k)Z^*(n,k) - \beta^{L_X}\mathbf{X}(n-L_X,k)Z^*(n-L_X,k),$$
(14)

where $Z(n,k) = \hat{\mathbf{F}}^H(n-1,k)\mathbf{X}(n,k)$. We also define,

$$P_Z(n,k) = \beta P_Z(n-1,k) + Z(n,k)Z^*(n,k) - \beta^{L_X}|Z(n-L_X,k)|^2.$$
(15)

Using (14) and (15), we obtain the quadratic cost function given as

$$J'[\hat{\mathbf{F}}(n,k)] = tr[\mathbf{C}^H\hat{\mathbf{R}}_X(n,k)\mathbf{C}] - 2tr[\hat{\mathbf{F}}^H(n,k)\mathbf{r}_{XZ}(n,k)]$$
$$+ tr[P_Z(n,k)\hat{\mathbf{F}}^H(n,k)\hat{\mathbf{R}}_N(n,k)\hat{\mathbf{F}}(n,k)].$$
(16)

The minimizer of the above cost function is straightforward:

$$\hat{\mathbf{F}}(n,k) = \mathbf{Q}(n,k)\mathbf{r}_{XZ}(n,k)P_Z^{-1}(n,k),$$
(17)

where $\mathbf{Q}(n,k) = \hat{\mathbf{R}}_N^{-1}(n,k)$, which can be obtained by rewriting $\hat{\mathbf{R}}_N(n,k)$ as

$$\hat{\mathbf{R}}_N(n,k) = \alpha\hat{\mathbf{R}}_N(n-1,k) + \mathbf{N}(n,k)\mathbf{N}^H(n) - \alpha^{L_N}\mathbf{N}(n-L_N,k)\mathbf{N}^H(n-L_N,k)$$
$$= \alpha\hat{\mathbf{R}}_N(n-1,k) + \mathbf{N}_N(n,k)\begin{bmatrix}1 & 0 \\ 0 & -\alpha^{L_N}\end{bmatrix}\mathbf{N}_N^H(n,k),$$
(18)

where $\hat{\mathbf{R}}_N(n,k) = (\mathbf{N}(n,k), \mathbf{N}(n-L_N,k))$, and applying the matrix inversion lemma to (18). Thus, we have

$$\mathbf{Q}(n,k) = \hat{\mathbf{R}}_N^{-1}(n,k)$$
$$= \frac{1}{\alpha}\Big\{\mathbf{Q}(n-1,k) - \mathbf{Q}(n-1,k)\mathbf{N}_N(n,k)$$
$$\times \left(\begin{bmatrix}\alpha & 0 \\ 0 & \alpha^{1-L_N}\end{bmatrix} + \mathbf{N}_N^H(n,k)\mathbf{Q}(n-1,k)\mathbf{N}_N(n,k)\right)^{-1}\mathbf{N}_N^H(n,k)\mathbf{Q}(n-1,k)\Big\}.$$
(19)

Algorithm 1. Summary of the proposed Max-SNR beamformer

1: **for** all frame n **do**
2:     **for** all frequency bin k **do**
3:         **if** VAD statistic (20) accepts H_0 for $\mathbf{X}(n)$ **then**
4:             Update $\hat{\mathbf{R}}_N(n,k)$ and $\mathbf{Q}(n,k)$ by (18) and (19), respectively.
5:         **else**
6:             Update $\mathbf{r}_{XZ}(n,k)$ and $P_z(n,k)$ by (14) and (15), respectively
7:         **end if**
8:         Update $\hat{\mathbf{F}}(n,k)$ by (17).
9:         Obtain $Y(n,k)$ by (21).
10:     **end for**
11: **end for**

4.2   Proposed Max-SNR Beamformer

The generalized eigenvalue tracking algorithm described in the previous section can be incorporated in the Max-SNR beamformer in the following manner. The proposed Max-SNR algorithm is summarized in **Algorithm 1**.

First, determine whether or not microphone signal $\mathbf{X}(n,k)$ includes the target signal. In this paper, we use the VAD proposed in [11], [10], which is given as

$$\Lambda(n) = \frac{1}{L} \sum_{k=0}^{L-1} \left\{ \frac{|X_m(n,k)|^2}{\sigma_N^2(k)} - \log \frac{|X_m(n,k)|^2}{\sigma_N^2(k)} - 1 \right\} \underset{H_0}{\overset{H_1}{\gtrless}} \eta, \tag{20}$$

where either H_0 or H_1 denotes the condition that target signal is absent or present, $\sigma_N^2(n)$ denotes the noise variance and η denotes the decision threshold. If H_0 is accepted (target is absent), we update $\mathbf{Q}(n,k)$ by (19). On the other hand, if H_1 is accepted (target is present), we update $\mathbf{r}_{XZ}(n,k)$, $P_Z(n,k)$, and $\hat{\mathbf{F}}(n,k)$ by (14), (15), and (17). Finally, we obtain the output as

$$Y(n,k) = \hat{w}^*(n,k)\hat{\mathbf{F}}^H(n,k)\mathbf{X}(n,k), \tag{21}$$

where, $\hat{w}(n,k)$ is the filter normalization coefficient given as

$$\hat{w}(n,k) = \frac{\sqrt{\hat{\mathbf{F}}^H(n,k)\hat{\mathbf{R}}_N(n,k)\hat{\mathbf{R}}_N(n,k)\hat{\mathbf{F}}(N,k)/M}}{\hat{\mathbf{F}}^H(n,k)\hat{\mathbf{R}}_N(n,k)\hat{\mathbf{F}}(n,k)}. \tag{22}$$

This normalization is necessary, since the estimation of the filter is made in each frequency bin independently. For details of the normalization, see [14] for the notion of the normalization for $\mathbf{F}(n,k)$.

5   Simulation Results

We compare the performance of the proposed method (denoted by MWC-DW; MWC with decaying windows) to the conventional methods (SGA). We also evaluate the performance of well-known tracking methods such as MWC and power method-based fast GEV extraction (PM-FGE) [13]. We set the parameters for MWC-DW as follows: $\beta = 0.99$, $\alpha = 0.99$, $L_X = 50$, $L_N = 200$.

Table 1. Angular frequencies in π rad

	ω_1	ω_2	ω_3
$n < 1000$	0.62	0.46	0.74
$n \geq 1000$	0.39	0.5	0.2

Table 2. SNR in dB of each signal

Source	Mic.	SGA [14]	MWC [17]	PM-FGE [13]	MWC-DW
1		−0.5768	7.9564	9.0553	9.2743	10.1849
2		0.5411	8.9105	11.3434	14.1264	15.5154

5.1 Performance Evaluation for Artificial Signals

Artificial signals are used for the performance evaluation of the proposed algorithm. Our aim is to observe the convergence speed and the tracking ability when statistics of the signal change. For this purpose, input vectors are assumed to be stochastic sinusoids defined as $\mathbf{f}(n) = (f(n), \ldots, f(n-5))^T$ and $\mathbf{g}(n) = (g(n), \ldots, g(n-5))^T$, where

$$f(n) = \sqrt{2}\sin(\omega_1 t + \phi_1) + \nu_1(n), \tag{23}$$

$$g(n) = \sqrt{2}\sin(\omega_2 t + \phi_2) + \sqrt{2}\sin(\omega_3 t + \phi_3) + \nu_2(n), \tag{24}$$

where n is the sample index. This simulated signal model has the following parameters: angular frequencies ω_1, ω_2 and ω_3; initial phases ϕ_1, ϕ_2 and ϕ_3; independent Gaussian noises ν_1 and ν_2. The angular frequencies are deterministic as shown in Table 1 to simulate the sudden change of the signal model at time instance $n = 1000$. The initial phases are uniformly distributed random variables, and the Gaussian noises of ν_1 and ν_2 have zero-mean and the variances of 1 and 0.01, respectively.

We tracked the principal GEV $\mathbf{w}$ of the above model without knowledge of the correlation matrices, $\mathbf{R}_f = E[\mathbf{f}(n)\mathbf{f}^T(n)]$ and $\mathbf{R}_g = E[\mathbf{g}(n)\mathbf{g}^T(n)]$. For the evaluation, we used the similarity measure called direction cosine [17]

$$\text{direction cosine}(n) = \frac{|\mathbf{w}^T(n)\mathbf{v}|}{\|\mathbf{w}(n)\|\|\mathbf{v}\|}, \tag{25}$$

where $\mathbf{v}$ is the true principal GEV of $\mathbf{R}_f = E[\mathbf{f}(n)\mathbf{f}^T(n)]$ and $\mathbf{R}_g = E[\mathbf{g}(n)\mathbf{g}^T(n)]$. Direction cosine is unity if the estimated vector is parallel to the truth.

For each method, the evolution of $1 - \text{direction cosine}(n)$ averaged over 100 independent runs is illustrated in Figure 1. The vertical and horizontal axes denote the direction cosine and the time index, respectively. As shown in Figure 1, compared to the other methods, SGA exhibits slow convergence and significant vibration as well as lower accuracy of estimation at the steady state. As seen in the first segment ($0 \leq n \leq 999$), the convergence speed of MWC is slower than PM-FGE and the proposed MWC-DW, which show a similar speed. However, it can be confirmed in the second segment ($n \geq 1000$) that MWC-DW adapts the change of signal statistics more rapidly than the other methods.

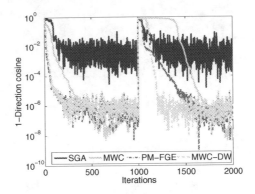

Fig. 1. The evolution of direction cosine

5.2   Comparison of Speech Extraction Performance

We examined the speech extraction with each method in a simulated room and evaluate the performance for changes of target signals position. The impulse response of the simulated room was calculated by a method described in [1]. The simulated room size was 6 m × 5 m × 3 m, and we arranged sources and microphones as shown in Figs. 2 and 3. The number of microphones M was 5, and the distance of each microphone was 4 cm. Sources 1 and 2 were female and male voices with a sampling frequency of 8 kHz [8], respectively, and each source length was 5 seconds. We used the Hann window ($L = 256$) for the short time Fourier transform and the overlap rate was 0.5. For the VAD, we estimated the noise variance $\sigma_N^2(k)$ from the pre-recorded noise.

We conducted the situation in the following way. One of two sources was only active and the source changes from 1 to 2 at time instance $t = 5$. We used recorded fan noise as the noise source, which appears during the experiment ($0 \leq t \leq 10$). We set the reflection coefficients of walls such that the reverberation time was 0.5 second. We adjusted signal levels such that the SNR of microphone signals is about 0 dB.

The result of VAD are illustrated in Figure 4. In Figure 4, the top panel shows the source signal, the middle panel shows the microphone signal ($m = 3$) and the bottom panel shows the result of VAD by binary representation of 1 (speech is on) and -1 (speech is off).

The waveforms of outputs are illustrated in Figure 5. Compared to the source signal shown in Figure 4, the output of SGA noise signal of the low frequency remains. Although the output of MWC seems more cleaner than that of SGA, the noise component still remains. The outputs of PM-FGE and MWC-DW seem the most cleanest among the output signals. It is seen that the noise level of MWC-DW is lower than PM-FGE from these figures. To see this objectively, in Table 2 we list the SNRs of one of the microphone signals ($m = 3$) and the output signals. In Table 2, the upper row shows the SNRs calculated from the output signal before changing the target signal (Source 1). The lower row shows the

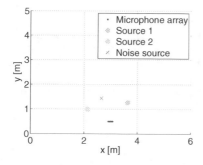

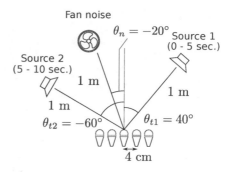

Fig. 2. Assignment of sources and microphones. Each source and microphone is located at the height of 1 m.

Fig. 3. Details for the assignment of sources and microphones.

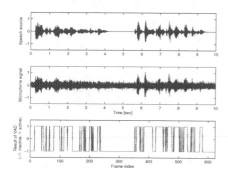

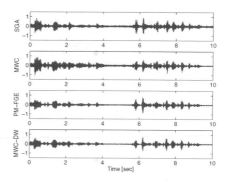

Fig. 4. The source signal, the microphone signal ($m = 3$), and the VAD result shown with active (1) and inactive (-1) frames (from top to bottom).

Fig. 5. Outputs of SGA, MWC, PM-FGE, and the proposed MWC-DW (from top to bottom).

SNRs calculated from the output signal after changing the target signal (Source 2). The SNR of output signal extracted by MWC-DW is the highest among the all methods. The SNR of Source 2 is 6.6049 dB higher than the conventional method, SGA. It seems that the decaying window contributes the improvement.

6 Conclusion

We have proposed a novel adaptive Max-SNR beamformer that enables fast and stable convergence. The main contribution is to establish an efficient tracking algorithm of the principal GEV that can be properly incorporated into Max-SNR beamforming. In our simulations, we have showed that the proposed method achieves faster convergence than the conventional method based on gradient ascent. Moreover, it has been shown that the proposed method can extract the target signal efficiently in the presence of a sudden position changes of targets.

References

1. Allen, J.B., Berkley, D.A.: Image method for efficiently simulating small-room acoustics. The Journal of the Acoustical Society of America 65, 943–950 (1979)
2. Fudge, G.L., Linebarger, D.A.: A calibrated generalized sidelobe canceller for wideband beamforming. IEEE Trans. Signal Process. 42(10), 2871–2875 (1994)
3. Greenberg, J.E., Zurek, P.M.: Evaluation of an adaptive beamforming method for hearing aids. The Journal of the Acoustical Society of America 91, 1662–1676 (1992)
4. Griffiths, L., Jim, C.: An alternative approach to linearly constrained adaptive beamforming. IEEE Trans. Antennas Propag. 30(1), 27–34 (1982)
5. Habets, E., Benesty, J., Cohen, I., Gannot, S., Dmochowski, J.: New insights into the MVDR beamformer in room acoustics. IEEE Trans. Audio, Speech, and Language Process. 18(1), 158–170 (2010)
6. Kolossa, D., Araki, S., Delcroix, M., Nakatani, T., Orglmeister, R., Makino, S.: Missing feature speech recognition in a meeting situation with maximum SNR beamforming. In: Proc. IEEE Int. Symp. Circuits Syst. (ISCAS 2008), pp. 3218–3221 (2008)
7. Kompis, M., Dillier, N.: Noise reduction for hearing aids: Combining directional microphones with an adaptive beamformer. The Journal of the Acoustical Society of America 96, 1910 (1994)
8. MacLean, K.: VoxForge Repository (2006),
   http://www.repository.voxforge1.org/downloads/SpeechCorpus/Trunk/
9. Seltzer, M.L., Raj, B., Stern, R.M.: Likelihood-maximizing beamforming for robust hands-free speech recognition. IEEE Trans. Speech Audio Process. 12(5), 489–498 (2004)
10. Sohn, J., Kim, N.S., Sung, W.: A statistical model-based voice activity detection. IEEE Signal Process. Lett. 6(1), 1–3 (1999)
11. Sohn, J., Sung, W.: A voice activity detector employing soft decision based noise spectrum adaptation. In: Proc. IEEE Int. Conf. Acoust., Speech, Signal Process. (ICASSP 1998), vol. 1, pp. 365–368 (1998)
12. Souden, M., Benesty, J., Affes, S.: A study of the lcmv and MVDR noise reduction filters. IEEE Trans. Signal Process. 58(9), 4925–4935 (2010)
13. Tanaka, T.: Fast generalized eigenvector tracking based on the power method. IEEE Signal Process. Lett. 16(11), 969–972 (2009)
14. Warsitz, E., Haeb-Umbach, R.: Blind acoustic beamforming based on generalized eigenvalue decomposition. IEEE Trans. Audio, Speech, and Language Process. 15(5), 1529–1539 (2007)
15. Yang, B.: Projection approximation subspace tracking. IEEE Trans. Signal Process. 43(1), 95–107 (1995)
16. Yang, J., Zhao, Y., Xi, H.: Weighted rule based adaptive algorithm for simultaneously extracting generalized eigenvectors. IEEE Transactions on Neural Networks 22(5), 800–806 (2011)
17. Yang, J., Zhao, Y., Xi, H.: Weighted rule based adaptive algorithm for simultaneously extraction generalized eigenvectors. IEEE Trans. Neural Netw. 22(5), 800–806 (2011)
18. Zhang, C., Florêncio, D., Ba, D.E., Zhang, Z.: Maximum likelihood sound source localization and beamforming for directional microphone arrays in distributed meetings. IEEE Trans. Multimedia 10(3), 538–548 (2008)
19. Zheng, Y., Goubran, R., El-Tanany, M., Shi, H.: A microphone array system for multimedia applications with near-field signal targets. IEEE Sensors Journal 5(6), 1395–1406 (2005)

Author Index